Catalysis: Science and Engineering

Catalysis: Science and Engineering

Edited by Ross Beckett

CLANRYE
INTERNATIONAL
www.clanryeinternational.com

Clanrye International,
750 Third Avenue, 9th Floor,
New York, NY 10017, USA

ISBN: 978-1-63240-629-3

Cataloging-in-Publication Data

Catalysis : science and engineering / edited by Ross Beckett.
    p. cm.
Includes bibliographical references and index.
ISBN 978-1-63240-629-3
1. Catalysis. 2.Catalysts. I. Beckett, Ross.
QD505 .C38 2017
541.395--dc23

For information on all Clanrye International publications
visit our website at www.clanryeinternational.com

Printed in the United States of America.

Contents

Preface

The rate of increase caused to a chemical reaction through the addition of another substance is known as catalysis. This book discusses the principles of the process of catalysis. It is a significant process in many industries such as the petroleum industry, chemical manufacturing and the food processing industries. This book unravels the recent studies in this field. Chapters in this book delve into the scientific processes as well as industrial applications of catalysis. It includes contributions of experts and scientists which will provide innovative insights into this field. The readers would gain knowledge that would broaden their perspective about catalysis.

The researches compiled throughout the book are authentic and of high quality, combining several disciplines and from very diverse regions from around the world. Drawing on the contributions of many researchers from diverse countries, the book's objective is to provide the readers with the latest achievements in the area of research. This book will surely be a source of knowledge to all interested and researching the field.

In the end, I would like to express my deep sense of gratitude to all the authors for meeting the set deadlines in completing and submitting their research chapters. I would also like to thank the publisher for the support offered to us throughout the course of the book. Finally, I extend my sincere thanks to my family for being a constant source of inspiration and encouragement.

Editor

A combined approach for deposition and characterization of atomically engineered catalyst nanoparticles

Q. Yang[1], D. E. Joyce[2], S. Saranu[2], G. M. Hughes[1], A. Varambhia[1], M. P. Moody[1] and P. A. J. Bagot[1]

[1]Department of Materials, University of Oxford, Parks Road, Oxford OX1 3PH, UK
[2]Mantis Deposition Ltd., Thame OX9 3RR, UK

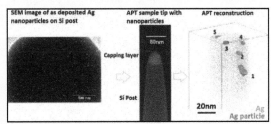

Abstract The structure and composition of catalytic silver nanoparticles (Ag-NPs) fabricated through a novel gas condensation process has been characterized by Scanning Electron Microscopy (SEM) and Atom Probe Tomography (APT). SEM was used to confirm the number density and spatial distribution of Ag-NPs deposited directly onto standard silicon microposts used for APT experiments. Depositing nanoparticles (NPs) directly by this method eliminates the requirement for focussed ion beam (FIB) liftout, significantly decreasing APT specimen preparation time and enabling far more NPs to be examined. Furthermore, by encapsulating deposited particles before final FIB sharpening, the APT reconstruction methodologies have been improved over prior attempts, as demonstrated by comparison to the SEM data. Progress in these areas is vital to enable large-scale catalyst research efforts using APT, a technique, which offers significant potential to examine the detailed atomic-scale chemistry in a wide variety of catalytic NPs.

Keywords Nanoparticles, Atom probe tomography, Heterogeneous catalysis

Introduction

Heterogeneous catalysts facilitate the production of a wide range chemicals that are critical to a broad spectrum of industries. In addition to their importance financially, they also often play a significant role for the environment, for example, by removing atmospheric pollutants from internal combustion engine exhausts.[1] Current research in the development of heterogeneous catalyst must balance two major, and at times, conflicting goals. These are to limit the total loading of expensive transition metals, while simultaneously improving catalyst performance in terms of selectivity and endurance. To achieve these demanding targets, a far more detailed understanding of the fundamental structure and chemistry of catalyst materials is required, and importantly how these may alter during reaction conditions. Such catalysts are most commonly employed in the form of nanoparticles (NPs), therefore advanced characterization methods are necessary to study

them in detail. A variety of experimental techniques have to date been applied to examine the structure and performance of numerous NPs, including transmission electron microscopy (TEM), X-ray diffraction (XRD), X-ray absorption spectroscopy (XAS), and Brunauer, Emmett and Teller (BET) isotherm tests, all of which supply valuable, complementary information. TEM is extensively used for determining the morphology and size distribution of particles, see, for example, Refs.[2–7] More recently, the application of Scanning TEM (STEM) has also enabled chemical analysis of individual NPs,[8–10] while XRD can also be employed to investigate crystallinity and particle size.[11,12] The local atomic structure and in particular chemical state of NPs are becoming increasingly accessible using XAS,[13,14] while finally BET isotherm tests are used to detect another vitally important parameter for catalysts, namely overall active surface area.[15,16]

The industrial production of NPs typically utilize wet chemical synthesis methods. A pressing issue is ensuring that all of the synthesis NPs are all of a similar size, shape, and with the desired structure, e.g. core-shell. Without this fine level of control, the efforts to design atomically engineered catalyst materials, to address the challenges outlined above, will be negated. A further difficulty is the risk that the high-resolution characterization tools and associated sample preparation methods used to characterize these NPs may yield unrepresentative data owing to the small numbers of NPs that can be realistically examined in the course of an experimental campaign.

In this study, a novel method to synthesize NPs of well-defined sizes by means of a cluster beam deposition method is described, together with the development of an approach to prepare these NPs in a form suitable for characterization by Atom Probe Tomography (APT). Atom probe tomography offers a unique combination of 3D atomic-scale spatial and chemical resolution. Indeed number of recent studies using this approach to examine catalytic NPs have been attempted.[17-19] These have highlighted the clear benefits of this method, particularly in the case of nano-engineered alloyed structures such as core-shell NPs; even the most advanced STEM instruments can struggle to resolve the atomic-scale chemical structure of these, particularly for elements of similar masses (and hence electron scattering factors). Despite the benefits, a current disadvantage for APT studies is that they require highly controlled field evaporation of ions from a smooth very sharp needle-shaped specimen to ensure maximum spatial resolution. Thus examining individual catalyst NPs, often dispersed over a porous support material remains a considerable challenge, and by no means routine. Therefore, a second goal of the current study is to demonstrate further developments in the approach to prepare specimen suitable for APT, using Focussed Ion Beam (FIB) methods to encapsulate a small number of NPs prepared by cluster beam deposition within a suitable sample geometry.

Experimental

Silver (Ag) NPs were prepared by means of terminated gas condensation, using a Mantis Nanogen 50 source. A schematic of the equipment and method is shown in Fig. 1. During sample preparation, a metallic vapor was generated in the condensation zone by magnetron sputtering of an Ag target under an argon environment. The number density of

atoms, ions, and clusters in this vapor was controlled by setting the appropriate magnetron power and gas flowrate conditions. A constriction in the outlet of the aggregation zone creates an elevated stagnation pressure inside the chamber on the order of a few tens of Pascals, equivalent to a mean free path of a few tens of micrometers. The large number of collisions in this region thermalizes the vapor, promoting the nucleation and subsequent coalescence of NPs. The magnetron is mounted on a linear translation arm, which provides a means to increase the dwell time of the NPs in the aggregation zone. An expansion zone (circled in red in Fig. 1) creates a rapid pressure drop where excess carrier gas is pumped away through the differential pumping port and a NP beam emerges. The sputtering power, aggregation length, inert gas flowrate, stagnation pressure, and chamber wall temperature can thus all strongly influence NP size and structure.

One of the features of this method is that a large fraction of the NPs become charged as a consequence by passing through a plasma in the aggregation zone. A quadrupole can therefore be used to filter NPs by their mass-to-charge ratio. The resulting charge state is a function of the experimental conditions, but generally a large proportion of the NPs (> 70% for Ag) have a single negative charge, meaning the total flux remains quite high even after mass filtering. Because of the associated charge, the NPs can also be accelerated toward a biased substrate to ensure good adhesion. The quadrupole mass filter is attached after the expansion zone (shown as a green rectangular in Fig. 1), and a mass spectrum can be generated by scanning the frequency of the voltage bias on the quadrupole rods. In order to mass-select nanoclusters, AC and DC voltage is applied to four straight metal rods inside the quadrupole. Any ionized NP passing through the rods will be forced into an oscillating path. For a given AC frequency and amplitude, only one mass (within the resolution of the instrument) will continue on a stable oscillating path and other masses will be rejected. In this way, particles having the mass-to-charge-ratio match with predetermined value will make their way through the quadrupole and on to the substrate. In this experiment, the threshold was set as 10 nm for the diameter of particles as an upper limit.

For APT experiments, the substrate onto which the Ag-NPs were deposited took the form of Si multi-tip coupon. These

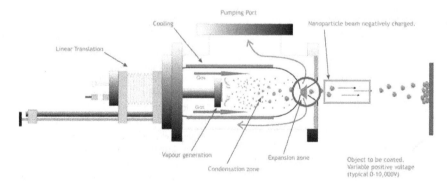

Figure 1 Schematic configuration of the Mantis Nanogen 50 source. The circled red area is the zone where pressure drop happens and particles are expelled. The rectangular green zoom indicates the mass spectrometer, used to filter charged particles (Courtesy Mantis Deposition Ltd. (Thame, UK))

coupons as used extensively to prepare conventional FIB liftout samples for APT studies (however FIB liftout was not required in this study). The APT sample preparation procedures in the current work are schematically depicted in Fig. 2. In Table 1, the operational parameters used in the Nanogen deposition are listed. The aim was to deposit a uniform layer of NPs across the array of flat-topped post-s on the Si coupon, within which the NPs were clearly separated yet still with relatively high number density on the surface. After particle deposition, the tops of the Si posts were covered in a platinum layer, using the e⁻ beam of a Zeiss Auriga Dual-Beam FIB microscope. The organic precursor used for this coating is $C_9H_{16}Pt$ (methylcyclopentadienly(trimethyl)platinum). The purpose of this step is to completely encapsulate the particles in a matrix, after which Ga ion beam milling can be used to mill the samples into their final needle shape suitable for APT. Overall, the goal of this sample preparation route serves to embed a small number of NPs near the apex of the APT needle. This should offer more controlled field evaporation process as the tip is analysed and hence improved reconstructions in comparison to previous electrophoresis-based specimen preparation methods that provided a layer of free-standing NPs completely exposed at the surface of the needle.[17,18] Prepared samples were then analyzed using a LEAP 3000 HR™, running with 200 kHz laser pulses at 0.2 nJ and a stage temperature of 50 K.

Results

Deposition

For the deposition process, the assumptions are that particles are in a single $(1+)$ charge state, spherical, and of known density then allows determination of their diameters, which are shown as a size distribution in Fig. 3. In this experiment, an upper limit of the NP size selection was set to 10 nm, so particles larger than this should be eliminated. However, inevitably some fraction of the produced particles are either neutral or doubly charged. Therefore, a small proportion of all particles deposited will lie out with the size range shown. Overall however, a narrow distribution of NPs is indeed observed, with the majority of particles deposited having a diameter ~ 6 nm.

Scanning Electron Microscopy

The quality of the deposition was examined after preparation by SEM. Figure 4 shows SEM images of as-deposited Ag-NPs on the apex of a single Si post. Bright spots on the image indicate individual NPs. The size of the NPs, as measured by SEM, is approximately 10 nm and these particles have a uniform size distribution. However, some particles larger than 20 nm can be observed, which could be the result of coalescence or non-singly charged ions. Overall, the SEM-images are in relatively close agreement with the results from the quadrupole mass filter (Fig. 3), indicating a good control of particle size. In addition, the higher magnification image of Fig. 4b shows an even coverage of well separated NPs, which is promising for the described APT specimen preparation route. To obtain further spatial information on the surface distribution of the NPs, a nearest neighbour (NN) distribution was calculated using an in-house developed image analysis algorithm based on MATLAB. The first stage in this analysis is to determine the centers of individual NPs and record the 2D coordinates (Fig. 5a). Following this, a NN distribution can be calculated (based on the 1-NN). The values obtained are plotted in Fig. 5b, which shows that most NPs lie within 10 nm of each other.

Atom probe tomography

Figure 6a shows an Ag-NP-coated post-along with a series of tests aimed at burying the NPs below a region of Pt deposited using the FIB electron beam (which appear as the bright protrusions in the picture). One of these Pt-based deposition protrusion was then subject to annular milling using the Ga + beam to produce the final sample ready for

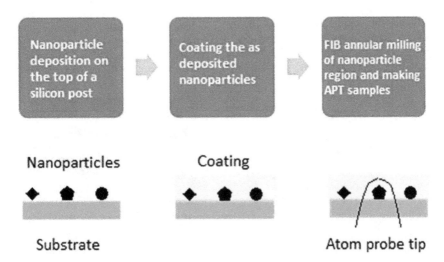

Figure 2 Simple schematic for producing atom probe specimens that incorporate the deposited NPs. Pink substrate represents the flat top of a silicon post. Black NPs are the products of cluster beam deposition. The yellow region is the Pt coating as deposited in the focussed ion beam (FIB)

Table 1   Operation parameters for the Nanogen 50

Running conditions	Parameters
Base pressure	4×10^{-6} mbar
Argon flowrate	100 sccm
Aggregation length	50 mm
Pulsed DC power	20 kHz, 1 μs, 100 W (0.34 A, 298 V)
MesoQ filter	10 nm
Substrate bias	0.5 kV
Deposition rate	60 ngm s^{-1}
Deposition time	45 s

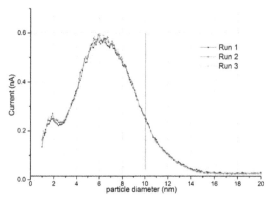

Figure 3   Size distribution of the produced NPs. The data is obtained from the mass-to-charge ratio detected from the quadrupole mass filter. The legends represent different scans during experiment, showing little change in the size distribution between different measurements

APT analysis as shown in Fig. 6b. In this, the Pt capping layer and silicon post have good contrast, which aids final sharpening; the Ag-NPs lie at the interface between the bright and dark regions. The prepared needle samples were then analyzed via APT, and the resulting atom map is shown in Fig. 6c.

In the reconstruction, the individual Ag-NPs, coating layer and supporting post-are clearly revealed upon application of the iso-concentration surface. The data set thus closely correlates with the SEM image of Fig. 6b. In the upper region, aside from Ag the major detected ions are O, Pt, and Ga. Oxygen is abundant on the outer surface of the dataset,

which is usual for samples exposed to the atmosphere. However, the oxygen layer is thin, indicating that the deposited Pt layer has preserved the embedded NPs. The precursor species used to form the Pt layer contains substantial amounts of C, which is also detected, along with Ga$^+$ from ion implantation during the FIB milling process.[20] Efforts to refine the quality of the deposited layers, utilizing higher purity dedicated coating instrumentation are actively being pursued for future refinements of this method.

To examine the distribution of the Ag atoms separate atom maps were generated, which are shown in Fig. 7. The first of these, Fig. 7a, shows that a significant quantity of the Ag atoms are distributed throughout the analysis volume, rather than all being contained within distinct NPs. The origin of this Ag distribution may be owing to a combination of effects. One cause may be trajectory aberrations in the evaporation of atoms during APT analysis, owing to the significant difference between the electric field strength required to evaporate Ag and Pt atoms respectively. Furthermore, the known high surface mobility of Ag under intense electric fields could further contribute to any trajectory aberration.[21] Another explanation is the possible presence of very small NPs/individual Ag atoms on the microtip coupon following deposition. Evidence for the latter is shown in the NN distribution of Fig. 5b, which indicates a substantial fraction of very small species (<1 nm) present. Refining deposition conditions (including substrate bias which may also be causing NP breakup on impact) and those of the FIB-capping layer would help to identify the origin of the spread in Ag distribution. Nevertheless, the atom maps demonstrate a considerable improvement on previous attempts using electrophoresis-based methods, with distinct intact Ag-NPs clearly resolvable in the data by means of an Ag iso-concentration surface (set at two Ag atoms/nm³). This is also shown alone in Fig. 7a, highlighting five isolated Ag-NPs detected, which are labeled from '1–5'.

The volume and calculated equivalent diameters of the five isolated particles are listed in Fig. 7b. The volume data is directly derived from isosurfaces, from which the equivalent diameters are then calculated assuming all particles are spherical. The calculated equivalent diameters are all below 10 nm, in agreement with the mass filtering and SEM results in Figs. 3 and 4. The calculated NN distances for the five NPs are also listed in Fig. 7b, and these, with the exception of NP, 5,

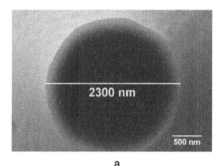

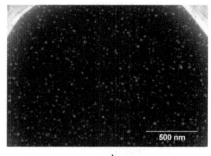

a	b

Figure 4   SEM images of the deposited NPs on the top of a silicon microtip post. a A full picture of the flat end of a silicon post, the diameter of which is about 2300 nm. b A magnified picture of the silicon post, bright dots on which are individual Ag-NPs

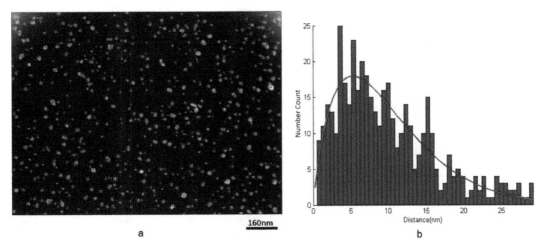

Figure 5 *a* SEM image with green dots shows the detected centers of deposited particles. Totally, 464 coordinates of those dots have been recorded. *b* Resulting nearest neighbour (NN) distibution of all detected particles. The distribution indicates that most particles are within 10 nm of each other

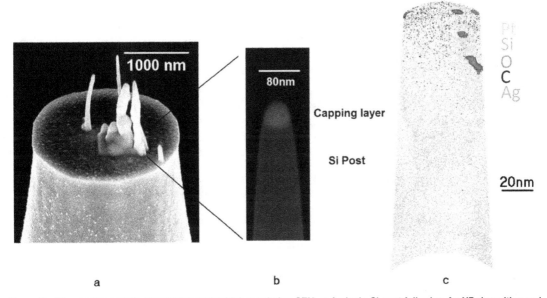

Figure 6 Atom probe sample preparation under high resolution SEM. *a* A single Si post-following Ag-NP deposition and subsequent capping by Pt using the electron beam. *b* Final sample produced by annular milling. The radius of the tip is around 50 nm. *c* Atom map reconstruction tip shown in *b*. The data set shows distinct regions: an upper layer consisting of Pt and C species, then a layer of Ag-NPs before the Si atoms in the post-are detected. (Ag particles are enclosed in blue color)

all lie within the broad distribution based on the SEM observations in Fig. 5*b*. Nanoparticle five resides just within the APT analysis field of view, and it may well be that further nearer NPs are located just outside the analysis volume.

A particle composition analysis shows there are a range of other elements enclosed within the isolated Ag particles. These are mostly Si, Pt, and Ga, which are most likely introduced through trajectory aberrations/Ag surface mobility along with sample preparation issues in the FIB. Alternatively, the existence of Si within identified Ag-NPs could be the result of the particle deposition process, whereby a

fraction of Ag particles impact on the surface and mix with Si atoms from the surface.[22] However, the accuracy of compositional analyses through this method could be easily enhanced by improvements in deposition/sample preparation conditions.

Discussion

In this study, we have demonstrated two important steps for the synthesis and accurate characterization of NPs. First, a novel method allowing the production of tightly controlled

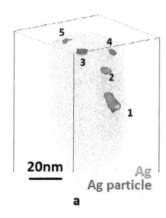

No.	Volume (nm³)	Equivalent diameter (nm)	Nearest Neighbour distance (nm)
1	332	8.6	18.4
2	80.7	5.4	15.0
3	66.0	5.0	18.4
4	30.4	3.9	15.0
5	14.7	3.0	33.6

a   **b**

Figure 7 *a* Ag atoms detected during APT analysis, along with distinct Ag-NPs as identified using an iso-concentration surface (2 atoms/nm³). Isolated NPs labeled '1–5'. *b* Dimensions of the isolated Ag-NPs. The volumes are derived from the isosurfaces and the equivalent diameters are based on a spherical volume. Nearest neighbor (NN) distances are calculated based on the central coordinates of those particles within the atom data reconstruction

distributions of NPs was presented. Importantly, this approach is highly suitable for the controlled deposition of deposition of NPs onto a substrate and hence can underpin a new specimen preparation route in providing much higher quality APT analyses.

The terminated gas condensation method yields reproducible batches of NPs with uniform sizes, but unlike conventional wet chemistry approaches, it does not result in the co-deposition of extensive carbon support materials/surfactants that can particularly complicate APT analysis.[23] Furthermore, terminated gas condensation also offers the capability of subsequent direct deposition onto either characterization substrates as shown here or onto support materials for catalysis applications, such as electrode assemblies for fuel cells, under the same deposition conditions. This presents an opportunity to assess using advanced electron microscopy, APT and other methods (e.g. AFM) whether the NPs engineered have been intended, if they have been deposited at the optimum number density to ensure maximum catalytic performance, and potentially even post-service analyses to assess particle stability during operation. An important additional benefit to APT is that entire Si microtip arrays can be coated, rather than just single tips as used previously.[18,19] This approach also avoids the traditionally used FIB liftout approach for APT sample preparation, which significantly reduces preparation time and hence increases sample throughput.[24] The latter of these benefits is particularly important to ensure representative data on catalyst particles can be obtained. Table 2 summarizes the pros and cons of the currently developed methods for preparing viable APT NP-containing samples. Note that in some cases sample form will also play a

significant factor; for example the electrophoresis method requires NPs to be suspended in a suitable solvent, while particle deposition methods require suitable target sources to be available.

The most promising finding in the characterization results obtained here is the identification of individual NPs within the APT analysis volume, with a size a spatial separation in good agreement with the Nanogen 50/SEM data. This clearly establishes the suitability of this approach, while also indicating areas where further attention needs to be paid to improve the method. For such small NPs, the use of TEM would provide higher resolution shape and size information that would aid the reconstructions, although the SEM examination is also valuable for providing accurate statistical data on number density and particle separation.

In terms of the chemical analysis of the particles, the current data shows them to contain high levels of Si as well as expected Ag. This indicates that the particles have perhaps penetrated the Si support. The mixing of these two elements within the reconstructed NP cores may be a result of either the known high surface mobility of Ag atoms under the intense electric field applied by the APT analysis,[21] intermixing of elements during NP deposition, or else trajectory aberrations in the field evaporation process resulting in a blurring of the spatial positions. This latter effect is a well known phenomenon, seen in other material systems, where there are significant differences in evaporation fields between analyzed species, for example Fe matrix atoms apparently located inside Y_2O_3 particles in oxide-dispersion strengthened steels.[26,27] Furthermore, a similar recent APT analysis on PtFe NPs revealed the same issue of matrix penetration into the particles.[22] However a lower substrate

Table 2 Comparison of different sample preparation methods for nano-particle analysis. There are four methods listed and the first three has been reported in various literatures. The fourth one is proposed in this work

Methods used for NP sample preparation	Electrophoresis method[17,18,23]	Liftout method[19,22,24]	*In-situ* growth method[25]	Particle deposition method
Reconstruction accuracy	Low	High	High	High
Cost	Low	High	Medium	Medium
Time for preparation	Fast	Slow	Slow	Medium

bias and/or the use of different techniques (e.g. sputter coaters) to embed the NPs in more closely matched matrix materials, in terms of the field required to evaporate, would greatly improve if not entirely eradicate these issues.

Preparing a closely evaporation field-matched matrix material that fully surrounds the NPs would unlock the full power of the APT technique.[22,28] Further to the model NP materials examined here, recent studies have already given strong indications of this potential, for example it has been possible using APT to correlate the thickness of Pd shell layers in Ag@Pd core-shell NPs to their efficacy as formic-acid cracking catalysts.[18] Modern catalysts are becoming increasingly complex in terms of their compositions, and APT can play a significant role in their characterization and hence further design, as a previous study of AuAg catalysts prepared for APT by FIB liftout has demonstrated.[19] Both these investigations have, however, been limited by the numbers of NPs that were able to be successfully analyzed, yet the current work suggests a clear approach to resolving many of these issues making routine analysis of catalytic NPs by APT a desirable and achievable goal.

Conclusions

A novel approach has been developed for preparing NPs using a terminated gas condensation method, which are tightly controlled in terms of size, shape, and number density as deposited on substrate materials. Following deposition, the clear benefits of using a multi-technique advanced characterization approach based on SEM/APT to examine deposited NPs in detail have been demonstrated. For APT analysis, deposition of NPs directly onto Si microtip coupons, followed by subsequent encapsulation/shaping using a FIB provides a new method for greatly increasing the yield and quality (for example in comparison to electrophoresis) of resulting APT data. This combined preparation/characterization approach offers great potential for producing nano-engineered catalysts with far improved understanding of the links between atomic-scale structure and catalytic performance.

Conflicts of interest

The authors declare that there are no conflicts of interest.

Acknowledgements

The authors want to thank M.D. Green from Mantis Deposition Ltd for the initial discussion of the results.

References

1. J. Kašpar, P. Fornasiero and N. Hickey: *Catal. Today*, 2003, **77**, (4), 419–449.
2. S. B. Simonsen, I. Chorkendorff, S. Dahl, M. Skoglundh, J. Sehested and S. Helveg: *J. Am. Chem. Soc.*, 2010, **132**, (23), 7968–7975.
3. R. He, Y.-C. Wang, X. Wang, Z. Wang, G. Liu, W. Zhou, L. Wen, Q. Li, X. Wang, X. Chen, J. Zeng and J. G. Hou: *Nat. Commun.*, 2014, **5**, 4327.
4. S. Helveg and P. L. Hansen: *Catal. Today*, 2006, **111**, (1-2), 68–73.
5. S.-J. Cho, J.-C. Idrobo, J. Olamit, K. Liu, N. D. Browning and S. M. Kauzlarich: *Chem. Mater.*, 2005, **17**, (12), 3181–3186.
6. S. Trasobares, M. L. ópez-Haro, M. Kociak, K. March and F. de: *Angew. Chem. Int. Ed. Engl*, 2011, **50**, (4), 868–872.
7. M. Lopez-Haro, L. Dubau, L. Guétaz, P. Bayle-Guillemaud, M. Chatenet, J. André, N. Caqué, E. Rossinot and F. Maillard: *Appl. Catal. B Environ.*, 2014, **152-153**, 300–308.
8. P. D. Nellist and S. J. Pennycook: *Science*, 1996, **274**, (5286), 413–415.
9. H. E. P. D. Nellist, S. Lozano-Perez and D. Ozkaya: *J. Phys. Conf. Ser.*, 2010, **241**, (1), 012067.
10. P. D. Nellist, S. Lozano-Perez and D. Ozkaya: *J. Phys. Conf. Ser.*, 2012, **371**, (1), 012027.
11. M. Min, J. Cho, K. Cho and H. Kim: *Electrochim. Acta*, 2000, **45**, (25-26), 4211–4217.
12. K. Uchino, E. Sadanaga and T. Hirose: *J. Am. Ceram. Soc.*, 1989, **72**, (8), 1555–1558.
13. R. Kaegi, A. Voegelin, B. Sinnet, S. Zuleeg, H. Hagendorfer, M. Burkhardt and H. Siegrist: *Environ. Sci. Technol.*, 2011, **45**, (9), 3902–3908.
14. J.-I. Park, M. G. Kim, Y. Jun, J. S. Lee, W. Lee and J. Cheon: *J. Am. Chem. Soc.*, 2004, **126**, (29), 9072–9078.
15. R. Mueller, L. Mädler and S. E. Pratsinis: *Chem. Eng. Sci.*, 2003, **58**, (10), 1969–1976.
16. S. Nakade, Y. Saito, W. Kubo, T. Kitamura, Y. Wada and S. Yanagida: *J. Phys. Chem. B*, 2003, **107**, (33), 8607–8611.
17. C. Eley, T. Li, F. Liao, S. M. Fairclough, J. M. Smith, G. Smith and S. C. E. Tsang: *Angew. Chem. Int. Ed. Engl.*, 2014, **53**, (30), 7838–7842.
18. K. Tedsree, T. Li, S. Jones, C. W. Chan, K. M. Yu, P. A. Bagot, E. A. Marquis, G. D. Smith and S. C. E. Tsang: *Nat. Nanotechnol.*, 2011, **6**, (5), 302–307.
19. P. Felfer, P. Benndorf, A. Masters, T. Maschmeyer and J. M. Cairney: *Angew. Chemie Int. Ed.*, 2014, **126**, (42), 11372–11375.
20. M. K. Miller, K. F. Russell, K. Thompson, R. Alvis and D. J. Larson: *Microsc. Microanal.*, 2007, **13**, (6), 428–436.
21. K. Moazed: *AIME*, 1964, **230**, 234.
22. E. Folcke, R. Lardé, J. M. Le Breton, M. Gruber, F. Vurpillot, J. E. Shield, X. Rui and M. M. Patterson: *J. Alloys Compd.*, 2012, **517**, 40–44.
23. T. Li, P. A. J. Bagot, E. Christian, B. R. C. Theobald, J. D. B. Sharman, D. Ozkaya, M. P. Moody, S. C. E. Tsang, G. D. W. Smith and A. C. S. Catal., 2014, **4**, (2), 695–702.
24. P. Felfer, T. Li, K. Eder, H. Galinski, A. Magyar, D. Bell, G. D. W. Smith, N. Kruse, S. P. Ringer and J. M. Cairney: *Ultramicroscopy*, 2015. doi:10.1016/j.ultramic.2015.04.014
25. O. Moutanabbir, D. Isheim, H. Blumtritt, S. Senz, E. Pippel and D. N. Seidman: *Nature*, 2013, **496**, (7443), 78–82.
26. D. Larson, P. Maziasz, I. -S. Kim and K. Miyahara: *Scr. Mater.*, 2001, **44**, (2), 359–364.
27. A. J. London, S. Lozano-Perez, M. P. Moody, S. Amirthapandian, B. K. Panigrahi, C. S. Sundar and C. R. M. Grovenor: *Ultramicroscopy*, 2015. doi:10.1016/j.ultramic.2015.02.013
28. D. J. Larson, A. D. Giddings, Y. Wu, M. A. Verheijen, T. J. Prosa, F. Roozeboom, K. P. Rice, W. M. M. Kessels, B. P. Geiser and T. F. Kelly: *Ultramicroscopy*, 2015. doi:10.1016/j.ultramic.2015.02.014

Determining surface structure and stability of ε-Fe₂C, χ-Fe₅C₂, θ-Fe₃C and Fe₄C phases under carburization environment from combined DFT and atomistic thermodynamic studies

Shu Zhao[1,2,3], Xing-Wu Liu[1,2,3], Chun-Fang Huo[1,2], Yong-Wang Li[1,2], Jianguo Wang[1] and Haijun Jiao*[1,4]

[1]State Key Laboratory of Coal Conversion, Institute of Coal Chemistry, Chinese Academy of Sciences, Taiyuan 030001 China

[2]National Energy Center for Coal to Liquids, Synfuels China Co., Ltd, Huairou District, Beijing 101400, China

[3]University of Chinese Academy of Sciences, No. 19A Yuquan Road, Beijing 100049, China

[4]Leibniz-Institut für Katalyse e.V. an der Universität Rostock, Albert-Einstein Strasse 29a, 18059 Rostock, Germany

Abstract The chemical–physical environment around iron based FTS catalysts under working conditions is used to estimate the influences of carbon containing gases on the surface structures and stability of ε-Fe₂C, χ-Fe₅C₂, θ-Fe₃C and Fe₄C from combined density functional theory and atomistic–thermodynamic studies. Higher carbon content gas has higher carburization ability; while higher temperature and lower pressure as well as higher H₂/CO ratio can suppress carburization ability. Under wide ranging gas environment, ε-Fe₂C, χ-Fe₅C₂ and θ-Fe₃C have different morphologies, and the most stable non-stoichiometric

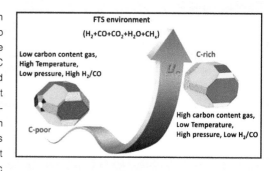

termination changes from carbon-poor to carbon-rich (varying surface Fe/C ratio) upon the increase in Δμ_C. The most stable surfaces of these carbides have similar surface bonding pattern, and their surface properties are related to some common phenomena of iron based catalysts. For these facets, χ-Fe₅C₂-(100)-2.25 is most favored for CO adsorption and CH₄ formation, followed by θ-Fe₃C-(010)-2.33, ε-Fe₂C-(1$\bar{2}$1)-2.00 and Fe₄C-(100)-3.00, in line with surface work function and the charge of the surface carbon atoms.

Keywords Iron carbides, Carburization, Morphology, Fischer–Tropsch synthesis, DFT

Introduction

Fischer–Tropsch synthesis (FTS) is an important technology in converting synthesis gas generated from coal, natural gas and biomass into oil and value added chemicals.[1] Almost a hundred years after its discovery, FTS has been attracting increasing interest worldwide due to the increasing oil prices. Despite of the large scale industrial applications and extensive studies on this important technology in the past decades, the detailed FTS mechanisms are still not fully understood and many explanations to the experimental

observations are premature and lack of scientific rationalization. One of these uncertainties is the surface structures, the corresponding active sites and their roles under FTS conditions.

Suitable FTS catalysts for industrial applications are iron or cobalt based, and iron based catalysts may become more dominant along with the expanding of FTS capacity due to the higher availability and lower cost of iron compared to cobalt. Freshly prepared iron based FTS catalysts are generally iron oxides [mainly hematite (α-Fe₂O₃) and also small amount of maghemite (γ-Fe₂O₃) or even ferrihydrate] and they have to be reduced before becoming FTS active. During the reduction, α-Fe₂O₃ is firstly reduced to magnetite

*Corresponding author, email haijun.jiao@catalysis.de

(Fe₃O₄) by using H₂, synthesis gas or CO and then partially transferred to metallic iron and iron carbides in varying proportions depending on the operating conditions.[2,3] Such multiple phases with very fine crystalline dimensions (several to tens of nanometers) make the characterization and identification of the active phases very difficult, and often such phases will change once the environment around changes as found in most *ex situ* analyses. This leads to the misunderstanding and misinterpretation of the experimentally observed phenomena from real FTS reaction tests.[4]

In Fe based FTS, ε-Fe₂C, χ-Fe₅C₂ and θ-Fe₃C phases have been detected experimentally.[3,5] Both ε-Fe₂C and χ-Fe₅C₂ phases have a hexagonally close packed structure[6,7] but differ in interstitial carbon sites. In ε-Fe₂C, the carbon atom is in the iron octahedral center, while in θ-Fe₃C and χ-Fe₅C₂, the carbon atom is in the iron trigonal prismatic center. Hexagonal carbide (ε-Fe₂C) has been identified as the carburization product of H₂ reduced iron and CO at low temperature,[8] and is the sole component up to 520 K and stable up to 600 K.[8,9] Not formed during FTS at low temperature (<575 K),[10] θ-Fe₃C is only found after carburization above 720 K.[11,12] Under both CO and synthesis gas, ε-Fe₂C is considered as χ-Fe₅C₂ precursor, which is subsequently transformed into θ-Fe₃C at high temperature.[8,13] Königer *et al.*[6] observed that ε-Fe₂C can be converted to χ-Fe₅C₂ after annealing at 423 K, and the χ-Fe₅C₂ phase starts to transform to θ-Fe₃C at 573 K, which is the dominant phase after annealing at 723 K.[6] The exact transformation temperature for the formation of the specific carbide phases depends on many factors, such as crystallite size, morphology, surface texture and promoters or inhibitors as well as the other environment conditions (pressure and gas composition). By using *ab initio* atomistic thermodynamics to investigate the stability of bulk carbide phases, de Smit *et al.*,[3] found that the stable carbide phases depend highly on carbon chemical potential (μc) imposed by gas phase surroundings and emphasized the importance of the controlling chemical–physical environment around the catalyst for forming an efficient FTS system.

Despite the fact that pretreatments can affect the catalytic performance of iron based catalysts, the corresponding studies of the surface properties along with the change of the gas environment (or chemical potential) are rare and most have focused on the defined stoichiometric terminations and the non-stoichiometric terminations of carbides have not been considered.[14,15] The stability and structure as well as electronic and magnetic properties of ε-Fe₂C[16–18] and θ-Fe₃C[14,19–22] have been investigated intensively. In addition, the adsorption and activation of CO and H₂ as well as CₓHᵧ formation on the (100), (001) and (010) surfaces of Fe₃C have been computed.[23–26] Although not directly detected under FTS conditions, Fe₄C can be formed by incorporating carbon atoms into the face centered cubic γ-iron lattices[27] and we included Fe₄C in our study for comparison. The properties of the (100), (110), (111) surfaces of Fe₄C[28] and the CO adsorption properties on these surfaces also have been investigated in our previous work.[29] Recently we found that pretreating conditions, such as temperature, pressure and H₂/CO ratios of an idealized and closed equilibrium system, have significant impact on the relative stability of the

χ-Fe₅C₂ facets in different Fe/C ratios.[30] However, the effects of non-idealized and wide varying operating environments on surface composition and stability of other iron carbides (ε-Fe₂C, θ-Fe₃C and Fe₄C) have not ever been considered. In fact, the real FTS chemical–physical environment may result in wide varying non-equilibrium nature for the catalytic system and the change trends of the catalyst phases will be driven by carbon chemical potential (μc) under conditions with complicated mechanisms. Although the gas environment may result in non-equilibrium, the catalysts can be considered to reach the steady state for a continuous flow of reactants and products at defined conditions. For a fundamental understanding into the FTS mechanisms, systematic studies of the relationship between catalyst surface structures and the thermodynamic parameters for pushing surface structure evolution on the basis of the FTS environment are highly desired.

In this work, the surface structure and stability of the ε-Fe₂C, θ-Fe₃C and Fe₄C as well as χ-Fe₅C₂ phases have been investigated on the basis of density functional theory (DFT) calculations and atomistic thermodynamics by considering the influence of temperature, pressure and H₂/CO ratio under simplified and wide varying non-equilibrium environment. The CO activation and the reactivity analysis on the obtained stable surfaces are also conducted aiming at approaching the overall landscape of Fe based FTS catalysts under real operating conditions.

Methodology

Structure calculation

The catalyst structures were calculated at the level of DFT with Vienna *ab initio* simulation package.[31,32] Electron exchange and correlation energy was treated within the generalized gradient approximation and the Perdew–Burke–Ernzerhof scheme (PBE).[33] Electron ion interaction was described by the projector augmented wave method.[34,35] Spin polarization was included in all calculations on the ferromagnetic iron carbide systems (ε-Fe₂C, θ-Fe₃C and Fe₄C) and this is essential for an accurate description of the magnetic properties. Iterative solutions of the Kohn–Sham equations were done using a plane wave basis with energy cutoff of 400 eV, and the samplings of the Brillouin zone were generated from the Monkhorst–Pack scheme. A second order Methfessel–Paxton[36] electron smearing with σ=0.2 eV was used to ensure accurate energies with errors due to smearing of less than 1 meV per unit cell. The convergence criteria for the force and electronic self-consistent iteration were set to 0.03 eV Å⁻¹ and 10⁻⁴ eV, respectively.

Catalyst models

The bulk structures and the corresponding Monkhorst–Pack grid of k points of the ε-Fe₂C, θ-Fe₃C and Fe₄C phases are listed in Table 1. The optimized lattice parameters agree well with those of the experiments and other calculations.[23,28,37–41] In calculating the ε-Fe₂C bulk structure, a 2 × 2 × 1 supercell was used, and the detailed information is given in the Appendix. For all surface calculations, symmetrical slab surface models were chosen. Each surface was represented by a slab in 10–15 Å thickness, enough to avoid significant influence on the surface energies from our benchmarks.[30] A

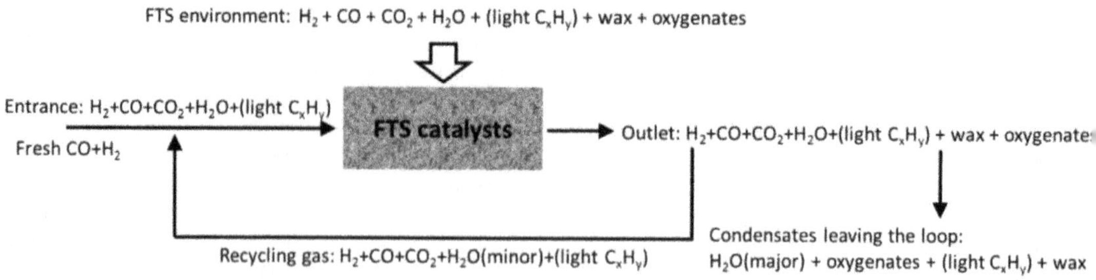

Figure 1 Overall scheme of complicated chemical environment of FTS reactions

vacuum layer of 15 Å was set to exclude the interactions among the periodic slabs, and all atoms were fully relaxed during the calculations. The Monkhorst–Pack grid of k points for each of the corresponding slab models is included in the Supplementary Material (Table S1).

Atomistic thermodynamics

The surface stability influenced by temperature, pressure and gas composition was investigated by using ab initio atomistic thermodynamics.[42,43] Since this is the same procedure used in our previous study on the surface composition and morphology of the χ-Fe$_5$C$_2$ phase,[30] detailed information can be found either in the Supplementary Material or in our previous work. Using the total energy of an isolated carbon atom (E_C) as reference for the variable μ_C, $\Delta\mu_C = \mu_C - E_C$, the minimum $\Delta\mu_C$ for the ε-Fe$_2$C, χ-Fe$_5$C$_2$,[30] θ-Fe$_3$C and Fe$_4$C phases is -7.83, -7.80, -7.80 and -8.01 eV, respectively. The total energies of gas phase molecules and carbon atom were calculated using a single k point (gamma point), where the periodic molecules were separated with 15 Å vacuum distances. These critical $\Delta\mu_C$ values indicate the lowest μ_C for the formation of stable carbides. Since the vibrational contribution to the Gibbs free energy of the χ-Fe$_5$C$_2$ slab is negligible[30] and this is also true for most solid matter, we used only the total energy (E_{slab}) as the predominant term obtained directly from DFT calculations.

CO adsorption

For systematic comparison of the surface properties, we computed CO adsorption on the most stable facets of these carbides. The adsorption energy per CO molecule (E_{ads}) is defined as $E_{ads} = [E_{CO/slab} - (nE_{CO} + E_{slab})]/n$; where $E_{CO/slab}$, E_{CO} and E_{slab} are the total energies of the slab with adsorbed CO on the surface, an isolated CO molecule and the slab of the clean surface, respectively, and n is the number of adsorbed CO molecules. The coverage (θ) is defined as the number of CO molecules over the number of the exposed layer iron atoms. The surface C atoms (C_s) of iron carbides can be considered as adatoms on the defective surfaces, and the binding energy of C_s can be obtained as $E_{ads}(C_s) = E_{slab} - E_{slab/defect} - E_C$, where E_{slab}, $E_{slab/defect}$ and E_C are the total energies of the slab, the defective slab and an isolated carbon atom, respectively. Since PBE functional can give reasonable optimized geometries but overestimates the adsorption energies,[44] we used PBE for structure optimization and RPBE single point energy for estimating the adsorption energy. The Bader charges are used for discussing the effects of charge transfer.[45-47]

Results and discussion

Carbon chemical potential (μ_C) models for real FTS environment

The chemical–physical environment of real FTS catalysts should be properly defined in terms of the operating conditions. Our focus is on the trend of the phase transition of iron carbides in FTS reaction, for which the iron based catalysts have been treated as iron carbides obtained from reduction steps.[3,8,11] For this purpose, the driving force of the phase transition of iron carbides is μ_C, which can be defined using ab initio atomistic thermodynamics. Due to the complicated phenomena of the carburization processes,[3,48]

Table 1 Bulk properties of iron carbides (experimental values in parentheses) and used Monkhorst–Pack grid of k points

Crystal	Cell parameters	μ_B (Fe)
ε-Fe$_2$C Hexagonal P6$_3$/mmc ($5 \times 5 \times 6$)	$a = 5.472$ Å (2×2.794 Å)* $b = 5.639$ Å (2×2.794 Å)* $c = 4.280$ Å (4.360 Å)* $\beta = 121.0°$ ($120.0°$)*	1.66 (1.70–1.72)§
θ-Fe$_3$C Orthorhombic Pnma ($7 \times 5 \times 9$)	$a = 5.025$ Å (5.080 Å)†	1.91 (1.72–1.79)§
Fe$_4$C Cubic Pm$\bar{3}$m ($7 \times 7 \times 7$)	$b = 6.726$ Å (6.730 Å)† $c = 4.471$ Å (4.510 Å)† $\beta = 90.0°$ ($90.0°$)† $a = 3.761$ Å (3.750 Å)‡	Fe(I): 3.04, Fe(II): 1.75

*Ref. 37.
†Ref. 38.
‡Ref. 39.
§Ref. 41.

many species in FTS system can either increase or decrease μ_C imposed over the iron based catalyst surfaces, as discussed in previous studies.[49–55] A comprehensive modeling of the chemical–physical environment of real FTS catalysts should consider most of the key factors, which will have normally non-linear type of contribution to the phase transition phenomena to be investigated. With this in mind, it is necessary to recall the schemes of a real FTS process as Fig. 1.

As given in Fig. 1, the chemical species involved in real FTS system include hydrogen, carbon monoxide, carbon dioxide, water and hydrocarbons within a wide carbon number distribution as well as small portions of oxygenates (mainly alcohols). The composition of these chemical species can easily be determined from industrial operation measurements and/or detailed material balance calculations for a real FTS system. This defines the rather accurate boundaries of the chemical–physical environment around working FTS catalysts, and the catalyst evolution trend may be predicted by using ab initio atomistic thermodynamics. For this goal, the key issue is to systematically develop the model to describe μ_C over the surfaces imposed by environment. However, this model is not very straight forward because of the wide varying and non-equilibrium nature of real FTS systems, namely the theoretical trends of the change in catalyst structures will largely depend on the kinetic factors of all the events related to the chemical species in the catalyst phases, and these events cover both carburization and decarburization reactions. Table 2 lists an overall summary of the events relating to the possible reactions between the catalyst phases (C_s) and the chemical species.

It should be noted that the reactions in Table 2 may occur thermodynamically under typical FTS conditions and affect μ_C. In fact the exact behaviors of μ_C will be also highly related to the rates of these reactions under FTS conditions. However, one can always study the thermodynamic trends by using energetics on the basis of the data from ab initio atomistic thermodynamics.

Obviously, CO is the most potential carburization agent in FTS and can easily deposit carbon atoms on the iron surface from the Boudouard reaction ($2CO \rightarrow C + CO_2$, reaction (2)). Gas phase molecular hydrogen and the adsorbed hydrogen which plays important roles in the transition of catalyst phases are in equilibrium.[56] Oxygen removal from the surface is rate limiting for carbide formation in pure CO, but this step becomes rapid in the presence of hydrogen, therefore addition of H_2 to CO can accelerate carbon deposition ($CO + H_2 \rightarrow C_s + H_2O$, reaction (3)).[57,58] However, the most important role of H_2 is the hydrogenation of surface carbon atoms ($C_s + H_2 \rightarrow -CH_2-$, reaction (1)) resulting in hydrocarbons as the primary products of FTS.[50,56] On the other hand, the light C_xH_y can also be transferred into surface carbons ($-CH_2 \rightarrow C_s + H_2$, reaction (7),[57,59] and $2(-CH_2-) + CO_2 \rightarrow 3C_s + 2H_2O$, reaction (8)). In addition, CO_2 can also consume hydrogen ($CO_2 + 2H_2 \rightarrow C_s + 2H_2O$, reaction (5)), and the reaction extent is limited by water content and temperature.[50,58] Otherwise, CO_2 and H_2O as byproducts can act as decarburizing agents ($C_s + CO_2 \rightarrow 2CO$, reaction (4); and $C_s + H_2O \rightarrow CO + H_2$, reaction (6)). The presence of CO_2 even in small quantities requires high CO concentration to

balance this decarburizing reaction at elevated temperature.[48]

It is suggested that reaction (3) has the fastest kinetics on the basis of the high metal dusting rates in CO/H_2 environment,[48,55,57,59] while reaction (2) is also rapid and the rate of carbon deposition decreases with the increasing CO_2 content.[53] Olsson and Turkdogan[53] showed that in CO–H_2 mixtures reaction (2) is most important for H_2 content less than 50%, while the contribution of reaction (3) to the total rate is dominant for more than 50% H_2. In their study, H_2O has great influence on the rate of carbon deposition. When H_2O is added into CO–H_2 mixture, the rate of carbon deposition decreases with the increasing water vapor content, and this is due to reaction (6) (the reverse of reaction (3)). On the other hand, under CO condition, the rate of reaction (2) increases with the increasing H_2O content.[53] Koeken et al.,[60] found that increasing the total pressure can increase carbon deposition rate for a H_2/CO ratio of 1 : 1, but when H_2/CO ratio is higher than 4 : 1, higher total pressure can suppress the carbon deposition, and increasing H_2/CO ratio can also decrease the rate of carbon deposition. Ando and Kimura[61] also found that the amount of deposited carbon on iron apparently increases by adding small amount of H_2 to pure CO, while an excessive H_2 retards carbon deposition. These results imply that the carbon deposition rate is sensitive to the operating conditions (temperature, pressure and gas composition).

Apart from reactions (2) and (3), we also considered the carbon transfer from light hydrocarbons (C_xH_y) in reaction (7) ($C_xH_y \rightarrow xC_{(Fe)} + y/2H_2$) to estimate their carburization ability. In this case, μ_C $\left[\mu_C = 1/x \left(\mu_{C_xH_y} - y/2\mu_{H_2} \right) \right]$ is determined by decomposition of light hydrocarbon. The influences of temperature, pressure and H_2/C_xH_y ratio on $\Delta\mu_C$ are given in Fig. 2. As temperature increases from 450 to 650 K at 30 atm with a 15% molar percentage of C_xH_y (Fig. 2a), $\Delta\mu_C$ changes hardly under C_2H_4 and C_2H_6 as gas reservoirs, while slightly decreases under C_2H_2 and increases under

Table 2 List of possible reactions between catalyst phases (C_s) and FTS species

Species	Events	Effect§
H_2	$C_s + H_2 \rightarrow -CH_2-$	(1)*	—
CO	$2CO \rightarrow C_s + CO_2$	(2)†	+
	$CO + H_2 \rightarrow C_s + H_2O$	(3)‡	+
CO_2	$C_s + CO_2 \rightarrow 2CO$	(4)†	—
	$CO_2 + 2H_2 \rightarrow C_s + 2H_2O$	(5)	+
H_2O	$C_s + H_2O \rightarrow CO + H_2$	(6)‡	—
$-CH_2-$	$-CH_2- \rightarrow C_s + H_2$	(7)	+
	$2(-CH_2-) + CO_2 \rightarrow 3C_s + 2H_2O$	(8)	+

*Molecular hydrogen may undergo rapid decomposition ($H_2 \rightarrow 2H_s$) on catalyst surfaces and surface hydrogen atoms may hydrogenate surface carbon atoms (FTS key steps).
†Under CO–CO_2 mixture, the carburizing mechanism consists of two elementary reactions: $CO \rightarrow C_s + O_s$ and $CO + O_s \rightarrow CO_2$, the later one was found to be rate limiting.
‡Hydrogen has been shown to be an accelerator of CO decomposition over iron based catalysts, while H_2O has been found to both accelerate and retard CO decomposition.[75] The rates of reactions (2) and (3) depend on the rates of the reactions of CO and H_2 with an oxygen atom to produce CO_2 and H_2O, respectively.[76]
§Positive effect (+) for carburization, and negative effect (–) for decarburization.

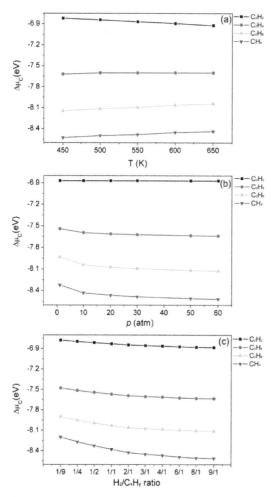

Figure 2 Relationship of carbon chemical potential ($\Delta\mu_C$) to *a* temperature (450–650 K) at 30 atm and C_xH_y=15%; *b* total pressure (1–60 atm) at 550 K and C_xH_y=15%; *c* H_2/C_xH_y ratio (1/9 to 9/1) at 550 K and 30 atm under hydrocarbons

CH_4. As the pressure rises from 1 to 60 atm at 550 K with a 15% molar percentage of C_xH_y (Fig. 2*b*), $\Delta\mu_C$ slightly decreases under C_2H_4, C_2H_6 and CH_4, while does not change under C_2H_2. As expected, increasing the H_2/C_xH_y ratio from 1/9 to 9/1 at 550 K and 30 atm lowers $\Delta\mu_C$ in some extent (Fig. 2*c*).

Figure 2 shows that the carburization ability of light hydrocarbons decreases with the decrease in carbon content from acetylene to saturated hydrocarbons, i.e. $C_2H_2 > C_2H_4 > C_2H_6 > CH_4$. It is noted that $\Delta\mu_C$ under saturated hydrocarbons (C_2H_6 and CH_4) becomes lower than the critical values for stable iron carbide phases, i.e. −7.83 eV for ε-Fe_2C, −7.80 eV for χ-Fe_5C_2[30] and θ-Fe_3C as well as −8.01 eV for Fe_4C, and consequently the carbide phases will transform to metallic iron phase. Therefore, we used ethylene as light hydrocarbon model for our discussion and comparison.

The μ_C to real FTS catalysts can be estimated under different conditions relating to the possible modes, e.g. hydrogen rich modes for starting-up and shutdown of the process, the normal modes typically for oil production, and

CO rich mode due to some unexpected reasons in the whole process. In this work, we try to understand the tendency of the change of the iron carbide phases in the above major situations. As discussed above, in CO/H_2 mixture, only reactions (2), (3) and (7) can contribute to carbon deposition at different levels (Table 2) and their reversible reactions can give overall description of the major physical and chemical relations to carbon deposition on real FTS catalysts. By considering only reaction (3) ($CO + H_2 \rightarrow C + H_2O$), the $\Delta\mu_C$ is much higher than the minimum of carbides (−7.80 eV). It is also true by raising the H_2/CO ratio from 1/1 to 100/1, the $\Delta\mu_C$ (from −6.76 to −6.91 eV) is still far away from the minimum. This would mean that the H_2/CO ratio could not affect the $\Delta\mu_C$, and the carbon-rich facets would remain stable at very high H_2/CO ratio. Obviously, this disagrees with the experimental results, because high H_2 partial pressure will retard carbon deposition and even reduce iron carbide into metallic iron. Instead of using only single reaction to estimate the changes of the $\Delta\mu_C$, we combined different reactions. However, it should be noted that these independent reactions impose chemical force to the phase transition of catalysts and the extent of each reaction to carbon balance of the catalyst has not been well defined. Therefore, we supposed three different extents of each reaction and discuss these situations respectively

Scheme A: (2) + (3) + (7)

$$1/4C_2H_4 + 3/4CO \rightarrow C_s + 1/4H_2 + 1/4CO_2 + 1/4H_2O$$
$$\mu_C = 3/4\mu_{CO} + 1/4\mu_{C_2H_4} - 1/4\mu_{H_2} - 1/4\mu_{H_2O} - 1/4\mu_{CO_2}$$

Scheme B: (2) × 2 + (3) + (7)

$$1/5C_2H_4 + CO \rightarrow C_s + 1/5H_2 + 2/5CO_2 + 1/5H_2O$$
$$\mu_C = \mu_{CO} + 1/5\mu_{C_2H_4} - 1/5\mu_{H_2} - 1/5\mu_{H_2O} - 2/5\mu_{CO_2}$$

Scheme C: (2) + (3) + (7) × 2

$$1/3C_2H_4 + 1/2CO \rightarrow C_s + 1/6H_2 + 1/6CO_2 + 1/6H_2O$$
$$\mu_C = 1/2\mu_{CO} - 1/3\mu_{C_2H_4} - 1/6\mu_{H_2} - 1/6\mu_{H_2O} - 1/6\mu_{CO_2}$$

In a typical Fe based FTS process, the most important parameters influencing the catalyst performance are the chemical composition of fluids surrounding the catalyst, temperature and pressure. In this study, the gas environment is designed in composition with $CO + H_2$ (varying H_2/CO ratio), −CH_2− (light hydrocarbons), CO_2 and H_2O of 75, 10, 12 and 3%, respectively, representing the typical industrial conditions. Since we did not consider the contribution of the condensed heavy hydrocarbons and oxygenates, the ratios of carbon, oxygen and hydrogen are not stoichiometric. For comparison, we also included a gas phase free from CO with H_2 and hydrocarbons, which simulates the weaker carburization environment for the iron catalyst as tested in fundamental studies in FTS.[62]

Under real CO involved environment, the influences of temperature and pressure on $\Delta\mu_C$ are evaluated for H_2/CO ratio of 2, 4 and 8 with other gas compositions presented in Table 3, and the main results are shown in Fig. 3. At a total pressure fixed at 30 atm (Fig. 3*a*), it is found that higher temperature leads to lower (more negative) $\Delta\mu_C$ for all three schemes with different H_2/CO ratios, in consistent with the results of de Smit *et al.*[63] At a given temperature the μ_C

determined in Scheme B is the highest, followed by those in Scheme A and Scheme C. This trend implies that CO has stronger carburization ability than C_2H_4.

Figure 3b shows the results with varying pressure at 550 K. For Scheme B and Scheme A, higher pressure leads to higher (less negative) $\Delta\mu_C$, which indicates lower pressure can retard carbon deposition. In the whole pressure range (1 to 60 atm), the $\Delta\mu_C$ determined by Scheme C keeps constant.

Figure 3c presents the H_2/CO ratio influence on $\Delta\mu_C$ at 550 K and 30 atm, with the gas composition presented in Table 4. As H_2/CO ratio increases from 1/1 to 20/1, the $\Delta\mu_C$ determined by all three schemes decreases and this implies that excess hydrogen would retard carbon deposition. However, it should be noted that even at extremely high H_2/CO ratio (20/1) $\Delta\mu_C$ does not become lower than the critical value for stable iron carbide phases, i.e. -7.83 eV for ε-Fe₂C, -7.80 eV for χ-Fe₅C₂ and θ-Fe₃C as well as -8.01 eV for Fe₄C. This reveals the thermodynamic possibility for carbon deposition at very high H_2/CO ratios beyond stable iron carbide phases under extended FTS operation conditions. Such carbon deposition destroys the mechanical structure of catalysts as observed in industrial practices and deactivates the catalysts.[57] It is noted that the same trends in Schemes A–C have been found for using CH₄ as light hydrocarbon model (Supplementary Material).

How to keep the activity of the catalysts by adjusting the chemical and physical parameters remains to be a headache problem and challenging. Rising temperature can retard carbon deposition thermodynamically, but accelerate carburization kinetically. Lowering pressure is thermodynamically and kinetically promising, but reduces significantly the process productivity. Increasing H_2/CO ratio is therefore a reasonable choice with thermodynamic and kinetic advantages as well as controllable process productivity. It should be noticed that our current calculations depend very strongly on three different carburization schemes that hopefully can cover all possibilities in real FTS systems. The exact schemes should be determined by experimental studies, which show the chemical extents of different carbon formations steps, namely (2), (3) and (7) in Table 2. This should be done by experiments with well defined techniques and operation conditions. Such studies provide a precise thermodynamic basis for the fundamental investigation along with ab initio atomistic thermodynamics developed in this work.

In FTS reaction system, the influence of CO_2 and H_2O contents on the $\Delta\mu_C$ is presented in Fig. 4, with the gas composition listed in Tables 5 and 6. It shows clearly that increasing the content of CO_2 (5 to 25%) and H_2O (1 to 15%) lowers $\Delta\mu_C$ at very low degree. The results indicate that CO_2

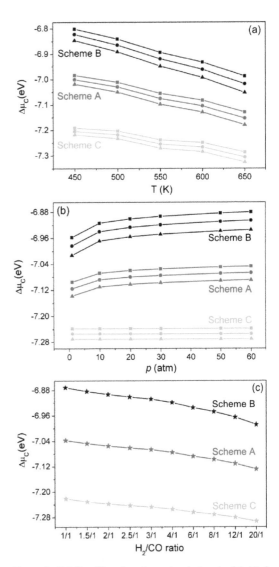

Figure 3 Relationship of carbon chemical potential ($\Delta\mu_C$) to a temperature (450–650 K) at 30 atm and H_2/CO ratio of 2, 4 and 8; b pressure (1–60 atm) at 550 K and H_2/CO ratio of 2, 4 and 8; c H_2/CO ratio (1 to 20) at 550 K and 30 atm (■ for $H_2/CO=2$; • for $H_2/CO=4$ and ▲ for $H_2/CO=8$)

Table 3 Gas composition under H_2/CO ratios of 2, 4 and 8

H_2/CO	2	4	8
H₂ (%)	50.00	60.00	66.67
CO (%)	25.00	15.00	8.33
CH₄ (%)	10.00	10.00	10.00
CO₂ (%)	12.00	12.00	12.00
H₂O (%)	3.00	3.00	3.00

Table 4 Gas composition under different H_2/CO ratios at 550 K and 30 atm

H_2/CO	H₂/%	CO/%	CH₄/%	CO₂/%	H₂O/%
1.0	37.50	37.50	10.00	12.00	3.00
1.5	45.00	30.00	10.00	12.00	3.00
2.0	50.00	25.00	10.00	12.00	3.00
2.5	53.57	21.43	10.00	12.00	3.00
3.0	56.25	18.75	10.00	12.00	3.00
4.0	60.00	15.00	10.00	12.00	3.00
6.0	64.29	10.71	10.00	12.00	3.00
8.0	66.67	8.33	10.00	12.00	3.00
12.0	69.23	5.77	10.00	12.00	3.00
20.0	71.43	3.57	10.00	12.00	3.00

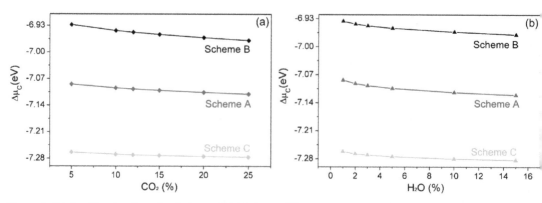

Figure 4 Relationship of carbon chemical potential ($\Delta\mu_C$) to *a* CO_2 content and *b* H_2O content at 550 K, 30 atm and H_2/CO=8

and H_2O, the byproduct in FTS, play only subordinate role in controlling phase transition process and should be removed from the process as usual (Fig. 1) for achieving other process benefits.

These compared results in Schemes A–C imply that we may use different unsaturated hydrocarbons and CO to optimize the environment for getting stable catalyst phases, especially for initializing the FTS process. It has been proved experimentally that unsaturated light hydrocarbons and H_2, instead of CO and H_2, can conduct chain growth reactions over iron based FTS catalysts.[62] In order to conduct efficient FTS reactions it is necessary to optimize the carburization ability of the chemical–physical environments (temperature, pressure and H_2/CO ratio) in terms of μ_C. The insight behind μ_C is the change of the stable iron carbide phases as well as the surface structure and composition.

Surface stability

To get the equilibrium shapes of ε-Fe_2C, θ-Fe_3C and Fe_4C under different operation conditions, we studied both low and high Miller index facets of these carbides, which contain all low Miller index surfaces and the characteristic peaks in X-ray diffraction.[37,64,65] All calculated surfaces and the equivalent Miller index are listed in the Supplementary Material (Table S3). Because of their complex bulk structures, each surface has several terminations (including both stoichiometric and non-stoichiometric terminations), e.g. five terminations for each of the (101), (102) and (103) surfaces of ε-Fe_2C; 16 terminations for each of the (111), (113), (133) and (131) surfaces of θ-Fe_3C; and each facet of Fe_4C has two terminations. Here we used the surface Fe/C ratio ($\alpha=n_{Fe}/n_C$) to distinguish these terminations as discussed previously.[30] In the following discussion, the number following the Miller index indicates the surface Fe/C ratio. The surface free

energies of these terminations within the $\Delta\mu_C$ range from −8.50 to −6.00 eV are given in the Supplementary Material (Fig. S3) for comparison, and only the results of the most stable termination of each facet are used for discussion.

Figures 5–7 show the relationship between surface free energy ($\gamma(T,p)$) of the most stable facets of ε-Fe_2C, θ-Fe_3C and Fe_4C and $\Delta\mu_C$. Similar to Fe_5C_2,[30] carbon-rich termination with lower α value becomes more stable at higher (less negative) $\Delta\mu_C$ for all iron carbides, while the carbon-poor terminations are more favorable at lower (more negative) $\Delta\mu_C$. With the increasing $\Delta\mu_C$, the most stable termination changes from carbon-poor (higher Fe/C ratio) to carbon-rich (lower Fe/C ratio), and the turn points represent the change of the stable termination and they differ from facet to facet. By combining theory and *in situ* XPS studies de Smit *et al.*[63] also found that body centered cubic Fe and surface/subsurface carbon are more stable at high temperature (low μ_C), while the carbon-rich χ-Fe_5C_2 (100) surface becomes thermodynamically more stable upon lowering the temperature (high μ_C). Since excessive carbon deposition will deactivate the catalysts and lower the catalytic performance,[57] avoiding carbon deposition can be improved by using proper temperature, pressure and gas environment.

For ε-Fe_2C (Fig. 5), ($1\bar{2}1$)-2.00 and (101)-1.50 are the most stable facets, followed by ($2\bar{2}1$)-2.67/1.33, and ($0\bar{1}1$)-2.00/1.33. The least stable surfaces are ($\bar{2}01$)-2.67/1.33, and (103)-2.50/1.00. The other surfaces, ($0\bar{1}3$)-2.67/1.50, (001)-4.00/1.00, (110)-2.00/1.33, (112)-2.00/1.00, (100)-2.00/1.00, ($0\bar{1}2$)-4.00/1.00, (102)-4.00/2.00, (111)-3.00/1.33, have intermediate stability. It is also noted that for the ($1\bar{2}0$), (010), ($1\bar{2}1$) and ($1\bar{2}2$) facets, the stoichiometric terminations are most stable (Fig. S3).

Table 5 Gas composition at 550 K and 30 atm with H_2/CO ratio of 8 for different CO_2 contents

CO_2/%	H_2/%	CO/%	CH_4/%	H_2O/%
5.00	72.89	9.11	10.00	3.00
10.00	68.44	8.56	10.00	3.00
12.00	66.67	8.33	10.00	3.00
15.00	64.00	8.00	10.00	3.00
20.00	59.56	7.44	10.00	3.00
25.00	55.11	6.89	10.00	3.00

Table 6 Gas composition at 550 K and 30 atm with H_2/CO ratio of for different H_2O contents

H_2O/%	H_2/%	CO/%	CH_4/%	CO_2/%
1.00	68.44	8.56	10.00	12.00
2.00	67.56	8.44	10.00	12.00
3.00	66.67	8.33	10.00	12.00
5.00	64.89	8.11	10.00	12.00
10.00	60.44	7.56	10.00	12.00
15.00	56.00	7.00	10.00	12.00

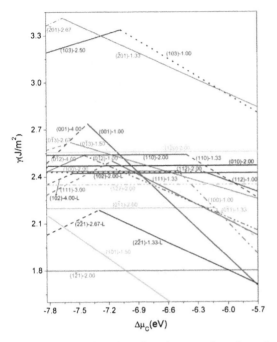

Figure 5 Relationship of surface free energies of most stable facets of ε-Fe₂C to $\Delta\mu_C$ (indices given in parentheses indicates corresponding Miller index, and second term of indices provides corresponding surface α=Fe/C ratio)

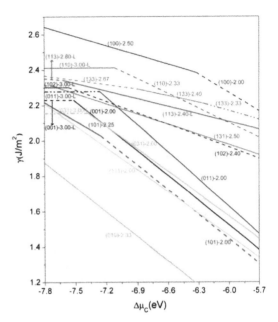

Figure 6 Relationship of surface free energies of most stable facets of θ-Fe₃C to $\Delta\mu_C$ (indices given in parentheses indicates corresponding Miller index, and second term of indices provides corresponding surface α=Fe/C ratio)

For θ-Fe₃C (Fig. 6), (010)-2.33 is the most stable one. The least stable surface is (100) with α=2.50 or 2.00. The other surfaces, (110)-3.00/2.33, (133)-2.67/2.40/2.33, (113)-2.80/2.40, (131)-2.50, (102)-3.00/2.40, (011)-3.00/2.00, (001)-3.00/2.00, (101)-2.25/2.00, (111)-2.00, (031)-3.00/2.00, have intermediate stability.

For Fe₄C (Fig. 7), (100) with α=3.00 is the most stable facet. The surface free energies of other facets are significantly higher than (100), and the least stable surface is (133)-6.00/3.00, followed by (110)-3.00, (111)-4.00/2.00, (131)-6.00/2.67 and (210)-5.00/3.33, have intermediate stability.

For Fe₅C₂[30] as reported previously, the (100) termination is most stable with α=2.25, followed by (111)-2.17/1.75, (510)-2.50, and (110)-2.40/2.00. The least stable surfaces are (10$\bar{1}$)-2.75/2.25, (001)-2.50; (11$\bar{3}$)-2.50, (113)-2.00 and (101)-1.50. In addition, (110)-2.40/2.00, (010)-2.50; (133)-1.75; (11$\bar{1}$)-2.50, (511)-2.25, (221)-3.00, ($\bar{4}$11)-2.50, (011)-2.40/2.20 have stability in between.

Crystallite morphology

In order to estimate the crystallite morphology of these iron carbides, it is necessary to determine the equilibrium crystal shape by using the standard Wulff construction.[66] In the standard Wulff construction, the surface free energy for a given closed volume is minimized and the exposure of a facet depends not only on surface free energy but also on orientation in crystal.[67] Since the surface free energy of each facet is a function of μ_C, the crystal shape should also be a function of the $\Delta\mu_C$ that corresponds to different experimental conditions of temperature, pressure and atmosphere. Figure 8 presents the morphology of the ε-Fe₂C, θ-Fe₃C and Fe₄C crystals at different $\Delta\mu_C$, corresponding to different gas

compositions at 550 K and 30 atm, respectively, along with that of χ-Fe₅C₂ (slightly modified form compared to our previous report, where an incorrect default crystal parameter was used; however, this does not affect our conclusion). The proportions of exposed terminations of ε-Fe₂C, χ-Fe₅C₂ and θ-Fe₃C are listed in Tables 7, 8 and 9.

At lower $\Delta\mu_C$ (−7.60 eV), the crystallite of ε-Fe₂C has 11 exposed surface terminations in different Fe/C ratios, (100), (010), (001), (101), (0$\bar{1}$1), (110), (111), (1$\bar{2}$1), (102), (0$\bar{1}$2) and (2$\bar{2}$1). The (1$\bar{2}$1) termination has the largest portion (35.3%) of the surface area, followed by the (101) and (2$\bar{2}$1) terminations (27.7 and 16.7%, respectively), and they cover about 80% of the total surface area of the crystal. As the $\Delta\mu_C$ increases to −7.10 eV, (0$\bar{1}$1), (001) and (102) are

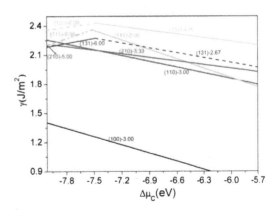

Figure 7 Relationship of surface free energies of most stable facets of Fe₄C to $\Delta\mu_C$ (indices given in parentheses indicates corresponding Miller index, and second term of indices provides corresponding surface α=Fe/C ratio)

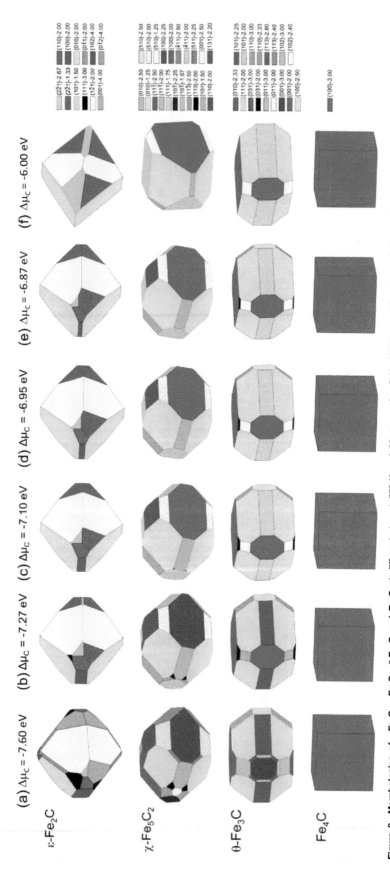

Figure 8 Morphologies of ε-Fe$_2$C, χ-Fe$_5$C$_2$, θ-Fe$_3$C and Fe$_4$C at different $\Delta\mu_C$ at 550 K and 30 atm: a for C$_2$H$_4$ (15%) condition; b for Scheme C with H$_2$/CO=8.0; c for Scheme A with H$_2$/CO=8.0; d for Scheme B with H$_2$/CO=8.0; e for C$_2$H$_2$ (15%) condition. Indices given in legend indicates corresponding Miller index, and second term of indices provides corresponding surface α=Fe/C ratio)

disappeared, and the carbon-rich $(2\bar{2}1)$-1.33 termination becomes more stable than the carbon-poor $(2\bar{2}1)$-2.67 termination. As the $\Delta\mu_C$ increases, the proportion of (100), (111), $(1\bar{2}1)$ and $(0\bar{1}2)$ decreases, while the area of (101) and (010) increases. When the $\Delta\mu_C$ reaches to -6.0 eV, the

crystallite of ε-Fe$_2$C has only four exposed surface terminations and they are (101), $(1\bar{2}1)$, (010) and $(2\bar{2}1)$. The (101) becomes the largest exposed surface (59.5%), and the facets $(1\bar{2}1)$, (101) and $(2\bar{2}1)$ still cover the most surface area (92%) of the crystal.

Table 7 Facets contributions (%) to total surface area in Wulff construction of ε-Fe$_2$C presented in Fig. 8

Facet	(a) C$_2$H$_4$	(b) Scheme C	(c) Scheme A	(d) Scheme B	(e) C$_2$H$_2$	(f) $\Delta\mu_C=-6.0$ eV
(100)-2.00	5.52	3.74	2.75	1.85	1.36	0.00
(010)-2.00	0.55	1.45	0.86	0.64	0.53	7.99
(001)-1.00	0.18	0.00	0.00	0.00	0.00	0.00
(101)-1.50	27.70	40.81	44.13	46.96	48.52	59.52
$(0\bar{1}1)$-2.00	4.14	0.00	0.00	0.00	0.00	0.00
(110)-2.00	3.43	4.96	3.96	3.01	2.51	0.00
(111)-1.33	...	...	0.03	0.06	0.07	0.00
(111)-3.00	4.91	0.28	...	...	...	...
$(1\bar{2}1)$-2.00	35.25	39.02	36.41	34.15	32.93	17.32
(102)-2.00	0.97	0.00	0.00	0.00	0.00	0.00
$(0\bar{1}2)$-1.00	0.65	0.03	0.00	0.00	0.00	0.00
$(2\bar{2}1)$-1.33	...	9.70	11.85	13.34	14.09	15.18
$(2\bar{2}1)$-2.67	16.71	...	...	...	...	...

Table 8 Facets contributions (%) to total surface area in Wulff construction of χ-Fe$_5$C$_2$ presented in Fig. 8

Facet	(a) C$_2$H$_4$	(b) Scheme C	(c) Scheme A	(d) Scheme B	(e) C$_2$H$_2$	(f) $\Delta\mu_C=-6.0$ eV
(010)-2.50	1.20	...	...	...	...	...
(010)-1.25	...	0.92	2.06	2.83	3.33	10.44
$(11\bar{1})$-2.50	11.72	...	...	...	...	...
$(11\bar{1})$-2.00	...	9.76	9.42	9.06	8.91	2.25
(001)-2.50	0.92	0.00	0.00	0.00	0.00	0.00
(510)-2.50	9.58	...	...	...	...	...
(510)-2.00	...	6.38	5.79	6.05	...	...
(510)-1.25	...	...	...	...	6.25	5.37
(113)-2.00	0.91	0.35	0.03	0.00	0.00	0.00
$(\bar{4}11)$-2.50	9.14	...	...	...	...	...
$(\bar{4}11)$-2.00	...	6.35	4.56	2.74	1.87	0.00
$(11\bar{3})$-2.50	3.33	1.02	0.00	0.00	0.00	0.00
(101)-1.50	4.53	4.92	4.65	4.24	4.00	1.42
(100)-2.25	16.63	20.24	...	...	...	...
(100)-2.00	...	...	21.67	23.07	23.76	31.84
(110)-2.00	10.49	12.90	11.87	10.54	9.64	0.00
(111)-1.75	23.81	30.41	32.60	33.65	34.22	39.99
(511)-2.25	1.19	0.00	0.00	0.00	0.00	0.00
(131)-2.20	3.34	0.68	0.20	0.00	0.00	0.00
$(10\bar{1})$-2.25	3.21	...	...	...	...	...
$(10\bar{1})$-1.67	...	6.07	7.35	7.82	8.02	8.69

Table 9 Facets contributions (%) to total surface area in Wulff construction of θ-Fe$_3$C presented in Fig. 8

	(a) C$_2$H$_4$	(b) Scheme C	(c) Scheme A	(d) Scheme B	(e) C$_2$H$_2$	(f) $\Delta\mu_C=-6.0$ eV
(001)-3.00	5.03	...	...	...	...	...
(001)-2.00	...	6.80	6.82	6.82	6.79	6.81
(010)-2.33	20.31	23.94	24.74	25.44	25.77	29.99
(100)-2.50	3.46	3.36	3.06	2.53	2.24	0.00
(101)-2.25	13.25	12.60	...	...	...	...
(101)-2.00	...	...	13.06	13.89	14.49	21.39
(110)-3.00	9.28	4.96	...	...	...	...
(110)-2.33	...	...	3. 04	2.38	2.04	0.00
(011)-3.00	5.60	2.36	...	...	...	...
(011)-2.00	...	...	2.28	2.42	2.50	2.92
(111)-2.00	35.88	44.38	45.75	45.63	45.45	38.88
(113)-2.80	1.21	...	...	...	...	...
(113)-2.40	...	0.00	0.00	0.00	0.00	0.00
(031)-3.00	3.92	...	...	...	...	...
(031)-2.00	...	1.61	1.25	0.89	0.71	0.00
(102)-3.00	0.85	...	...	...	...	...
(102)-2.40	...	0.00	0.00	0.00	0.00	0.00

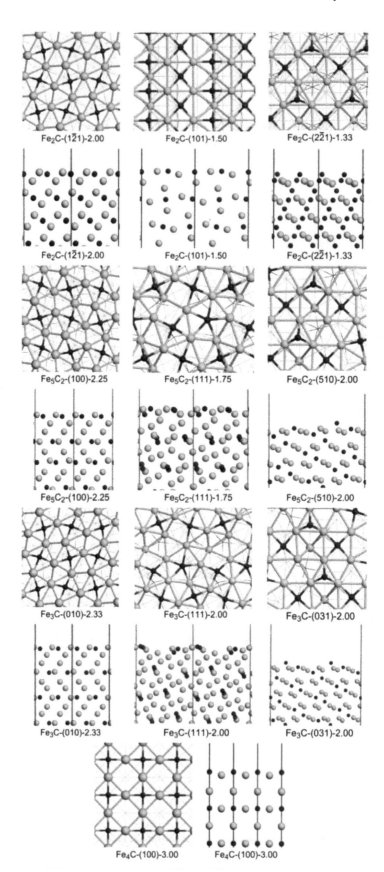

Fe₂C-(1̄21)-2.00   Fe₂C-(101)-1.50   Fe₂C-(2̄2̄1)-1.33

Fe₂C-(1̄21)-2.00   Fe₂C-(101)-1.50   Fe₂C-(2̄2̄1)-1.33

Fe₅C₂-(100)-2.25   Fe₅C₂-(111)-1.75   Fe₅C₂-(510)-2.00

Fe₅C₂-(100)-2.25   Fe₅C₂-(111)-1.75   Fe₅C₂-(510)-2.00

Fe₃C-(010)-2.33   Fe₃C-(111)-2.00   Fe₃C-(031)-2.00

Fe₃C-(010)-2.33   Fe₃C-(111)-2.00   Fe₃C-(031)-2.00

Fe₄C-(100)-3.00   Fe₄C-(100)-3.00

Figure 9 Surface structures of most stable surfaces and surfaces that have largest exposed surface area in Wulff construction of ε-Fe$_2$C, χ-Fe$_5$C$_2$, θ-Fe$_3$C and Fe$_4$C (indices given in parentheses indicates corresponding Miller index, and the third term of the indices provides the corresponding surface α=Fe/C ratio, Fe atoms are shown by blue balls, C atoms are shown by black balls)

At lower $\Delta\mu_C$ (−7.60 eV), the crystallite of χ-Fe$_5$C$_2$ has 14 exposed surface terminations in different Fe/C ratios, (010), (11$\bar{1}$), (001), (510), (113), ($\bar{4}$11), (11$\bar{3}$), (101), (100), (110), (111), (511), (131) and (10$\bar{1}$). The (111), (100) and (11$\bar{1}$) surfaces cover 52.1% of the total surface area of the crystal (23.8, 16.6 and 11.7%, respectively). When the $\Delta\mu_C$ increases to −6.95 eV, the facets (001), (511), (113), (11$\bar{3}$), and (131) are disappeared, and the carbon-rich termination of (010), (11$\bar{1}$), (10$\bar{1}$), (510), ($\bar{4}$11) and (100) are exposed. As the $\Delta\mu_C$ increases, the proportion of (100), (010), (111) and (10$\bar{1}$) increases, while the exposed areas of all the other surfaces decrease. At higher $\Delta\mu_C$ (−6.00 eV), only seven facets are still exposed, among which the (111) and (100) terminations cover as much as 71.8% of the surface area of the crystal.

When the $\Delta\mu_C$ is −7.60 eV, the crystallite of θ-Fe$_3$C has 11 exposed surface terminations, (001), (100), (010), (101), (110), (011), (111), (113), (131), (102) and (031). The (111) facet has the largest portion of the total surface area (35.9%), followed by the (010) (20.3%) and (101) (13.3%) facets. As the $\Delta\mu_C$ increases to −6.95 eV, the proportions of (111) and (010) increase to 45.6 and 25.4%, respectively. The carbon-rich (031), (011), (001), (101) and (110) terminations are exposed under higher $\Delta\mu_C$ compared with lower μ_C. There are five facets still exposed at higher $\Delta\mu_C$ (−6.00 eV), among them the (111), (010) and (101) terminations cover 90.3% of the total surface area of the crystal (38.9, 30.0 and 21.4%, respectively).

In the whole range of $\Delta\mu_C$ that we considered, the crystallite of Fe$_4$C only exposes the (100)-3.00 termination.

Surface property

In the most stable terminations of each iron carbides (Fig. 9), ε-Fe$_2$C-(1$\bar{2}$1)-2.00, χ-Fe$_5$C$_2$-(100)-2.25, θ-Fe$_3$C-(010)-2.33 and Fe$_4$C-(100)-3.00, each surface carbon atom coordinates with four surface iron atoms, and each surface iron atom coordinates with two surface carbon atoms. In addition, the most exposed surfaces, χ-Fe$_5$C$_2$-(111)-1.75 and θ-Fe$_3$C-(111)-2.00, have similar atom arrangement on partial surface structures. The third exposed facet of ε-Fe$_2$C, (2$\bar{2}$1)-1.33, has also some similar surface structure with χ-Fe$_5$C$_2$-(510)-2.00 as

well as θ-Fe$_3$C-(031)-2.00. The computed density of states of the surface layer atoms (Fig. 10) also revealed the similarity of these surface structures.

Since the pattern and density of carbonaceous deposit on surface can significantly influence the catalytic performance,[68,69] similar and unique catalytic activities of the carbides facets with the same atom arrangement on surface layer should be expected. At first we analyzed the surface properties, e.g. the charge and binding energy of the surface carbon atoms, as well as the surface work function (difference between the electrostatic potential energy in the vacuum region and the Fermi energy of the slab), which is an important electronic indicator of a surface, i.e. lower work function indicates the higher electron donating ability of the surface. In addition, we also computed the adsorption structure and energy of CO on these surfaces. As shown in Fig. 11, the most stable CO adsorption site is the Fe-top site

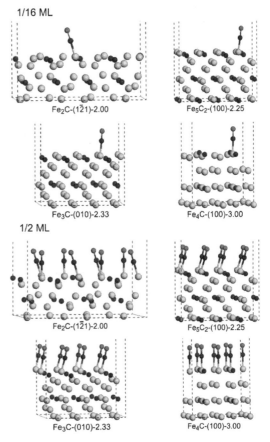

1/16 ML

Fe$_2$C-(1$\bar{2}$1)-2.00   Fe$_5$C$_2$-(100)-2.25

Fe$_3$C-(010)-2.33   Fe$_4$C-(100)-3.00

1/2 ML

Fe$_2$C-(1$\bar{2}$1)-2.00   Fe$_5$C$_2$-(100)-2.25

Fe$_3$C-(010)-2.33   Fe$_4$C-(100)-3.00

Figure 11 Adsorption of CO on most stable facet of ε-Fe$_2$C, χ-Fe$_5$C$_2$, θ-Fe$_3$C and Fe$_4$C with different coverage (indices given in parentheses indicates corresponding Miller index, and third term of indices provides corresponding surface α=Fe/C ratio, Fe atoms in blue, C atoms in black and O atoms in red)

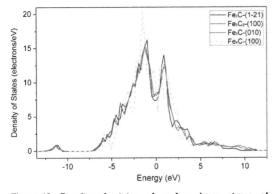

Figure 10 Density of states of surface layer atoms of most stable facets

on the ε-Fe$_2$C-(1$\bar{2}$1)-2.00, χ-Fe$_5$C$_2$-(100)-2.25 and θ-Fe$_3$C-(010)-2.33 surfaces. The Fe$_4$C-(100)-3.00 facet has hollow site on the surface, the most stable CO adsorption site is the 4-fold site. In order to compare with other three carbides, the less stable Fe-top site of CO adsorption is taken into account (adsorption energy of CO on the Fe-top site is only 0.054 eV higher than that of the 4-fold site).

At 1/16 and 1/2 ML (Table 10), as the surface work function increases in the order of χ-Fe$_5$C$_2$-(100)-2.25<θ-Fe$_3$C-(010)-2.33<ε-Fe$_2$C-(1$\bar{2}$1)-2.00<Fe$_4$C-(100)-3.00, both CO adsorption energies and C-O bond elongation as well as the net negative charge of the adsorbed CO molecules decrease, implying that CO favors to adsorb on the surface with lower work function. Consequently, the χ-Fe$_5$C$_2$-(100)-2.25 has the largest CO adsorption energy, followed by θ-Fe$_3$C-(010)-2.33, ε-Fe$_2$C-(1$\bar{2}$1)-2.00 and Fe$_4$C-(100)-3.00, respectively. The C-O bond activation degree is almost the same on χ-Fe$_5$C$_2$-(100)-2.25, θ-Fe$_3$C-(010)-2.33 and ε-Fe$_2$C-(1$\bar{2}$1)-2.00, while Fe$_4$C-(100)-3.00 has the weakest ability to activate the C-O bond. For each surface, when the coverage of CO increases from 1/16 to 1/2 ML, both CO adsorption energies and C-O bond elongation decrease.

On the basis of the computed binding energy of surface carbon atoms, it is evident that less negatively charged surface carbon atoms have weaker bonding to the surface. Since CH$_4$ formation energy exhibits a linear relationship with the charge of surface carbon atom,[24] one can expect that CH$_4$ formation is most favored thermodynamically on χ-Fe$_5$C$_2$-(100)-2.25, followed by θ-Fe$_3$C-(010)-2.33, ε-Fe$_2$C-(1$\bar{2}$1)-2.00 and Fe$_4$C-(100)-3.00.

Conclusion

In this work, we employed DFT calculations and *ab initio* atomistic thermodynamics to investigate the surface structure and stability of the low and high Miller index surfaces of the ε-Fe$_2$C, χ-Fe$_5$C$_2$, θ-Fe$_3$C and Fe$_4$C phases as well as their crystal shapes. The goal is to understand the effects of the FTS conditions on the structure and stability of iron carbides as FTS catalysts as well as their differences in surface properties.

The chemical–physical environment around iron based FTS catalysts under working conditions is described from thermodynamic aspect. With different carbon containing gas environments under real FTS operating conditions, it is found that the carburization ability depends mainly on the carbon content of the gas environments, i.e. the higher carbon content of C containing gases, the higher the carburization ability. It is also found that higher temperature, lower pressure and higher H$_2$/CO ratio can suppress carburization ability and retard carbon deposition.

The crystal shapes of ε-Fe$_2$C, χ-Fe$_5$C$_2$, θ-Fe$_3$C and Fe$_4$C have been determined by using the standard Wulff construction on the basis of the calculated surface free energies. Under different pretreatment conditions, the surface morphologies of ε-Fe$_2$C, χ-Fe$_5$C$_2$ and θ-Fe$_3$C are different in termination and proportion of each facet area and the most stable non-stoichiometric termination changes from carbon-poor to carbon-rich (varying surface Fe/C ratio) upon the increase in $\Delta\mu_C$. The surface structure and composition of the most stable terminations have similar atom arrangement on the surface layer and the catalytic activities of these facets have been investigated. It is found that lower work function of the surface leads to larger adsorption energy of CO. Less negatively charged surface carbon atoms have weaker binding energy on the surface. Among these four carbides, χ-Fe$_5$C$_2$-(100)-2.25 is most favored for CO adsorption and CH$_4$ formation, followed by θ-Fe$_3$C-(010)-2.33, ε-Fe$_2$C-(1$\bar{2}$1)-2.00 and Fe$_4$C-(100)-3.00, respectively. The activation degree of C-O bonds are almost same on χ-Fe$_5$C$_2$-(100)-2.25, θ-Fe$_3$C-(010)-2.33 and ε-Fe$_2$C-(1$\bar{2}$1)-2.00, while Fe$_4$C-(100)-3.00 has the weakest ability for activating the C-O bond.

Appendix

Configuration modeling of fractional site occupancy in ε-Fe$_2$C

Figure A1 shows the unit cell structure of ε-Fe$_2$C, and this unit cell has the occupancy of carbon atoms of only 0.5

Table 10   Calculated adsorption energies (E_{ads}, eV) per CO, bond lengths (d, Å), net charges (q, e) and CO stretching frequencies (ν, cm^{-1}) on carbide surfaces, as well as surface properties of each facet

Surface	χ-Fe$_5$C$_2$ (100)-2.25	θ-Fe$_3$C (010)-2.33	ε-Fe$_2$C (1$\bar{2}$1)-2.00	Fe$_4$C (100)-3.00
Work function/eV	3.847	3.991	4.009	4.713
q^*(C$_s$)	-1.062	-1.090	-1.098	-1.119
E_{ads}(C$_s$)	-8.442	-8.459	-8.580	-9.126
E_{ads}(1/16 ML)†	-1.464	-1.442	-1.345	-0.946
E_{ads}(1/16 ML)‡	-1.741	-1.719	-1.667	-1.263
d(C-O) (1/16 ML)	1.171	1.172	1.170	1.164
q^* (CO) (1/16 ML)	-0.273	-0.277	-0.254	-0.226
ν_{CO} (1/16 ML)	1937	1933	1939	1990
E_{ads}(1/2 ML)†	-1.142	-1.135	-1.133	-0.317
E_{ads}(1/2 ML)‡	-1.482	-1.483	-1.489	-0.758
d(C-O) (1/2 ML)	1.162	1.163	1.163	1.159
q^*(CO) (1/2 ML)	-0.179	-0.176	-0.178	-0.091
ν_{CO} (1/2 ML)	1976	1971	1973	2017

*From Bader charge analysis.
†RPBE energies.
‡PBE energies.

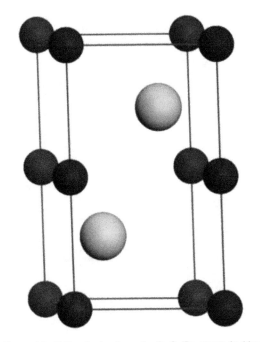

instead 1.0. This means that Fe atoms form a hexagonal close packed array and half of the octahedral interstitial sites are occupied by carbon atoms in a random way. Similar site-occupancy disorder structures are also found in β-Mo_2C[70] and Fe_2N.[71] According to this bulk structure, the Fe/C ratio is 1 to 1. In order to keep the 2 to 1 stoichiometry, the usually used practice is to delete half of carbon atoms from the bulk structure. However, it will generate several configurations and the number of the possible configurations increases dramatically with the supercell size. Since there are no systematic investigations into the bulk structure of ε-Fe_2C known, several structures of ε-Fe_2C were used in previous studies. Jack reported a ε-Fe_2C structure by deleting carbon atoms on the vertices of the unit cell (Fig. A1).[72] Jang et al.[17] and Fang et al.[18] calculated the ε-Fe_2C bulk structure with the space group $P6_322$, which is different from the experimental data ($P63/mmc$).[37] In this work, we used the unit cell of ε-Fe_2C with the space group $P63/mmc$ to generate the supercell by deleting half of the carbon atoms from the supercell. For supercell with different size, we calculated all possible structures in order to give a reasonable configuration of ε-Fe_2C.

Firstly, we generated the $2 \times 2 \times 1$ and $2 \times 2 \times 2$ supercells by deleting half of the carbon atoms from these

Figure A1 Unit cell structure of ε-Fe_2C (Fe atoms in blue balls, C atoms in black balls)

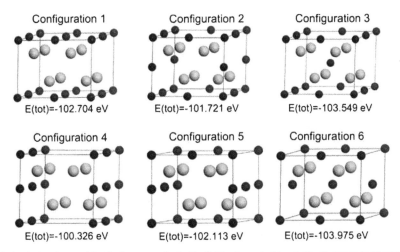

Configuration 1 — E(tot)=-102.704 eV
Configuration 2 — E(tot)=-101.721 eV
Configuration 3 — E(tot)=-103.549 eV
Configuration 4 — E(tot)=-100.326 eV
Configuration 5 — E(tot)=-102.113 eV
Configuration 6 — E(tot)=-103.975 eV

Figure A2 Optimized structures of $2 \times 2 \times 1$ supercell of ε-Fe_2C (Fe atoms in blue balls, C atoms in black balls)

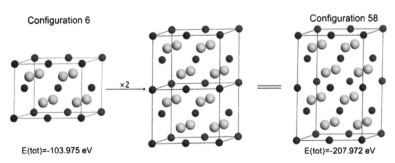

Configuration 6 — E(tot)=-103.975 eV   ×2   Configuration 58 — E(tot)=-207.972 eV

Figure A3 Optimized structures of $2 \times 2 \times 2$ supercell of ε-Fe_2C (Fe atoms are shown by blue balls, C atoms are shown by black balls)

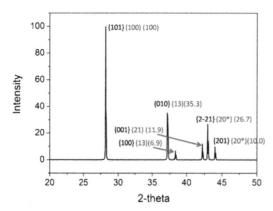

Figure A4 X-ray diffraction of ε-Fe₂C (detected intensities in red and calculated intensities in blue)

two structures. During this process, the site-occupancy disorder program was used to obtain all the supercell structures.[73] By taking the advantage of isometric transformation, site-occupancy disorder excludes the equivalent configurations and reduces the configurations only to the independent ones (Table A1). All the energies were obtained using the method described in the section on 'Structure calculation'. The occurrence probability of each configuration at temperature T can be obtained from equation (9)

$$P_n = \frac{1}{Z}\exp(-E_n/k_BT) \qquad (9)$$

where $k_B = 8.6173 \times 10^{-5}$ eV K^{-1}, E_n is the energy of that configuration and

$$Z = \sum_{n=1}^{N}\exp(-E_n/k_BT) \qquad (10)$$

The $2 \times 2 \times 1$ supercell has six independent configurations; the optimized structures and the corresponding total energies are shown in Fig. A2. Among these six structures, configuration 6 is most stable and its occurrence probability is 100% from 0 to 1000 K. After optimizing all 128 configurations of the $2 \times 2 \times 2$ supercell, we found configuration 58 to be most stable (Fig. A3). Its occurrence probability is also 100% from 0 to 1000 K. In addition, configuration 58 is two times of configuration 6, therefore they are the same structure. The simulated XRD spectrum is presented in Fig. A4, the relative intensity of the characteristic peaks agrees well with the experimental data.[37,74] This rationalizes our calculated structure of ε-Fe₂C.

Table A1 Number of independent configurations for a series of supercell of ε-Fe₂C

Supercell	N₀*	N†	M‡
2×2×1	96	70	6
2×2×2	192	12,870	122

*Number of symmetry operations.
†Total number of configurations.
‡Number of independent configurations.

Conflicts of interest

The authors declare no conflicts of interest.

Acknowledgements

This work was supported by National Natural Science Foundation of China (Grant No. 21273261), National Basic Research Program of China (Grant No. 2011CB201406), and Chinese Academy of Science and Synfuels China Co., Ltd.

Supporting Information Available: The atomic thermodynamic method used for calculating surface free energies and the calculated surface free energies are available. This material is available free of charge via the Internet at http://www.publicationethics.org.

References

1. F. Fischer and H. Tropsch: Brennstoff-Chem., 1923, **4**, 276–285.
2. F. H. Herbstein, J. Smuts and J. N. van Niekerk: Anal. Chem., 1960, **32**, 20–24.
3. E. de Smit, F. Xinquini, A. M. Beale, O. V. Safonova, W. van Beek, P. Sautet and B. M. Weckhuysen: J. Am. Chem. Soc., 2010, **132**, 14928–14941.
4. E. de Smit, I. Swart, J. F. Creemer, G. H. Hoveling, M. K. Gilles, T. Tyliszczak, P. J. Kooyman, H. W. Zandbergen, C. Morin, B. M. Weckhuysen and F. M. F. de Groot: Nature, 2008, **456**, 222–226.
5. A. K. Datye, Y. Jin, L. Mansker, R. T. Motjope, T. H. Dlamini and N. J. Coville: Stud. Surf. Sci. Catal., 2000, **130**, 1139–1144.
6. A. Königer, C. Hammerl, M. Zeitler and B. Rauschenbach: Phys. Rev. B, 1997, **55B**, 8143–8147.
7. D. L. Williamson, K. Nakazawa and G. A. Krauss: Metall. Trans. A, 1979, **10A**, 1351–1363.
8. M. Manes, A. D. Damick, M. Mentser, E. M. Cohn and L. J. E. Hofer: J. Am. Chem. Soc., 1952, **74**, 6207–6209.
9. E. M. Cohn and L. J. E. Hofer: J. Chem. Phys., 1953, **21**, 354–359.
10. J. F. Shultz, W. K. Hall, T. A. Dubs and R. B. Anderson: J. Am. Chem. Soc., 1955, **78**, 282–285.
11. L. J. E. Hofer and E. M. Cohn: Nature 1951, **167**, 977–978.
12. S. Nagakura: J. Phys. Soc. Jpn, 1959, **14**, 186–195.
13. H. Merkel and F. Weinrotter: Brennstoff-Chem., 1951, **32**, 289–297.
14. W. C. Chiou Jr and E. A. Carter: Surf. Sci., 2003, **530**, 88–100.
15. D. C. Sorescu: J. Phys. Chem. C, 2009, **113C**, 9256–9274.
16. Z. Q. Lv, S. H. Sun, P. Jiang, B. Z. Wang and W. T. Fu: Comput. Mater. Sci., 2008, **42**, 692–697.
17. J. H. Jang, I. G. Kim and H. K. D. H. Bhadeshia: Scr. Mater., 2010, **63**, 121–123.
18. C. M. Fang, M. A. van Huis and H. W. Zandbergen: Scr. Mater., 2010, **63**, 418–421.
19. H. I. Faraoun, Y. D. Zhang, C. Esling and H. Aourag: J. Appl. Phys., 2006, **99**, 093508–093515.
20. C. Jiang, S. G. Srinivasan, A. Caro and S. A. Maloy: J. Appl. Phys., 2008, **103**, 043502–043509.
21. B. Hallstedt, D. Djurovic, J. von Appen, R. Dronskowski, A. Dick, F. Körmann, T. Hickel and J. Neugebauer: CALPHAD: Comput. Coupling Phase Diagr. Thermochem., 2010, **34**, 129–133.
22. C. K. Ande and M. H. F. Sluiter: Metall. Mater. Trans. A, 2012, **43A**, 4436–4444.
23. L. J. Deng, C. F. Huo, X. W. Liu, X. H. Zhao, Y. W. Li, J. G. Wang and H. J. Jiao: J. Phys. Chem. C, 2010, **114C**, 21585–21592.
24. C. F. Huo, Y. W. Li, J. G. Wang and H. J. Jiao: J. Am. Chem. Soc., 2009, **131**, 14713–14721.
25. X. Y. Liao, D. B. Cao, S. G. Wang, Z. Y. Ma, Y. W. Li, J. G. Wang and H. J. Jiao: J. Mol. Catal. A: Chem., 2007, **269**, 169–178.
26. X. Y. Liao, S. G. Wang, Z. Y. Ma, Y. W. Li, J. G. Wang and H. J. Jiao: J. Mol. Catal. A: Chem., 2008, **292**, 14–20.
27. B. Q.Wei, M. Shima, R. Pati, S. K. Nayak, D. J. Singh, R. Ma, Y. Li, Y. Bando, S. Nasu and P. M. Ajayan: Small, 2006, **6**, 804–809.
28. C. M. Deng, C. F. Huo, L. L. Bao, X. R. Shi, Y. W. Li, J. G. Wang and H. J. Jiao: Chem. Phys. Lett., 2007, **448**, 83–87.
29. C. M. Deng, C. F. Huo, L. L. Bao, G. Feng, Y. W. Li, J. G. Wang and H. J. Jiao: J. Phys. Chem. C, 2008, **112C**, 19018–19029.
30. S. Zhao, X. W. Liu, C. F. Huo, Y. W. Li, J. G. Wang and H. J. Jiao: J. Catal., 2011, **294**, 47–53.
31. G. Kresse and J. Furthmüller: Comput. Mater. Sci. 1996, **6**, 15–50.
32. G. Kresse and J. Furthmüller: J. Phys. Rev. B, 1996, **54B**, 11169–11186.

33. J. P. Perdew, K. Burke and M. Ernzerhof: *Phys. Rev. Lett.* 1996, **77**, 3865–3868.

34. P. E. Blochl: *Phys. Rev. B*, 1994, **50B**, 17953–17979.

35. G. Kresse: *Phys. Rev. B*, 1999, **59B**, 1758–1775.

36. M. Methfessel and A. T. Paxton: *Phys. Rev. B*, 1989, **40B**, 3616–3621.

37. G. H. Barton and B. Gale: *Acta Cryst.*, 1964, **17**, 1460–1462.

38. I. G. Wood, L. Vocadlo, K. S. Knight, D. P. Dobson, W. G. Marshall, G. D. Price and J. Brodholt: *J. Appl. Cryst.*, 2004, **37**, 82–90.

39. A. V. dos Santos, M. I. da Costa and C. A. Kuhnen: *J. Magn. Magn. Mater.*, 1997, **166**, 223–230.

40. E. L. P. Y. Biancá, J. Desimoni and N. E. Christensen: *Phys. B*, 2004, **354B**, 341–344.

41. L. J. E. Hofer and E. M. Cohn: *J. Am. Chem. Soc.*, 1958, **81**, 1576–1582.

42. K. Reuter and M. Scheffler: *Phys. Rev. B*, 2001, **65B**, 035406–035416.

43. K. Reuter and M. Scheffler: *Phys. Rev. B*, 2003, **68B**, 045407–045417.

44. B. Hammer, L. B. Hansen and J. K. Norskov: *Phys. Rev. B*, 1999, **59B**, 7413–7421.

45. W. Tang, E. Sanville and G. Henkelman: *J. Phys.: Condens. Matter*, 2009, **21**, 084204–084210.

46. E. Sanville, S. D. Kenny, R. Smith and G. Henkelman: *J. Compos. Chem.*, 2007, **28**, 899–908.

47. G. Henkelman, A. Arnaldsson and H. Jónsson: *Comput. Mater. Sci.*, 2006, **36**, 254–360.

48. O. Karabelchtchikova: 'Fundamentals of mass transfer in gas carburizing', PhD thesis, Worcester Polytechnic Institute, Worcester, MA, USA, 2008.

49. H. M. T. Galvis, J. H. Bitter, T. Davidian, M. Ruitenbeek, A. L. Dugulan and K. P. de Jong: *J. Am. Chem. Soc.*, 2012, **134**, 16207–16215.

50. M. D. Shroff, D. S. Kalakkad, K. E. Coulter, S. D. Kohler, M. S. Harrington, N. B. Jackson, A. G. Sault and A. K. Datye: *J. Catal.*, 1995, **156**, 185–207.

51. R. J. O'Brien, L. Xu, R. L. Spicer and B. H. Davis: *Eng. Fuels*, 1996, **10**, 921–926.

52. P. Sautet and F. Cinquini: *Chem. Cat. Chem.*, 2010, **2**, 636–639.

53. R. G. Olsson and E. T. Turkdogan: *Metall. Trans.*, 1974, **5**, 21–26.

54. R. Asano, Y. Sasaki and K. Ishii: *ISIJ Int.*, 2002, **42**, 121–126.

55. J. Q. Zhang, A. Schneider and G. Inden: *Corros. Sci.*, 2003, **45**, 281–299.

56. M. Watanabe and P. Wissmann: *Catal. Lett.*, 1991, **1**, 15–26.

57. C. M. Chun, T. A. Ramamarayanan and J. D. Mumford: *Mater. Corros.*, 1999, **50**, 634–639.

58. J. B. Butt: *Catal. Lett.*, 1990, **7**, 83–106.

59. H. J. Grabke: *Mater. Corros.*, 1998, **49**, 303–308.

60. A. C. J. Koeken, H. M. T. Galvis, T. Davidian, M. Ruitenbeek and K. P. de Jong: *Angew. Chem. Ind. Ed.*, 2012, **51**, 7190–7193.

61. S. Ando and H. Kimura: *Scr. Metall.*, 1989, **23**, 1767–1772.

62. L. P. Zhou: 'Kinetic study of the Fischer-Tropsch synthesis over an industrial iron-based catalyst', PhD thesis, Institute of Coal Chemistry, Chinese Academy of Sciences, China, 2011.

63. E. de Smit, M. M. van Schooneveld, F. Cinquini, H. Bluhm, P. Sautet, F. M. F. de Groot and B. M. Weckhuysen: *Angew. Chem. Int. Ed.*, 2011, **50**, 1584–1588.

64. E. J. Fasiska and G. A. Jeffrey: *Acta. Crystallogr.*, 1965, **19**, 463–471.

65. D. Fruchart, P. Chaudouet, R. Fruchart, A. Rouault and J. P. Senateur: *J. Solid. State. Chem.*, 1984, **51**, 246–252.

66. K. Momma and F. Izumi: *J. Appl. Crystallogr.*, 2011, **44**, 1272–1276.

67. S. V. Khare, S. Kodambaka, D. D. Johnson, I. Petrov and J. E. Greene: *Surf. Sci.*, 2003, **522**, 75–83.

68. N. D. Spencer, R. C. Schoonmaker and G. A. Somorjai: *J. Catal.* 1982, **74**, 129–135.

69. V. R. Stamenkovic, B. Fowler, B. S. Mun, G. F. Wang, P. N. Ross, C. A. Lucas and N. M. Markovic: *Science*, 2007, **315**, 493–497.

70. J. Dubois, T. Epicier, C. Esnouf, G. Fantozzi and P. Convert: *Acta Metall.*, 1988, **36**, 1891–1901.

71. D. H. Jack and K. H. Jack: *Mater. Sci. Eng.*, 1973, **11**, 1–27.

72. K. H. Jack: *Acta Cryst.* 1950, **3**, 392–394.

73. R. G. Crespo, S. Hamad, C. R. A. Catlow and N. H. de Leeuw: *J. Phys.: Condens. Matter*, 2007, **19**, 256201–256216.

74. L. J. E. Hofer, E. M. Cohn and W. C. Peebles: *J. Am. Chem. Soc.*, 1949, **71**, 189–195.

75. D. C. Gardner and C. H. Bartholomew: *Ind. Eng. Chem. Prod. Res. Dev.*, 1981, **20**, 80–87.

76. P. L. Walker Jr, J. F. Rakszawski and G. R. Imperial: *J. Phys. Chem.*, 1959, **63**, 140–149.

3

Effect of acidity and texture of micro-, mesoporous, and hybrid micromesoporous materials on the synthesis of paramenthanic diol exhibiting anti-Parkinson activity

A. Torozova[1,2], P. Mäki-Arvela[1], N. D. Shcherban[3], N. Kumar[1], A. Aho[1], M. Stekrova[1], K. Maduna Valkaj[1], P. Sinitsyna[2,3,4], S. M. Filonenko[3], P. S. Yaremov V. G. Ilyin[3], K. P. Volcho[5], N. F. Salakhutdinov[5] and D. Yu. Murzin[1]*

[1]Johan Gadolin Process Chemistry Centre, Åbo Akademi University, 20500 Turku/Åbo, Finland
[2]Tver State Technical University, Tver 170026, Russia
[3]L.V. Pisarzhevskii Institute of Physical Chemistry, National Academy of Sciences, 03028 Kiev, Ukraine
[4]St. Petersburg State Institute of Technology (Technical University), St. Petersburg 190013, Russia
[5]N. N. Vorozhtsov Institute of Organic Chemistry, Russian Academy of Sciences, Novosibirsk 630090, Russia

Abstract Microporous, mesoporous, and new hybrid materials were studied in verbenol oxide isomerization for the synthesis of biologically active substance with anti-Parkinson activity. H-Si-MCM-41, H-Al-MCM-41, H-Al-MCM-48, H-Beta-25, and H-Beta-300 were compared with hybrid materials. The latter with a zeolite-like micro-mesoporous structure were characterized and evaluated

Verbenol oxide   Diol   Ketone   Oxetane

for their catalytic activity for the first time. The approach of dual templating for synthesis of new materials was applied in this work to combine properties of Beta-zeolites and mesoporous cellular foams (MCF). The selectivity to the target product was the highest over microporous mild acidic H-Beta-300 and hybrid ZF-100, with also mild acidity and even absence of strong acid sites. Selectivity at 97 and 99% of conversion was 61 and 59% for H-Beta-300 and hybrid ZF-100, respectively.

Keywords Zeolites, Hybrid micro-mesoporous materials, MCM-41, MCM-48, Beta zeolites, Verbenol oxide isomerization

Introduction

Zeolites are crystalline microporous materials that have different by nature active adsorption, ion exchange, acid and catalytic sites. Because of these properties, zeolites are widely used in industry – in the process of catalysis, adsorption of small molecules, separation of multi-component mixtures. However, microporous nature of these materials limits their use in the processes of catalytic transformation of bulk organic molecules and viscous liquids, which is necessary in fine chemicals, pharmaceutical industry, etc. It was expected that this drawback can be avoided after the synthesis of mesoporous molecular sieves (MMS), which are characterized by uniform mesoporosity, developed surface, and spatial ordering and periodicity in the nanometer range distances. However, the amorphous state

of a substance of MMS walls leads to poor hydrolytic stability and acidity, which significantly limits the use of mesoporous materials in catalysis.

In the recent years, creation of mesoporous or hierarchical zeolites containing at least two levels of porosity is attracting much attention.[1-6] Mesoporous zeolites exhibit an increased accessibility to reagents molecules, which leads to an increased catalytic activity and/or productivity in the reactions which are hindered in zeolites with diffusive or steric limitations or do not occur at all. In addition, secondary porosity makes possible the introduction of additional active phases and functionalization with organic groups. The presence of secondary porosity provides mesoporous zeolite with new functions and, consequently, opens new application areas.

Mesoporosity in zeolites can be created as intracrystalline cavities in zeolite single crystals or as intercrystallite pores in aggregates of zeolite nanoparticles.[6]

Many approaches were developed for generation of secondary porosity in zeolites. Various methods of synthesis of mesoporous zeolites can be divided into six categories: removal of the part of the framework atoms; template synthesis involving surfactants; hard template synthesis; zeolitization of MMS aluminosilicate and approaches using organosilanes as building units; mixed approaches.[5]

There are data in the literature about the formation of micro-mesoporous aluminosilicate during zeolitization in non-aqueous solvents of MMS and xerogels-precursors of zeolites obtained by vacuum drying of sol-precursors of zeolites,[7] or during thermal steam treatment of carbon–silica composites containing a molecular template for the formation of zeolite.[8] Mesoporous zeolite ZSM-5 was obtained using thermal steam treatment of xerogel achieved after drying of aluminosilicate gel containing the molecular template (TPA$^+$) and a template for mesopore formation (Pluronic F-127).[9] The use of a similar approach (quasi-solid phase conversion) to form hierarchical zeolites with oriented nanocrystals was described.[10]

Micro-mesoporous materials can also be formed by the 'core-shell' approach, when zeolite nanoparticles (Y-, ZSM-5-type) can be used as a core and mesoporous aluminosilicate as a shell.[11,12] Nanosized micro-mesoporous composites containing nanoparticles with a mean radius of about 90 nm and 140 nm were prepared via a hydrothermal reaction of a zeolite Beta seeds solution and a mesoporous precursor solution MCM-41[13] and MCM-48,[14] respectively. Composite micro-mesoporous materials based on Beta-type zeolite nanoparticles were obtained without adding structure-directing agents.[15]

Method of synthesis of hierarchical mesoporous zeolite Y through design of cationic surfactant cetyltrimethylammonium bromide (CTAB) micelle with a co-solvent tertbutyl alcohol (TBA) and the 1,3,5-trimethylbenzene (TMB) additive was reported.[16] Formation of hydrothermally stable aluminosilicate foams and SBA-15 from zeolite Y seeds was demonstrated.[17]

Ordered cubic Ia3d and hexagonal p6mm mesoporous aluminosilicates were synthesized via hydrothermal treatment of the previously obtained zeolite BEA precursor with CTAB surfactant through a two-step crystallization procedure.[18]

The approach of dual templating used in this paper as such allows adjusting the structure and sorption properties of the samples. An attempt was made to obtain hybrid materials that combine properties of Beta zeolite, an important industrial catalyst, and mesoporous cellular foams (MCF), which unlike other mesoporous silica (MCM-41, SBA-15) have large pores (more than 20 nm).

Catalytic isomerization of verbenol oxide was used as a model reaction (Scheme 1).

The main product is (1R, 2R, 6S)-3-methyl-6-(prop-1-en-2-yl)cyclohex-3-ene-1,2-diol, which exhibits anti-Parkinson activity.[19] Verbenol oxide is an intermediate, which can be produced via epoxidation of verbenone followed by hydrogenation. The raw material, (-)-verbenone is a naturally occurring monoterpenoid extract of *Rosmarinus officinalis* and rosemary essential oils.[20] It is also present in *Pinus Sylvestris*.[21] Verbenol oxide isomerization has been very scarcely investigated. It was demonstrated for the first time by medicinal chemists using a large excess of K10 clay to promote the reaction.[19] Catalytic isomerization of verbenol oxide was demonstrated for the first time over Fe-Beta-300 and Ce-MCM-41.[22] In addition, H-USY and ZSM-5 type zeolites with varying SiO_2 to Al_2O_3 ratio were systematically investigated as catalysts in verbenol oxide isomerization.[23] The results showed that about 47–50% selectivity toward diol was obtained with these catalysts. The conversion over medium pore size zeolite ZSM-5 was, however, only 78% in 3 h owing to pore blockage.

Thus, the purpose of the work to explore template synthesis of micro-mesoporous materials from primary products of zeolite-formation (sols-precursors) Beta-type and the MCF, determine their structure and properties as well as establish catalytic activity in verbenol oxide isomerization. It is important to understand if acidity or mesoporosity plays a key role in this reaction, and to compare the performances of micro-mesoporous hybrid materials, mesoporous ordered MCM-41 and MCM-48 and microporous Beta materials. Zeolite Beta with 12-membered rings has channel dimensions of 0.77×0.66 nm and 0.56×0.56 nm.[24–26] Mesoporous MCM-41 and MCM-48 have cavity sizes of 2.0–3.0 nm[27] and 3.1–3.3 nm,[28] respectively. The purpose of selecting these catalysts was to study influence of the structure, porosity, and acidity.

Experimental

Catalyst preparation

Ammonium forms of Beta-25 and Beta-300 (in which the number denotes SiO_2 to Al_2O_3 ratio) were purchased from

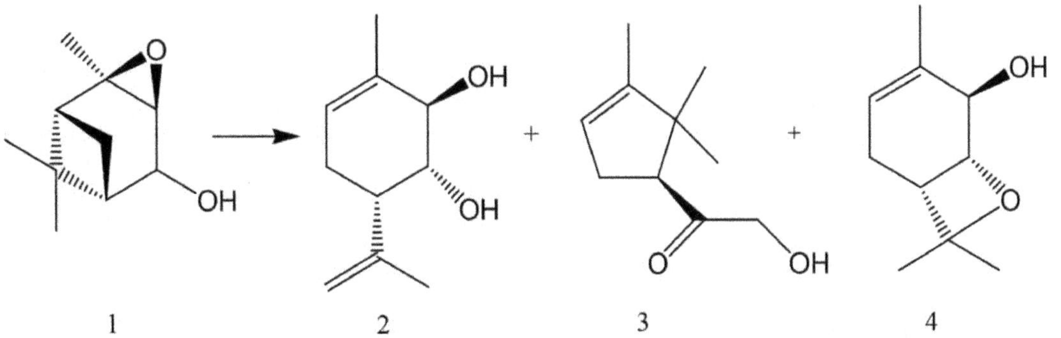

Scheme 1 Isomerization of verbenol oxide (1) and production of *trans*-diol (2), cyclopentylhydroxyketone (3), and oxetane (4)

Zeolyst International. These were transformed to proton forms in an oven at 450°C with a step calcination method. Na-MCM-41 mesoporous material was synthesized according to a slightly modified method published in.[29,30] Na-MCM-41 was ion-exchanged with ammonium nitrate and subsequently H-MCM-41 was obtained via calcination of NH_4-MCM-41 at 450°C. Al-MCM-48 was synthesized using a modified method of ref.[31,32] Cetyltrimethylammonium chloride (CTMACl) and NaOH were added to distilled water. Thereafter aluminium isopropoxide (AIP) was added into the solution followed by stirring for 15 min to hydrolyze AIP. Tetraethylorthosilicate was added finally to the solution at room temperature for 1 h to facilitate hydrolysis of TEOS. The synthesis was performed in an autoclave at 100°C for 75 h. The obtained Na-MCM-48 was ion-exchanged with ammonium chloride for 48 h, washed with distilled water, dried, and calcined.

Zeolite-like mesocellular foams were obtained via dual template synthesis from sol-precursor of Beta zeolite in the presence of micellar template which is used for synthesis of MCF. The gel of Beta was aged at 140°C for 20 h under static conditions, yielding the zeolite Beta precursor. Then obtained zeolite Beta precursor was added to the solution which was necessary for synthesis of MCF (P123 was dissolved in the solution of concentrated HCl–water solution and distilled water at 40°C, then 1,3,5-trimethylbenzene (TMB) was added). Then the reagent mixture was transferred into a teflon-lined stainless steel autoclave and aged at 100°C (ZF-100) or 120°C (ZF-120) for 24 h under static conditions. The sample ZF-100-3d was heated at 100°C for 72 h. After cooling down to room temperature, the products were filtered, washed with deionized water repeatedly, and dried in air at 100°C. The template was eliminated by calcination in air at 550°C (heating rate 2°C min^{-1}) for 5 h. The proton forms of the samples were obtained by ion-exchange of NH_4NO_3, followed by calcination at 550°C for 4 h.

Characterization

Acidity measurements

The acidity of the catalysts was determined by pyridine adsorption/desorption method using FTIR (ATI Mattson). Pyridine (Sigma Aldrich, ≥99.5%) was used as a probe molecule. Thin pellets of 10–20 mg were pressed from the catalysts. Before the measurement, the pellets were heated up to 450°C for 1 h. Pyridine was adsorbed at 100°C for 30 min. The desorption was performed at 250, 350, and 450°C, respectively, in order to determine weak, medium, and strong acid sites. Brønsted and Lewis acid sites were calculated from the integrated peaks of 1545 and 1455 cm^{-1}, respectively. The quantification of the acid sites was made by applying the extinction factor published by Emeis.[33]

The acidic properties of the synthesized samples were studied with temperature-programmed desorption of ammonia (TPDA) with a thermal conductivity detector. For TPDA, the following procedure was used:[34] activation of 0.2 g of the sample in a glass reactor for 1 h in a flow of helium at 550°C, followed by cooling to 100°C, saturation with ammonia, venting of the residual NH_3 with helium. Thereafter temperature-programmed desorption of NH_3 in the temperature range of 100–650°C at a heating rate of 15°C min^{-1} was performed; the amount of adsorbed ammonia was determined by titrating with 2.5×10^{-3} M solution of hydrochloric acid.

Nitrogen adsorption

Nitrogen adsorption measurements were performed with Sorptometer 1900 (Carlo Erba Instrument). The samples were evacuated before measurement at 150°C for 3 h. The specific surface areas were calculated with Dubinin's method. For the hybrid materials, the total surface area was estimated by Brunauer–Emmett–Teller equation,[35] the mesopore size was determined with the desorption branch of the isotherm, using the method of Barrett–Joyner–Halenda,[36] the micropore size was calculated by the equation of Saito–Foley.[37]

XRD

XRD measurements were performed with Philips X'Pert pro MPD operating with monochromated CuKα radiation at 40 kV/50 mM. The primary X-ray beam was collimated with 0.25° divergency slit and a 20 mm mask. The measurements were performed in a range of 1.15°–90° using a scanning speed of 0.04°4/2 s. The diffractograms were analyzed with Philips HighScore and MAUD programs[38] and compared with those retrieved from the International Zeolite Association website.[39]

The phase composition and spatial organization of the hybrid materials were analyzed using X-ray diffractometer (Bruker AXS D8 Advance) with CuKα-radiation in the 2θ range of 0.3–60°. The degree of crystallinity was estimated by changes in the ratio of the intensities of the characteristic reflections at $2\theta = 7.7°$; 21.4°; 22.4° of the investigated samples and BEA. For recording IR spectra the samples were tabletted with KBr (1:100, 400–4000 cm^{-1}), the Fourier spectrometer (Perkin-Elmer Spectrum One) was also used.

SEM and TEM

Scanning and transmission electron microscopies using Zeiss Leo Gemini 1530 and JEM 1400 plus, respectively, were applied to study the morphology of the catalysts. The acceleration voltage of 120 kV and the resolution of 0.98 nm for Quemsa II MPix bottom mounted digital camera were used for transmission electron microscopy.

Micrographs of the synthesized samples were obtained with a scanning electron microscope (SEM) (JEOL JSM-6610LV).

Liquid phase isomerization of verbenol oxide

Verbenol oxide (synthesized at Novosibirsk Institute of Organic Chemistry with a procedure published in ref.[19] NMR purity of 87.5%) was isomerized in a four-necked batch reactor equipped with a condenser and a motor stirrer. The total liquid phase volume was 100 mL and the initial VO concentration was 0.016 mol/L. Isomerization was performed using dimethylacetamide (anhydrous, ≥99.9%) as a solvent under argon atmosphere. In order to avoid internal and external mass transfer limitations, the catalyst particle size below 90 μm and efficient stirring, 390 rpm, were, respectively, applied. The catalyst was pre-dried in the reactor before the experiment by heating it under argon flow to 250°C for 30 min.

The samples were taken from the reactor and analyzed with a GC equipped with a FID detector and a DB Petro column (100 m, 250 μm internal diameter, 0.50 μm film thickness) using the following temperature programme: injection at 100°C, followed by a ramp of 3°C min^{-1} until 150°C, where it was maintained for 3 min, then a ramp of 2°C min^{-1} till 195°C (maintained for 10 min). The products were confirmed by GC-MS.

Results and discussion

Catalysts characterization results

X-ray powder diffraction

The diffraction patterns of the samples obtained by template method from BEA sol-precursors are shown in Figs. 1a and b. Increasing of the temperature of hydrothermal treatment of bitemplate reaction mixture of BEA and micellar template solution from 100 to 120°C leads to generating of the structure (sample ZF-120) with an increased degree of crystallinity (about 0.4, Fig. 1a). Following increase of the temperature and duration of hydrothermal treatment did not result in more crystalline samples, which obviously can be explained by unfavorable pH of the reaction mixture. For example heating the bi-template mixture for 72 h (sample

ZF-100-3d) instead of 24 h (sample ZF-100) did not result in higher crystallinity of the final product.

According to XRD data in the low-angle region, the obtained samples have disordered mesostructure, which is possibly stipulated by the formation of the framework from relatively large particles (Fig. 1b). The initial sol-precursors of beta-zeolite obtained via hydrothermal treatment at 140°C for 20 h were characterized with X-ray powder diffractometer. The synthesized material exhibited an amorphous phase. Before recording XRD-patterns, the sol-precursors were dried at room temperature, washed with water, and dried again. XRD patterns of Beta zeolite,[40] MCM-41,[30] and MCM-48[28,41] resembled those published in the literature.

Scanning electron microscopy

Scanning electron micrographs (Figures S1a and b) of the synthesized catalysts indicate an absence of the zeolite single phase, or probably a very small crystal size. The synthesized samples have the spongy structure typical for MCF. The ratio of silicon to aluminium in the obtained structures is equal to 48–49 (Figure S1c and d).

Transmission electron microscopy

The transmission electron micrographs (Fig. 2) of samples with the low crystallinity (ZF-100) have a structure similar to MCF (there are pores larger than 20 nm) and particles of zeolite structure about 200 nm. Increasing of the degree of crystallization (sample ZF-120) leads to the formation of slightly larger zeolite seeds (up to 300 nm) (Fig. 2). Transmission electron micrographs of MCM-type materials and H-Beta-25 zeolite clearly show their uniform channel structures (Figures S3a-d).

IR spectroscopy

The IR spectra of sol-precursors possess absorption band at 570 cm^{-1}, which can be attributed to the asymmetric stretching vibration of aluminosiloxane bonds of five-membered rings of $Si(Al)O_{4/2}$ tetrahedra of zeolite. Thus, the presence of this band in the IR spectra of the framework indicates that sol-precursors of Beta contain primary elements of the structure of zeolite, which under adding of the micellar template can form mesostructure. In the IR spectrum (Fig. 3) of the sample ZF-120 (degree of crystallinity is ~0.4), there are absorption bands at 620, 575, and 520 cm^{-1}, which are characteristic for fully crystallized Beta owing to

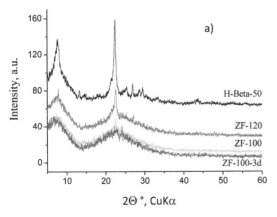

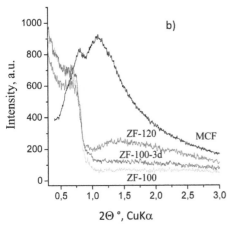

Figure 1 a High-angle XRD patterns and b small-angle XRD patterns of calcined samples obtained from sol-precursors of zeolite H-Beta-50

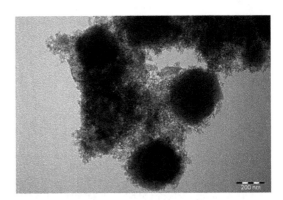

Figure 2 Transmission electron micrograph of ZF-100

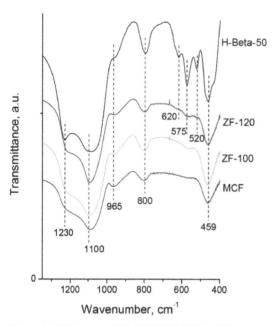

Figure 3 FTIR spectra of ZF-100, ZF-120, MCF, and H-Beta-50

the presence of 5- and 6-membered rings of $Si(Al)O_{4/2}$ tetrahedra in its structure. Absorption band at $1230\,cm^{-1}$ corresponds to asymmetric vibrations of siloxane bonds on the surface of the pore walls. The intensity of this band may be owing to the presence of certain siloxane bridges that have the same length of the bond across the whole framework. The absorption at $1100\,cm^{-1}$ is a superposition of different absorption bands of silicates, among which the largest contribution belongs to asymmetric stretching vibrations of the bonds $Si(Al)-O$ in tetrahedra. All IR spectra (Fig. 3) contain a broad absorption band at $1100\,cm^{-1}$ and a band at $1230\,cm^{-1}$, and the intensity of the latest one correlates with the degree of crystallinity of the synthesized samples. The presence of the absorption band at $1230\,cm^{-1}$ indicates that the structure of the obtained samples is formed by mainly siloxane bonds, close to bonds in zeolite Beta.

Nitrogen ad(de)sorption

The adsorption isotherms of zeolite-containing mesoporous foams and the isotherms of Beta and MCF are shown in Fig. 4. The synthesized mesoporous foams on the basis of the products of partial crystallization are predominantly mesoporous materials ($V_{meso} \sim 1.1\,cm^3\,g^{-1}$, $S_{meso} = 150-330\,m^2\,g^{-1}$), which also include the micropores with a lower volume (0.13 and $0.15\,cm^3\,g^{-1}$ for ZF-100, and ZF-120, respectively). The detailed characterization for these new materials is given in Table 1. The total pore volume for samples ZF-100 and ZF-120 ($\sim 1.5\,cm^3\,g^{-1}$) is smaller compared to a typical mesoporous sample MCF ($\sim 1.7\,cm^3\,g^{-1}$, Table 1), this can be related to a decrease of the spatial ordering degree of mesostructure ZF-100 and ZF-120 as a result of their self-assembly from particles of larger size than those for MCF. This effect is pronounced less with the temperature increase of the hydrothermal treatment (sample ZF-120). The obtained hybrid materials are characterized with a slightly higher specific surface area (in particular for ZF-120 with S_{BET} of $690\,m^2\,g^{-1}$) than that for MCF ($\sim 500\,m^2\,g^{-1}$, Table 1) because of the appearance of inclusions of Beta zeolite phase into the walls of cellular foams. Accordingly, higher micropore volume is obtained. Hybrid material ZF-120 exhibited 1.4 fold higher specific surface area than its counterpart ZF-100, aged at a lower temperature (Table 1).

Formation of mesoporous structure from the products of partial crystallization was observed because of the presence of micellar template Pluronic P123. Based on mesopore size distribution, ZF-100, ZF-100-3d, and ZF-120 have relatively homogeneous mesoporosity. The average mesopore size is between 22 and 68 nm. It can be suggested that self-organization of mesoporous samples from sol-precursor of zeolite in the presence of the micellar template and 1,3,5-trimethylbenzene is accompanied by an increase of the mesopore size and decrease of their geometric uniformity. The complexity of spatially ordered AlSi-framework formation around the micellar structures of P123 from the particles of the sol-precursor of zeolite is obvious, considering their large pores and possible heterogeneity.

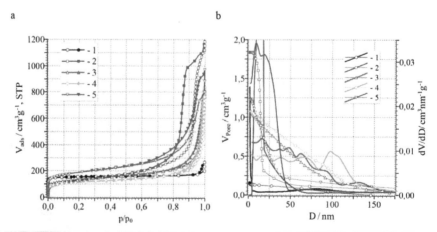

Figure 4 *a* The isotherms of nitrogen ad(de)sorption (77 K), *b* mesopore size distribution for samples H-Beta-50 (1), MCF (2), ZF-100 (3), ZF-100-3d (4), ZF-120 (5)

Table 1 Textural properties of the hybrid materials

Sample	V_{micro}^a (cm^{3}g$^{(1)}$)	D_{micro}^b (nm)	E_o^d (kJ mol^{-1})	V_{meso}^d (cm^{3}g^{-1})	S_{meso}^a (m^{2}g^{-1})	D_{meso}^e (nm)	$|\Delta\mu_o|^f$ (kJ mol^{-1})	S_{BET}^g (m^{2}g^{-1})	V_t^h (cm^{3}g^{-1})
ZF-100	0.13	0.91 (1.38)^c	3.0	1.11	190	52.0^c	46.7	490	1.24
ZF-100-3d	0.11	0.88 (1.30)^c	3.1	1.12	150	68.0^c	72.2	410	1.23
ZF-120	0.15	0.86 (1.36)^c	3.0	1.33	330	22.0^c	42.6	690	1.48
Beta-50	0.23	0.86 (0.94)^c	4.5	0.08	20	–	–	615	0.31
MCF	0.04	–	2.3	1.70	360	14.5 ± 1.5	13.5	470	1.74

^{a}V_{micro}, S_{meso}: micropore volume and mesopore specific surface area by t-plot method; ^{b}D_{micro}: micropore diameter (maximums of micropore size distribution); ^{c}D_{micro}: micropore diameter (mean); ^{d}E_o: the characteristic energy of adsorption; ^{d}V_{meso}: mesopore volume ($V_t - V_{micro}$); ^{e}D_{meso}: mesopore diameter (^{c}D_{meso} – mean diameter); ^{f}$|\Delta\mu_o|$: initial potential for adsorption; ^{g}S_{BET}: total specific surface area; ^{h}V_t: total pore volume.

As a comparison to hybrid materials, the nitrogen adsorption results for MCM and zeolite materials are given in Table 2. The high specific surface area of Al-H-MCM-41 is owing to its high micropore volume compared to that of H-Beta-zeolite. The specific surface area of Al-H-MCM-48 was also slightly lower than those of zeolites despite of its relative high micropore volume.

Acidity

Acid site concentration and strength were characterized by ammonia TPD and pyridine adsorption desorption techniques (Table 3). According to TPD of ammonia, ZF-120 (the degree S4–S6 of crystallinity ca. 0.4) contains sites of average strength (230°C), moreover the sample has strong acid sites (420°C). The total concentration of acid sites for this sample is 0.37 mmol g^{-1}. The Sample ZF-100 predominantly contains acid sites of an average strength as the shoulder on the ammonia TPD curve appears at ~360°C (not shown here) and the total concentration of acid sites for this catalyst is 0.17 mmol g^{-1}. The maximum of thermal desorption of ammonia at ~190°C is owing to interactions of ammonia with surface silanol groups forming hydrogen bonds (weakly bound NH$_3$).

IR spectroscopy allows revealing the nature of the hydroxyl groups in the studied materials. The most intense

Table 2 Specific surface area and micropore volume of the catalysts

Catalyst	Specific surface area (m^{2}g^{-1})	Micropore volume (cm^{3}g^{-1})	Ref.
H-Beta-25	680	0.24	
H-Beta-300	805	0.28	
H-Si-MCM-41	848	0.30	
H-Al-MCM-48	718	0.71	42
H-Al-MCM-41	808	0.29	

absorption bands in the IR spectra of the samples recorded after evacuation at 400°C were observed at 3745 and 3735 cm^{-1} (Fig. S4). These bands correspond to the stretching vibrations O–H and the non-acidic silanol groups, respectively, on the external surface and in the areas of structural defects caused by incomplete condensation of the framework or removal of lattice atoms under calcination. The IR spectra of zeolite Beta also possess the absorption band at 3610 cm^{-1} (Fig. S4), which corresponds to the stretching vibrations of O–H in acidic bridging hydroxyl groups SiOHAl (Brønsted acid sites). In addition, the IR spectrum of X-ray amorphous sample ZF-100 has weakly expressed shoulder at 3680–3550 cm^{-1}, which also can be attributed to the vibrations of this type. As a result of the interactions of pyridine with Brønsted acid sites, the absorption band at 3610 cm^{-1} disappears. At the same time, the IR spectra of all samples recorded after desorption of pyridine at 150°C have the absorption band at 1548 and 1456 cm^{-1}, corresponding to Brønsted and Lewis acid sites, respectively (Fig. S5). The intensity of the signal, corresponding to Brønsted acid sites (1548 cm^{-1}), decreases with a decrease of the degree of crystallinity.

Strength of the Brønsted acid sites in the samples with a higher degree of crystallinity (ZF-120) is close to the zeolite Beta: absorption band at 1548 cm^{-1} remains after desorption of pyridine at 400°C, which also indicates the relative strength of these sites. Brønsted acid sites of low crystalline sample ZF-100 are stable up to 300°C (Fig. S6, Table 3). The highest amount of Brønsted acid sites among all catalysts was observed for H-Beta-25 followed by hybrid ZF-120, whereas H-Beta-300, H-Al-MCM-48, and H-Al-MCM-41 exhibited only a relatively low amount of Brønsted acid sites (Table 3). The amount of Lewis acid sites was also the highest for H-Beta-25, but the second largest Lewis acidity was in H-Al-MCM-48.

Table 3 Brønsted and Lewis acidity of the catalysts determined by pyridine adsorption/desorption method

Catalyst	Brønsted acidity (μmol g^{-1})			Lewis acidity (μmol g^{-1})			Ref.
	250°C W + M + S	350°C M + S	450°C S	250°C W + M + S	350°C M + S	450°C S	
H-Beta-25	219	187	125	82	43	25	43
H-Beta-300	54	49	23	28	9	4	43
H-Si-MCM-41	0	0	0	15	4	1	
H-Al-MCM-48	59	18	2	63	25	7	43
H-Al-MCM-41	26	11	3	40	20	12	43
ZF-120	89	69	33	57	23	3	
ZF-100	30	15	0	35	15	0	

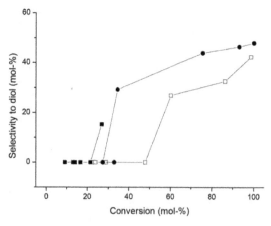

Figure 5 Selectivity to diol as a function of conversion in the isomerization of verbenol oxide over mesoporous catalysts at 140°C in dimethylacetamide. Symbols: (■) H-Si-MCM-41, (□) H-Al-MCM-48, and (•) H-Al-MCM-41

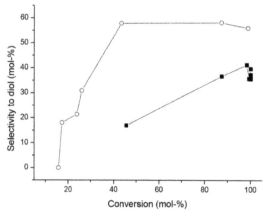

Figure 6 Selectivity to diol as a function of conversion in the isomerization of verbenol oxide over mesoporous catalysts at 140°C in dimethylacetamide. Symbols: (■) ZF-120 and (○) ZF-100

H-Si-MCM-41 mesoporous material exhibited only Lewis acidity which was the lowest among the tested catalysts.

It should be stated here that the aging temperature affected significantly the catalyst acidity, since ZF-120 contained nearly threefold more Brønsted acid sites than ZF-100. Qualitatively ammonia TPD and pyridine desorption method gave analogous trends for the acidities of ZF-120

and ZF-100, i.e. the former one exhibited stronger acidity than the latter one.

Comparison of the obtained acidity–crystallinity results for hybrid materials indicates significant deviations from additivity for characteristics of the synthesized sample (suggesting mechanical mixtures of zeolite and a mesoporous material). Thus, at a relatively steady increase of the degree of crystallinity, the obtained values of acidity (concentration of acid sites) are significantly higher than the values obtained under conditions when only the zeolite phase contains acid sites (concentration of acid sites in zeolite (0.396) is multiplied with crystallinity of the samples (0.08 for ZF-100 and 0.38 for ZF-120)). For low-crystalline sample ZF-100, this deviation is the highest being fourfold (Fig. S7). Based on the obtained results, it can be assumed that the presence of zeolite precursors, which probably have a high surface energy (and as a result have a higher reactivity, possibly owing to structural defects), causes the increase of the concentration of acid sites which does not correspond to the degree of crystallinity and consequently also to the catalytic activity of the samples.

Evaluation of catalytic properties results in liquid phase isomerization of verbenol oxide

Initial reaction rates

The highest initial rates were observed for mildly acidic catalysts, i.e. hybrid ZF-100, zeolite H-Beta-300, and mesoporous H-Al-MCM-41 and H-Al-MCM-48, which contained Brønsted acidity (Table 3). The lowest rate as expected was achieved with a non-acidic H-Si-MCM-41 catalyst (Figure S8a). On the other hand, with highly acidic catalyst the initial deactivation caused a lower initial rate for H-Beta-25. For ZF-100, the reaction proceeded rapidly (Figures S9a and S10a, Table 4). Thus, it can be stated that an optimum acidity gives the highest initial isomerization rate.

Conversions after prolonged reaction times

Conversions after 60-min reaction time decreased in the following order: ZF-120 > H-Beta-25 > H-Al-MCM-41 > H-Beta-300 > ZF-100 > H-Al-MCM-48 >> H-Si-MCM-41. This order shows that some Brønsted acidity is required for isomerization of verbenol oxide, since H-Si-MCM-41 was not active (Table 4). Furthermore, the most acidic catalysts ZF-120 and H-Beta-25 exhibited the highest rates after 60 min, showing that even Beta zeolite is quite suitable catalyst from the structural point of view. The third and fourth active catalysts contained also some strong acid

Table 4 Catalytic test results in verbenol oxide isomerization

Catalyst	Initial rate (mmol min^{-1}/g$_{cat}$)	Conversion after 60 min (mol.-%) in parenthesis after 180 min (mol.-%)	Selectivity to diol at 50% (90%) conversion (%)	Selectivity to ketone at 50% (90%) conversion (%)	Selectivity to oxetane at 50% (90%) conversion (%)
H-Beta-25	7.7	81 (100)	46 (46)	21 (22)	13 (10)
H-Beta-300	14.4	69 (97)	46 (58)	32 (25)	5 (4)
H-Si-MCM-41	6.2	17 (27)	17*	31*	14*
H-Al-MCM-48	13.0	60 (99)	3 (33)	22 (21)	13 (11)
H-Al-MCM-41	15.9	75 (100)	32 (43)	20 (25)	12 (5)
ZF-120	25.9	100 (100)	16 (36)	25 (43)	12 (12)
ZF-100	7.2	56 (99)	(56)	26 (18)	8 (13)

*At 27% of conversion.

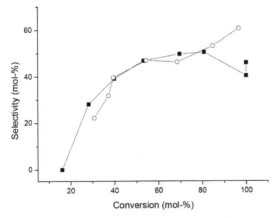

Figure 7 Selectivity to diol as a function of conversion in the isomerization of verbenol oxide over mesoporous catalysts at 140°C in dimethylacetamide. Symbols: (■) H-Beta-25 and (○) H-Beta-300

sites, whereas hybrid ZF-100 had no strong acid sites being thus not so active.

Product selectivity

The most selective catalyst for diol formation was H-Beta-300 giving 58% selectivity at 90% conversion (Table 4). Nearly the same selectivity was achieved with a mildly acidic, hybrid ZF-100. However, no diol was initially formed over mildly acidic mesoporous catalysts, H-Al-MCM-41, H-Al-MCM-48, and H-Si-MCM-41 (Fig. 5), whereas selectivity to diol increased rapidly with conversion using less acidic, mesoporous ZF-100 hybrid materials as a catalyst (Fig. 6). On the other hand, when using more acidic mesoporous hybrid material ZF-120 as a catalyst, selectivity was increasing gradually, reaching maximally only about 36 mol.-% level, thereafter decreasing. Among MCM mesoporous materials, the highest selectivity to paramenthanic diol was obtained over H-Al-MCM-41 and the second highest selectivity over H-Al-MCM-48. Performance of H-Al-MCM-41 is owing to the presence of mild Brønsted and Lewis acid sites (Table 3). For large pore zeolites, the selectivities to diol were quite similar as a function of conversion, except at high conversion levels, when diol reacted further to other compounds over strongly acidic H-Beta-25 (Fig. 7). Analogous results were obtained with a highly acidic H-USY-12 with SiO_2 to Al_2O_3 ratio 12, for which the diol selectivity decreased with increasing conversion, whereas diol selectivity continuously increased with increasing conversion over H-USY-80.[23] When comparing the selectivity to diol at 50% conversion among all the studied catalysts, namely H-USY and H-ZSM-5 with different SiO_2 to Al_2O_3 ratio in ref.[23] as well as in the current work, it can be observed that the highest selectivity was achieved with ZF-100 being 58%, whereas it was 43% for mildly acidic H-Beta-300 (Figs. 6 and 7). On the other hand, almost the same selectivity for ZF-100 and H-Beta-300, being 56 and 58%, respectively, at 90% conversion level, was achieved. These results indicate that mildly acidic catalysts are more selective toward diol than highly acidic ones and both microporous and hybrid materials give about the same selectivities.

Another major product was cyclopentyl ketone. Kinetic results revealed that the C5 ketone concentration as a function of diol was nearly independent on acidity for mesoporous MCM-type catalysts (Figure S8b), whereas over larger mesoporous hybrid materials C5 ketone formation was favored over more acidic ZF-120 compared to less acidic hybrid foam catalysts (Figure S9b). For zeolites the ratio between the concentrations of C5 ketone to diol was constant over H-Beta-25, whereas over mildly acidic H-Beta-300 C5 ketone reacted further to diol, which might be caused by diffusional limitations present with Beta zeolite, where the reaction proceeded more slowly owing to lower amounts of Brønsted and Lewis acid sites than with H-Beta-25 catalyst.

The third major product was a bicyclic oxetane. Its concentration increased with increasing diol concentration over MCM-type catalysts (Figure S8c). On the other hand, oxetane reacted further to diol over a more acid catalyst, such as ZF-120 (Figure S9c) and H-Beta 25 (Fig. S10c). According to molecular modeling, the stability of verbenol oxide is lower than that of oxetane followed by diol, which could explain the fact that a relatively strong acidity is needed for oxetane isomerization to diol. As a comparison to recently obtained results,[23] it can however, be stated that with very acidic H-USY catalysts, such as H-USY-15 and H-USY-12, oxetane reacted not to diol, but to other unknown products, indicating that an optimum acid strength is needed to transform oxetane to diol.

When comparing diol selectivity as a function of Brønsted and Lewis acidity it can be seen from Fig. 8 that the selectivity at 90% conversion toward diol decreased with increasing amounts of Brønsted and Lewis acid sites for microporous zeolites, whereas for mesoporous materials the selectivity decreased with increasing Lewis acid site concentration. It is noteworthy to mention here, thus the H-Al-MCM-48 mesoporous material did not exhibit higher selectivity to diol than H-Al-MCM-41, although it has larger amount of Brønsted acid sites (59 $\mu mol/g_{cat}$) than H-Al-MCM-41 catalyst (26 $\mu mol/g_{cat}$). A possible explanation of this behavior of H-Al-MCM-48 catalyst could be the structural

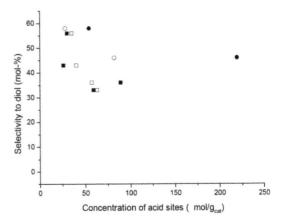

Figure 8 Selectivity to diol at 90% conversion of verbenol oxide as a function of total acid site concentration. Symbols: solid symbol concentration of Brnsted acid sites, open symbol concentration of Lewis acid sites, (•) microporous zeolites, (■) mesoporous materials

differences, pore sizes, and thickness of the wall of this mesoporous material compared to H-Al-MCM-41.

Conclusions

Verbenol oxide isomerization over micro-, meso-, and hybrid micro-mesoporous materials was studied. Well characterized H-Beta-25 and H-Beta-300 zeolites and MCM-41 and MCM-48 mesoporous structures were chosen for comparison with new synthesized materials. Zeolite-like mesoporous materials with spongy structure possessing high textural (S_{BET} up to $690\,m^2\,g^{-1}$, V_{total} up to $1.5\,cm^3\,g^{-1}$, $D_{pore} = 21-23\,nm$) and acid (total concentration of Brønsted and Lewis acid sites is $0.13-0.26\,mmol\,g^{-1}$) characteristics were obtained via the template method using the effect of solubilization (1,3,5-trimethylbenzene).

The final conversion of verbenol oxide was almost 100% over all studied catalysts, except H-Si-MCM-41, which exhibited a very low conversion (17%) owing to lack of Brønsted acid sites. Therefore, the acidic properties of the catalysts such as the type of acid sites, amount, and strength played a very important role in verbenol oxide isomerization. The concentration of Brønsted acid sites should be not too high to obtain a high selectivity, and not too low to achieve reasonable conversion. The highest yields were obtained over H-Beta-300 and ZF-100 catalysts, being 59 and 58%, respectively. It can be stated that mildly acidic catalysts are more selective toward diol than highly acidic ones. The influence of the structure of the catalyst is minimal. The highest selectivity to the desired product of diol was obtained over H-Beta-300 zeolite catalyst.

Conflict of interest

The authors declare that they have no conflict of interest.

Acknowledgement

This work is a part of the activities at Åbo Akademi University Johan Gadolin Process Chemistry Centre. Scanning electron micrographs for hybrid materials were performed at the University of Oviedo (Spain). The work was also performed as a part of the state contract awarded on the basis of a grant of the Government of the Russian Federation for support of scientific research conducted under the supervision of leading scientists at Russian institutions of higher education, research institutions of State Academies of Sciences and state research centres of the Russian Federation on March 19, 2014, no. 14.Z50.31.0013.

References

1. L. H. Chen, X. Y. Li, J. C. Rooke, Y. H. Zhang, X. Y. Yang, T. Tang, F.-S. Xiao and B. L. Su: *J. Mater. Chem.*, 2012, **22**, 17381–17403.
2. Z. Le Hua, J. Zhou and J. L. Shi: *Chem. Commun.*, 2011, **47**, 10536–10547.
3. K. Na, M. Choi and R. Ryoo: *Microp. Mesop. Mater.*, 2013, **166**, 3–19.
4. J. Zhu, X. Meng and F. Xiao: *Front. Chem. Sci. Eng.*, 2013, **7**, 233–248.
5. D. P. Serrano and P. Pizarro: *Chem. Soc. Rev.*, 2013, **42**, 4004–4035.
6. K. Möller and T. Bein: *Chem. Soc. Rev.*, 2013, **42**, 3689–3707.
7. W. Han, B. Y. Jia, G. Xiong and W. Yang: *Sci. Technol. Adv. Mater.*, 2007, **8**, 101–105.
8. C. Xue, F. Zhang, L. Wu and D. Zhao: *Micropor. Mesopor. Mater.*, 2012, **151**, 495–500.
9. J. Zhou, Z. Hua, Z. Liu, W. Wu, Y. Zhu and J. Shi: *ACS Catal.*, 2011, **1**, 287–291.
10. X. Wang, Y. Li, C. Luo, J. Liu and B. Chen: *RSC Adv.*, 2013, **3**, 6295–6298.
11. Y. Lv, X. Qian, B. Tu and D. Zhao: *Catal. Today*, 2013, **204**, 2–7.
12. J. Zhao, Z. Hua, Z. Liu, Y. Li, L. Guo, W. Bu, X. Cui, M. Ruan, H. Chen and J. Shi: *Chem. Commun.*, 2009, 7578–7580.
13. P. Prokešová, S. Mintova, J. Čejka and T. Bein: *Mater. Sci. Eng. C*, 2003, **23**, 1001–1005.
14. P. Prokešová, S. Mintova, J. Čejka and T. Bein: *Micropor. Mesopor. Mater.*, 2003, **64**, 165–174.
15. C. J. Van Oers, W. J. J. Stevens, E. Bruijn, M. Mertens, O. I. Lebedev, G. Van Tendeloo, V. Meynen and P. Cool: *Micropor. Mesopor. Mater.*, 2009, **120**, 29–34.
16. F. N. Gu, F. Wei, J. Y. Yang, N. Lin, W. G. Lin, Y. Wang and J. H. Zhu: *Chem. Mater.*, 2010, **22**, 2442–2450.
17. Y. Liu and T. J. Pinnavaia: *Chem. Mater.*, 2002, **14**, 3–5.
18. J. Huang, G. Li, S. Wu, H. Wang, L. Xing, K. Song, T. Wu and Q. Kan: *J. Mater. Chem.*, 2005, **15**, 1055–1060.
19. O. V. Ardashov, A. V. Pavlova, I. V. Il'ina, E. A. Morozova, D. V. Korchagina, E. V. Karpova, K. P. Volcho, T. G. Tolstikova and N. F. Salakhutdinov: *J. Med. Chem.*, 2011, **54**, (11), 3866–3874.
20. I. Gainer, M. Zorca, *Analele Universitatii din Bucaresti-chimie*, 2005, **I-II**, 287–290.
21. R. Szmigielski, M. Cieslak, K. J. Rudziński and B. Maciejewska: *Environ. Sci. Pollut. Res.*, 2011, **19**, (7), 2860–2869.
22. M. Stekrova, N. Kumar, P. Mäki-Arvela, A. Aho, K. P. Volcho, N. F. Salakhutdinov and D. Yu Murzin: *React. Kin. Mech. Catal.*, 2013, **110**, 449–458.
23. A. Torozova, P. Mäki-Arvela, N. Kumar, A. Aho, A. Smeds, M. Peurla, R. Sjöholm, I. Heinmaa, K. Volcho, N. F. Salakhutdinov and D. Yu Murzin: *React. Kinet. Mech. Catal.*, 2015, **116**, 299–314.
24. W. M. Meier and D. H. Olson: 'Ch. baerlocher, atlas of zeolite structures', 4th edn; 1996, Amsterdam, Elsevier.
25. Available at: hppt://www.iza-structure.org/database (accessed 10 June 2015)
26. R. A. Sheldon, I. Arends and U. Hanefeld: 'Green Chemistry and Catalysis'; 2007, Weinheim, Wiley-VCH.
27. R. Schmidt, E. W. Hansen, M. Stöcker, D. Akroriae and O. H. Ellestad: *J. Am. Chem. Soc.*, 1995, **17**, 4049–4056.
28. K. Schumacher, P. I. Ravikovitch, A. Du Chesne, A. V. Neimark and K. K. Unger: *Langmuir*, 2000, **16**, 4648–4654.
29. C. T. Kresge, M. E. Leonowicz, W. J. Roth, J. C. Vartuli, Synthetic mesoporous crystalline material, US Patent 5,098,684; 1992.
30. A. Bernas, P. Laukkanen, N. Kumar, P. M. äki-Arvela, J. Väyrynen, E. Laine, B. Holmbom, T. Salmi and D. Yu Murzin: *J. Catal*, 2002, **210**, 354–366.
31. S. B. Pu, J. B. Kim, M. Seno and T. Inui: *Microp. Mater*, 1997, **10**, 25–33.
32. M. Käldström, N. Kumar and D. Yu Murzin: *Catal. Today*, 2011, **167**, 91–95.
33. C. A. Emeis: *J. Catal*, 1993, **141**, (2), 374–375.
34. P. Hudec, A. Smieskova, Z. Zidek, P. Schneider and O. Solcova: *Stud. Surf. Sci. Catal*, 2002, **142**, 1587–1594.
35. S. G. Gregg and K. S. W. Sing: 'Adsorption, Surface Area and Porosity', 94; 1982, New York, NY, Acad. Press.
36. E. P. Barrett, L. G. Joyner and P. P. Halenda: *J. Am. Chem. Soc*, 1951, **73**, 373–380.
37. A. Saito and C. Foley: *AIChE J*, 1991, **3**, 429–436.
38. MAUD program Available at: http://www.ing.unitn.it/maid/ [Cited 2015 June 10]
39. W. Guo, C. Xiong, L. Huang and Q. Li: *J. Mater. Chem*, 2001, **11**, 1886–1890.
40. M. Käldström, N. Kumar, T. Heikkilä, M. Tiitta, T. Salmi and D. Yu Murzin: *ChemCatChem*, 2010, **2**, 539–549.
41. A. Torozova, P. M. äki-Arvela, A. Aho, N. Kumar, A. Smeds, M. Peurla, R. Sjöholm, I. Heinmaa, D. V. Korchagina, K. P. Volcho, N. F. Salakhutdinov and D. Yu Murzin: *J. Mol. Catal. A Chem*, 2015, **397**, 48–55.
42. M. Käldström, N. Kumar, T. Heikkilä, M. Tiitta, T. Salmi and D. Yu Murzin: *Biomass Bioenergy*, 2011, **35**, 1967–1976.
43. A. Aho, N. Kumar, K. Eränen, T. Salmi, M. Hupa and D. Yu Murzin: *IChemE*, 2007, **85**, 473–480.

4

Catalytic degradation of Acid Orange 7 by H_2O_2 as promoted by either bare or V-loaded titania under UV light, in dark conditions, and after incubating the catalysts in ascorbic acid

Marco Piumetti[1,†], Francesca S. Freyria[1,2,†], Marco Armandi[1], Guido Saracco[1], Edoardo Garrone[1], Giovanny Esteban Gonzalez[1], and Barbara Bonelli[1]* ⓘ

[1]Department of Applied Science and Technology and INSTM Unit of Torino-Politecnico, Politecnico di Torino, Corso Duca degli Abruzzi, 24, I-10129 Turin, Italy
[2]Department of Chemistry, Massachusetts Institute of Technology, 77 Massachusetts Ave, 02139 Cambridge, MA, USA

Abstract Pure and V-loaded mesoporous titania (with 2.5 wt-% V) were prepared by template-assisted synthesis and compared to commercial titania (Degussa P25), both as such and after vanadium loading. Mesoporous TiO_2 occurred as pure anatase nanoparticles with higher surface area (SSA = 150 m^2 g^{-1}) than P25 (SSA = 56 m^2 g^{-1}). Degradation of the azo dye Acid Orange 7 by H_2O_2 was used as a test reaction: under UV light, no difference emerged between mesoporous TiO_2 and P25, whereas in dark conditions, higher SSA of the mesoporous sample resulted in

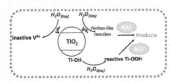

higher conversions. Under UV illumination, surface V^{5+} species inhibited photocatalytic activity, by forming inactive V^{4+} species. Similarly, in dark conditions, V^{5+} surface species reacted with H_2O_2, likely yielding $\cdot O_2H$ radicals and reducing to V^{4+}. On the contrary, V-containing catalysts were very active after pretreatment with ascorbic acid, which reduced V^{5+} species to V^{3+} species, the latter promoting very lively a Fenton-like reaction.

Keywords Mesoporous titania, Azo dye degradation, Hydrogen peroxide, Ascorbic acid, Fenton reaction

Introduction

Titania, anatase in particular, is both a very popular photocatalyst and catalytic support,[1-7] not only because of its (reasonably small) band gap, but also for the possibility of obtaining it in nanoporous and/or nanoparticle form, with increased specific surface area (SSA).[8-12]

One current application of TiO_2 is the degradation of organic pollutants, including azo dyes,[4] a class of organic molecules widely applied in the photographic industry (and bearing an environmental impact once released in the surroundings),[13] which can be removed by photocatalytic degradation with TiO_2, usually in the presence of an oxidizing agent, like H_2O_2.[2]

P25 is one of the most common commercial forms of TiO_2, occurring as a mixture of mainly anatase and rutile with SSA = 50 ± 15 m^2 g^{-1}.[14]

In the present paper, high SSA mesoporous TiO_2 (MT) was obtained by template-assisted synthesis, and calcined at 450 °C to avoid phase transition to rutile. Anatase has, indeed, a larger adsorptive affinity than rutile for organic molecules[15] and is generally regarded as the most photocatalytically active phase of TiO_2.

Since vanadium has been shown to improve the catalytic activity of TiO_2, extending its absorption the Vis-range,[16] a sample of MT with a 2.5 wt- % V loading was prepared; for comparison, two samples were obtained by impregnating P25 with (i) same V content and (ii) 1/3 V content to get, in principle, the same vanadium dispersion as for V2.5-MT.

The degradation of a model azo dye (Acid Orange 7, AO7)[13] was studied either under UV illumination or not (hereafter referred to as "dark conditions"). The effects of H_2O_2 concentration in the reaction and the presence of ascorbic acid (H_2Asc) were also studied. H_2Asc, especially in biochemical processes, plays different roles as an antioxidant, a "mild" reductant of both metal ions and organic moieties, and a radical scavenger.[17] It is able, *inter alia*, to reduce transition metal ions, including vanadium, in biological systems.[18,19] Nonetheless, reports are available on the effects of an oxidant–reductant system (H_2O_2 and H_2Asc, for instance) in the catalytic activity of V-containing catalysts.[20,21]

*Corresponding author, email barbara.bonelli@polito.it
†Both authors contributed equally to this work.

Table 1 Textural properties as obtained by XRD diffraction, N_2 physisorption at -196 °C, and EDX chemical analysis

Sample	Crystallites size[a] (nm)	S_{BET} (m² g⁻¹)	Metal content (wt. %)[b]	Surface metal density (MO$_x$ nm⁻²)[c]
MT	14 ± 3	150	–	–
V2.5-MT	17 ± 5	122	2.5	2.3
P25	19 ± 3 (anatase) 23 ± 4 (rutile)	56	–	–
V2.5-CT	–	38	2.8	8.7
V0.80-CT	–	59	0.77	2.6

[a]As calculated by applying the Debye–Scherrer formula.
[b]As determined by EDX analysis.
[c]As obtained by applying the formula: (metal content)/(SSA × AM), where metal content is the metal weight percentage as measured by EDX, SSA is the sample specific surface area, and AM is the atomic mass of the metal.

Experimental

Catalysts preparation

All reagents were ACS grade chemicals from Sigma-Aldrich.

The synthesis of mesoporous titania (MT) and of V-containing MT with a vanadium nominal content of 2.5 wt-% (V2.5-MT) is detailed in Ref. 23. Vanadium resulted present mainly at the surface of V2.5-MT.[22] The samples were calcined at 450 °C in air to remove the template and to avoid phase transition to rutile.

The commercial titania (CT) was Degussa P25 (TiO$_2$ content ≥99.5%); two samples with nominal V content of either 2.5 or 0.80 wt-% (V2.5-CT and V0.80-CT) were obtained by impregnation with NH$_4$VO$_3$ solution, followed by drying at 60 °C and calcination in air at 450 °C for 4 h.

Catalysts characterization

Powder X-ray diffraction patterns were collected on a X'Pert Philips PW3040 diffractometer using Cu Kα radiation (2θ range = 20°–85°; step = 0.05° 2θ; time per step = 0.2 s) and indexed according to the Powder Data File database (PDF 2000, International Centre of Diffraction Data, Pennsylvania). Crystallites average size (D) was determined by using the Debye–Scherrer formula, $D = 0.9\ \lambda/b\cos\theta$, where λ is the wavelength of the Cu Kα radiation, b is the full width at half-maximum (in radians), 0.9 is the shape factor for spherical particles, and θ is the angle of diffraction peaks. The anatase content was evaluated by the full-profile Rietveld method applied to diffraction patterns using the GSAS-EXPGUI free software. XRD background was modeled by a 10-term cosine polynomial function, and pseudo-Voigt functions were adopted for peaks curve fitting.

The SSA and total porous volume (V_p) were measured by N$_2$ sorption isotherms at -196 °C (Quantachrome Autosorb 1C) on powders outgassed at 150 °C for 4 h to remove water and other atmospheric contaminants; the SSA was determined according to the Brunauer–Emmett–Teller method.

The metal content was determined by (semi-quantitative) chemical analysis carried out by means of an energy-dispersive X-ray probe (low-vacuum Scanning Electron Microscope Quanta inspect 200) on 10–50-nm diameter spots. For each sample, about 10 measurements were carried out in different spots of the sample, from which an average metal content was calculated, as reported in Table 1.

Diffuse reflectance (DR) UV–vis spectra of powder samples dehydrated at 150 °C were measured on a Cary 5000 UV–vis–NIR spectrophotometer (Varian instruments) equipped with a DR sphere.

X-ray photoelectron spectroscopy (XPS) measurements were obtained on an XPS PHI 5000 Versa probe apparatus. The C 1s peak at 284.6 eV was used as a reference for charge correction. To study spent catalysts, powders recovered by centrifugation were dried in air and pressed in order to obtain pellets that were further outgassed under vacuum at room temperature before XPS analysis.

Catalytic tests

A total volume of 50 mL of 0.67 mM aqueous solution (natural pH of 6.80) of AO7 (Fluka) and an amount of catalyst corresponding to 1.0 g L⁻¹ concentration were used systematically for catalytic tests, during which the suspension was stirred by means of a magnetic stirrer, operated at 600 rpm. Preliminary blank experiments were run in dark conditions by mixing 0.67 mM AO7 with either 0.030 M or 0.80 M H$_2$O$_2$ without any catalyst.

Concerning the amount of H$_2$O$_2$ used during experiments, the reaction leading to the complete degradation of the dye is:

$$C_{16}H_{11}N_2SO_4Na + 42H_2O_2 \rightarrow 16CO_2 + 46H_2O + 2HNO_3 + NaHSO_4$$

(1)

According to reaction (1), a 0.030 M H$_2$O$_2$ concentration roughly corresponds to the stoichiometric amount of H$_2$O$_2$ necessary for the complete AO7 degradation, and 0.80 M H$_2$O$_2$ to an excess of hydrogen peroxide.

A first set of catalytic tests was performed under UV light with 0.80 M H$_2$O$_2$ by illuminating with a medium-pressure Hg lamp (Hamamatsu, LC3; light intensity of 55 mWcm⁻²).

A second set of experiments was carried out in dark conditions by adding either 0.030 M or 0.80 M H$_2$O$_2$ to the suspension containing AO7 and the solid. In a third set, 0.40 M H$_2$Asc was added to the suspension containing AO7 and the solid, and after 20 min, 0.80 M H$_2$O$_2$ was added in dark conditions.

In summary, a typical catalytic test involves 3.35 × 10⁻⁵ moles of AO7, either 2.45 × 10⁻⁵ or 8.17 × 10⁻⁶ moles of V, either 1.50 × 10⁻³ or 40.0 × 10⁻³ moles of H$_2$O$_2$. In the third set of experiments, 20.0 × 10⁻³ moles of H$_2$Asc were also present. Moreover, since the mixture was not deaerated, ca. 12.3 × 10⁻³ moles of atmospheric O$_2$ were always present.

Aliquots of the suspension were collected at regular intervals of time, the supernatant fraction was separated by centrifugation (ALC centrifuge PK110, at 4000 rpm for 2 min) and the UV–vis spectrum was measured in the 190–800 nm range on a Cary 5000 UV–vis–NIR spectrophotometer (Varian instruments), using a quartz cell with 1-mm path length. The concentration of AO7 was evaluated by the intensity of its band at 484 nm (vide infra), after a proper calibration procedure.

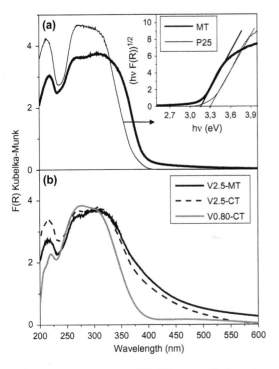

Figure 1 DR UV–vis spectra (200–600 nm range) of samples outgassed at 150 °C. Part *a*: MT (bold curve) and P25; the inset reports the Tauc's plot of MT (bold curve) and P25. Part *b*: V2.5-MT (bold black curve), V2.5-CT (dashed black curve), and V0.80-CT (bold gray curve)

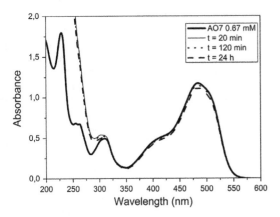

Figure 2 UV–vis spectra of the solutions obtained 20 min, 120 min, and 24 h after addition of 0.80 M H_2O_2 to the starting 0.67 mM AO7 solution (bold curve)

Results and discussion

Survey of the main physicochemical features of the catalysts

The XRD analysis (not reported) showed that anatase is the only phase present in MT samples (99.8%, as obtained by Rietveld refinement), whereas P25 is a mixture of anatase (88.8%) and rutile (11.2%). The size of MT crystallites was around 15 nm, as calculated according to the Debye–Scherrer

formula (Table 1), in agreement with previous transmission electron microscopy observation.[22]

Type IV N_2 isotherms were measured at −196 °C on MT samples, with high SSA values due to both intra- and inter-particle mesopores; a limited decrease of SSA was observed with V2.5-MT (Table 1).

Fig. 1 reports DR UV–vis spectra of samples outgassed at 150 °C to remove water and other atmospheric contaminants: comparison between MT and P25 shows that the former absorbs in a broader range of wavelengths (Fig. 1*a*). The spectrum of MT is shifted toward higher wavelengths with respect to P25 (arrow): accordingly, the corresponding Tauc's plots, in the version for indirect band gap (E_g) semiconductors (inset to Fig. 1*a*), yield $E_g \approx 3.30$ eV for P25 and $E_g \approx 3.15$ eV for MT. The redshift observed in MT absorption edge is likely due to slightly different optical and electrical properties of the mesoporous material with respect to bulk TiO_2, as already observed with porous films of titania.[23]

Fig. 1*b* compares the UV–vis spectra of V-containing catalysts, characterized by a pale yellow color: with respect to parent MT, V2.5-MT shows a small redshift of the absorption band and an increased absorption above 375 nm. Both isolated VO_x and oligomeric V_xO_y species ($y = 2x + 1$ in dehydrated powders), respectively, absorbing at 270 and 325 nm,[24,25] are likely present at the surface, though their occurrence is masked by TiO_2 absorption. V2.5-CT has similar UV–vis spectrum, whereas absorption related to VO_x species is less intense with V0.80-CT, in agreement with the lower metal content.

Preliminary considerations about catalytic tests

Scheme 1 reports the possible structures of AO7 in water: the hydrazone form (B), stable in the solid phase, undergoes, in water, azo-hydrazone tautomerism via intramolecular proton transfer, so that both hydrazone (B) and azo-form (A) are simultaneously present in solution.

The UV–Visible spectrum of 0.67 mM AO7 in water is reported as the bold curve in Fig. 2: the two peaks at 310 and 230 nm and the shoulder at 256 nm are due to aromatic ring absorptions. The peak at 484 nm is due to the n–π* transition involving the lone pair on the N atoms and the conjugated system extending over the two aromatic moieties and encompassing the $N–N$ group of the hydrazone form.[26] The shoulder at 403 nm has a similar nature, involving the $N–N$ group of the azo-form.[26]

The (non-catalytic) reaction between AO7 and H_2O_2 (either 0.030 M or 0.80 M) was preliminarily studied. The UV–vis spectra taken with 0.80 M H_2O_2 are reported in Fig. 2: the region below 300 nm is dominated by H_2O_2 absorption; small changes are observed in the AO7 bands, for which a limited decrease in intensity is observed after 24 h, as H_2O_2 is, per se, able to attack aromatic rings.[27] This shows that without catalyst, a limited conversion of AO7 is obtained, even in the presence of excess H_2O_2.

Catalytic tests under illumination

Fig. 3 compares results obtained after 20-min UV illumination without H_2O_2: in such conditions, the oxidizing agent is dissolved oxygen. UV irradiation, in turn, generates electron (e⁻)/hole (h⁺) pairs in TiO_2. Dissolved O_2 most probably acts as

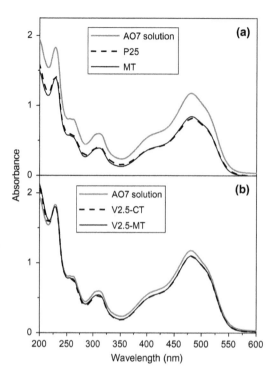

Figure 3 UV–vis spectra obtained under UV light in the absence of H_2O_2. Part a: starting 0.67 mM AO7 solution (gray curve), supernatant solution recovered after 20-min contact with P25 (dashed curve) and MT (bold curve). Part b: starting 0.67 mM AO7 solution (gray curve), supernatant solution recovered after 20-min contact with V2.5-CT (dashed curve) and V2.5-MT (bold curve)

an acceptor of e^- (equation (2)) and the powerful oxidant h^+ reacts with OH^- ions, generating radical $HO\cdot$ species (equation (3)) taking part in the AO7 degradation:

$$O_2 + e^- \rightarrow O_2^-$$ (2)

$$OH\text{-} + h^+ \rightarrow HO$$ (3)

The MT and P25 had the same photocatalytic activity (Fig. 3a), whereas V2.5-CT and V2.5-MT (Fig. 3b) were practically inactive. The same behavior of the bare TiO_2 samples, notwithstanding the different SSA, may be surprising. It has to be recalled, however, that P25, an optimized commercial sample, contains a substantial fraction of rutile, which is recognized to be able to prevent the fast electron–hole recombination occurring in anatase, due to a positive synergic effect between anatase and rutile nanoparticles in P25.[14,28]

The inactivity of V-containing TiO_2 has another explanation: surface V^{5+} centers may trap electrons, giving rise to reduced (and probably inactive) V^{4+} species. Inhibition of TiO_2 reactivity by surface V species is probably related to the fact that the latter act as recombination centers for electron–hole pairs, eventually suppressing the intrinsic photocatalytic activity of the support. This is in agreement with previous literature results, showing that vanadium has a detrimental effect on the

photocatalytic activity of V-doped TiO_2 under UV light, though extending absorption in the Vis range.[16]

Results obtained under UV light with 0.030 M H_2O_2 are reported in Fig. 4: as expected, addition of H_2O_2 to the reaction mixture had a positive effect on the photocatalytic activity of all the samples. With pure TiO_2 (Fig. 4a), H_2O_2 most probably acts as an acceptor of photogenerated electrons (e^-), according to reaction (4):

$$H_2O_2 + e^- \rightarrow OH^- + HO$$ (4)

forming hydroxyl radicals for the degradation of AO7. Again, MT and P25 show the same activity as before, and V has almost no effect on photocatalytic activity (Fig. 4b), which seems only due to TiO_2.

Catalytic behavior in dark conditions

Successive tests were run without UV irradiation to exploit the mere surface properties of the catalysts.

Fig. 5a reports, as an example, spectra obtained after contacting AO7 solution with MT sample and 0.030 M H_2O_2 in dark conditions; Fig. 5b reports results obtained in dark conditions with the same catalyst and 0.80 M H_2O_2. In both cases, bands of AO7 decreased in intensity: no new band, related to any decomposition product, however, appeared. As shown in Fig. 2, the spectral region below 300 nm is dominated by H_2O_2 absorption at higher concentration of the reactant: this feature was observed with all catalysts, so that the presence, if any, of reaction products absorbing in this region was not detectable. On the other hand, the intensity of the absorption below 340 nm does not decrease, so showing that the 0.80 M solution provides excess H_2O_2, in agreement with previous stoichiometric considerations.

Fig. 6 reports conversions obtained with all the studied catalysts in the same conditions (i.e. dark conditions and 0.80 M H_2O_2): MT at any reaction time exhibits conversion values about double than P25, in agreement with the higher SSA. The same explanation holds for the higher activity of V2.5-MT, with respect to both V2.5-CT and V0.80-CT samples, which showed similar curves, independently of the V content, indicating a worse vanadium dispersion attained with P25.

Table 2 reports values of final conversion (Fig. 6) and initial rate obtained in dark conditions with either 0.030 M or 0.80 M H_2O_2: as expected, final conversion always increases with H_2O_2 concentration and in the presence of a catalyst. Conversely, higher initial rates are observed at lower H_2O_2 concentration. This corresponds to a negative order as it concerns H_2O_2: as acknowledged by the literature, at high H_2O_2 concentrations, hydroxyl radicals are scavenged by H_2O_2 molecules with formation of $\cdot O_2H$ radicals, characterized by a lower oxidizing power than $HO\cdot$,[29] and so reaction (5) may slow down AO7 depletion:

$$H_2O_2 + HO\cdot \rightarrow H_2O + \cdot O_2H$$ (5)

With respect to P25, conversion reached with MT is higher, in agreement with the higher SSA of the latter catalyst; concerning initial rate, the positive effect of MT SSA may be appreciated at 0.80 M H_2O_2 concentration (Table 2).

Reactivity of Ti with H_2O_2 is well documented in the literature as it concerns Ti-Silicalite, and involves formation of

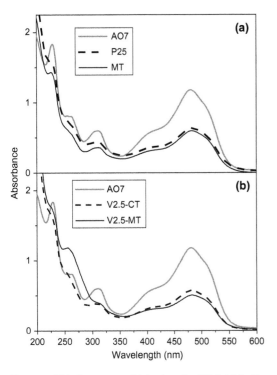

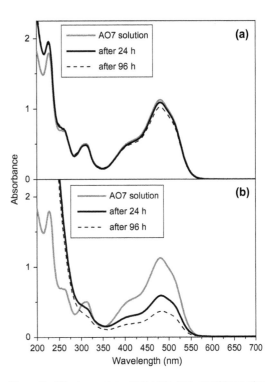

Figure 4 UV–vis spectra obtained under UV light in the presence of 0.030 M H$_2$O$_2$. Part a: starting 0.67 mM AO7 solution (gray curve), supernatant solution recovered after 20-min contact with P25 (dashed curve) and MT (bold curve). Part b: starting 0.67 mM AO7 solution (gray curve), supernatant solution recovered after 20-min contact with V2.5-CT (dashed curve) and V2.5-MT (bold curve)

Figure 5 UV–vis spectra obtained in dark conditions with either 0.030 M H$_2$O$_2$ (part a) or 0.80 M H$_2$O$_2$ (part b) in the presence of MT catalyst. Legend as follows: starting 0.67 mM AO7 solution (gray curve), supernatant solution recovered after 24 h (bold curve) and 96 h (dashed curve)

peroxo bridges with expansion of the coordination sphere of Ti.[30] Concerning TiO$_2$, in the presence of H$_2$O$_2$, surface Ti–OH groups are partially converted into Ti–OOH species, more reactive than H$_2$O$_2$ in partial oxidation reactions.[31]

Fig. 7 reports XPS spectra of MT before and after reaction with 0.80 M H$_2$O$_2$, for Ti 2p (section a) and O 1s range (section b). Before reaction, with MT, two peaks are seen at 464.02 and 458.28 eV, respectively, assigned to the 2p$_{1/2}$ and 2p$_{3/2}$ lines of Ti^{4+},[32] the spectrum of P25 being very similar. After reaction, MT XPS spectrum shifts to higher BE values: the same behavior, observed for anatase treated with H$_2$O$_2$,[32] was ascribed to the formation of surface Ti–OOH groups, as the peroxyl group has an electron-withdrawing effect and causes a shift to higher BE values.[32]

The O 1s region of MT and P25 XPS spectra is similar before reaction: that of MT shifts to higher BE values after reaction. This could be ascribed to a more extensive formation of Ti–OOH species in MT, due to the higher SSA. Curve fits of the O 1s range are reported for MT in Fig. 7c and d: before reaction, a satisfactory curve fit was obtained with two peaks at 529.53 and 530.90 eV, due to O^{2-} related to Ti^{4+} and to OH$^-$ species, respectively. Conversely, after reaction, a satisfactory curve fit was obtained with three peaks, the additional peak at 532.33 eV being ascribed to Ti–OOH species.[32] Although based on delicate curve fitting procedures, such result is in agreement with previous literature reports on H$_2$O$_2$-treated

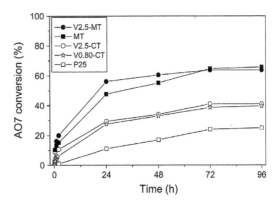

Figure 6 Comparison of AO7 catalytic conversions as obtained in dark conditions in the presence of 0.80 M H$_2$O$_2$

titania[31,32] supporting the idea of the formation of reactive Ti–OOH groups, which are responsible for the catalytic activity of MT.

The "quenching effect" of H$_2$O$_2$ observed at higher concentration (Table 2) is in agreement with the hypothesis of a radical mechanism, in which reactive OH· species are, in this case, due to decomposition of Ti–OOH.

Table 2 shows that with 0.030 M H$_2$O$_2$, initial rates are higher in the presence of V2.5-MT and V2.5-CT catalysts, whereas with 0.80 M H$_2$O$_2$ V-containing samples are very active at the

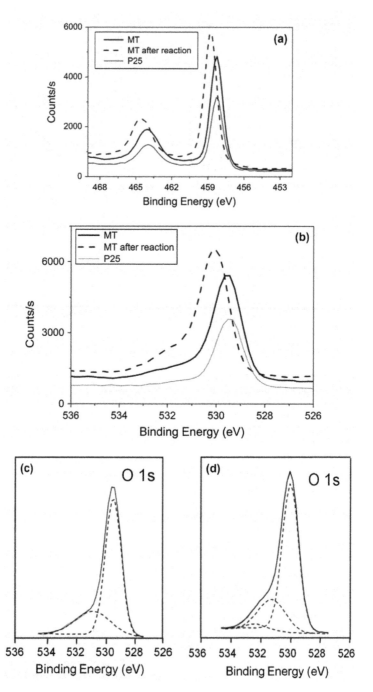

Figure 7 Part *a*: high-resolution XP spectra in the Ti2p BE of P25, MT before reaction (bold curve) and MT after reaction (dotted curve). Part *b*: the same in the O 1s BE range. Part *c*: O 1s XP spectrum of MT before reaction with curve fits. Part *d*: O 1s XP spectrum of MT after reaction with curve fits

beginning of the reaction, then conversion reaches a plateau (Fig. 6), indicating that they are subjected to deactivation. The surface density of heteroatoms (Table 1) is low; therefore, catalytically active patches of TiO_2 still occur at the surface. The curve *conversion vs. time* of V-containing samples is describable as the superposition of the corresponding undoped TiO_2

and a sudden initial increase and so the activity of doped surfaces should thus be not only to metal sites, but also to such patches.

In agreement with the literature,[33,34] we propose that reduction of V^{5+} to V^{4+} species occurs according to the reactions (6) and (7):

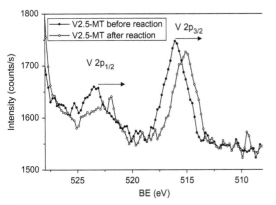

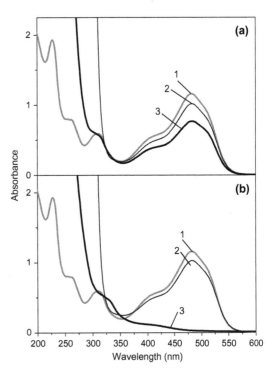

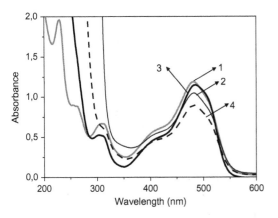

Figure 8   High resolution XP spectra of V2.5-MT catalyst both before and after reaction (V 2p BE range)

Figure 10   UV–vis spectra obtained in dark conditions in the presence of either MT (part *a*) and V2.5-MT (part *b*). Legend as follows: starting 0.67 mM AO7 solution (curves 1), supernatant solution contacted with the catalyst and 0.40 M H$_2$Asc (curves 2), and the same 20 min after addition of 0.80 M H$_2$O$_2$ (curves 3)

Figure 9   UV–vis spectra concerning blank experiments in the absence of any catalysts: starting 0.67 mM AO7 solution (curve 1), the same after 90 min in the presence of 0.40 M H$_2$Asc (curve 2), 0.80 M H$_2$O$_2$ (curve 3), and both 0.40 M H$_2$Asc and 0.80 M H$_2$O$_2$ (curve 4)

$$\equiv O_3V = O + H_2O_2 \rightarrow \equiv O_3V(OOH)OH \qquad (6)$$

$$\equiv O_3V\text{-}(OOH)OH \rightarrow \equiv O_3V\text{-}OH + HO2 \qquad (7)$$

$$HO_2 \cdot + H_2O_2 \rightarrow HO \cdot + H_2O + O_2 \qquad (8)$$

V^{4+} is formed together with a hydroperoxyl radical HO$_2$· than can react with excess H$_2$O$_2$ forming more reactive HO· (equation (8)). Such process is not catalytic, but stoichiometric, and indeed vanadium effect is only observed at the beginning of the reaction, the final conversion being almost the same as with bare TiO$_2$. If the MT conversion curve is subtracted from that of V2.5-MT, indeed, only 10% AO7 results to be converted by surface vanadium species, indicating that a limited fraction of surface V sites is reactive.

Formation of surface V^{4+} species is confirmed by XPS measurement run V2.5-MT: before reaction, two peaks are seen at

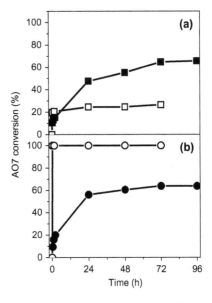

Figure 11   Comparison of the catalytic activity in dark conditions obtained with: (i) 0.80 M H$_2$O$_2$ without H$_2$Asc (full symbols) and (ii) 0.80 M H$_2$O$_2$ in the presence of 0.40 M H$_2$Asc (hollow symbols) for MT (part *a*) and V2.5-MT (part *b*)

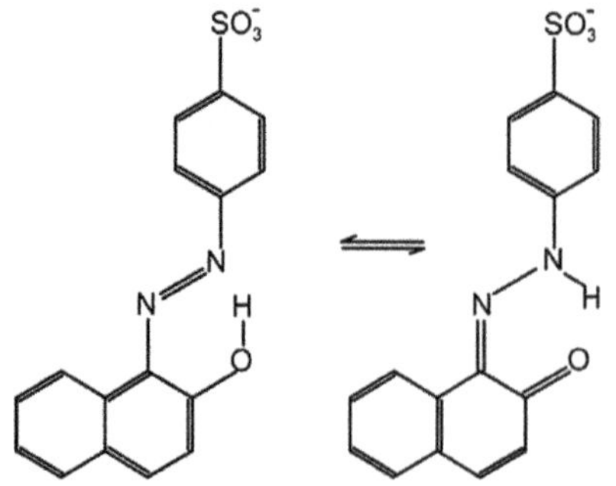

Scheme 1   Azo- and hydrazone form of AO7

515.9 and 523.4 eV, due to pentavalent V $2p_{3/2}$ and V $2p_{1/2}$ lines, respectively (Fig. 8).[35] The shift in peaks position after reaction shows that V^{5+} species underwent reduction to V^{4+}, in agreement with the literature reports on V-doped TiO_2,[36–38] supporting the mechanism proposed. Moreover, in both cases V $2p_{3/2}$ lines were satisfactorily curve-fitted with one peak, so showing that practically all the surface V^{5+} species present before reaction were reduced to V^{4+}.

Effect of ascorbic acid on catalytic activity in dark conditions

H_2Asc, a well-known reducing agent, was used to reduce vanadium to V^{3+} species, which are active in the Fenton-like reaction (vide infra).

Blank experiments without any catalyst were firstly run by contacting 0.67 mM AO7 solution with either 0.40 M H_2Asc or 0.80 M H_2O_2, and with both chemicals (0.40 M H_2Asc and 0.80 M H_2O_2). Fig. 9 compares the related UV–vis spectra: curve 2 is obtained with 0.40 M H_2Asc after 120-min contact, showing that some reaction occurs between AO7 and H_2Asc, whereas the strong absorption below 320 nm is due to H_2Asc itself.[39] AO7 conversion after 90 min with 0.80 M H_2O_2 is ca. 5.0% (curve 3), the intense absorption below 280 nm being due to H_2O_2. When both H_2A and H_2O_2 are present (curve 4), the band of H_2Asc decreases, due to reaction with H_2O_2.

Fig. 10 reports UV–vis spectra obtained with MT a and V2.5-MT b in the presence of 0.40 M H_2Asc (curves 2) and 20 min after addition of 0.80 M in the presence of 0.40 M H_2Asc, whereas Fig. 11 compares catalytic activities in dark conditions with and without H_2Asc. H_2Asc depresses the activity of MT; the conversion remaining constant to a value well below that obtained without H_2Asc (Fig. 11a). In agreement with the

literature,[40] H_2Asc is mostly adsorbed at the surface of titania through the reaction:

$$Ti-OH + H_2A \rightarrow Ti-OAH + H_2O \qquad (9)$$

decreasing the amount of surface Ti–OH, so that less sites are available for the formation of Ti–OOH species by interaction with H_2O_2.

With V2.5-MT, 100% conversion is reached in few minutes (Fig. 11b); the same effect was observed with both V2.5-CT and V0.80-CT.

V^{5+} surface species are likely reduced by H_2A to V^{3+} species, which may undergo Fenton-like[41,42] reaction switching to V^{4+}:

$$V^{3+} + H_2O_2 \rightarrow V^{4+} + OH^- + HO\cdot \qquad (10)$$

$$V^{4+} + H_2O_2 \rightarrow V^{3+} + H^+ + HO_2\cdot \qquad (11)$$

$$HO\cdot + H_2O_2 \rightarrow HO_2\cdot + H_2O \qquad (12)$$

$$HO_2\cdot + H_2O_2 \rightarrow HO\cdot + H_2O + O_2 \qquad (13)$$

$$AO7 + HO\cdot \rightarrow \text{degradation products} \qquad (14)$$

In this case, the presence of an excess of H_2Asc ensured reduction of all surface species and allowed several redox cycles, in which very reactive V^{3+} species were regenerated, avoiding catalyst deactivation due to formation of V^{4+}.

Conclusions

In dark conditions, MT is more active than P25 in the degradation of AO7, likely due to its larger SSA, which favors the formation of reactive Ti–OOH species in the presence of H_2O_2; under UV illumination, the two types of TiO_2 showed comparable activity, due to a balance between higher SSA of MT and mixed crystalline composition of P25.

Absorption of UV light gives rise to the formation of inactive V^{4+} species, and therefore, surface doping with V may be detrimental to photocatalytic activity. In dark conditions, V^{5+} species are reduced by H_2O_2 to the same inactive V^{4+} ions, but after contact with H_2Asc, reduction to V^{3+} species occurs, the latter species being, instead, very active in a Fenton-like mechanism.

Acknowledgments

Authors thank: Prof. Marco Sangermano (DISAT, Politecnico di Torino) for borrowing the UV lamp and Dr Edvige Celasco (DIFI, Università degli Studi di Genova) for XPS measurements.

Table 2   Initial rate (M s⁻¹) and final conversion, as obtained in dark conditions with both 0.030 M and 0.80 M H_2O_2

Catalyst	Initial rate with 0.030 M H_2O_2 (M s⁻¹)	Final conversion with 0.030 M H_2O_2 (%)	Initial rate with 0.80 M H_2O_2 (M s⁻¹)	Final conversion with 0.80 M H_2O_2 (%)
None	–	–	9.54×10^{-9}	11
MT	1.8×10^{-7}	9.0	5.8×10^{-8}	66
V2.5-MT	4.2×10^{-7}	49	5.3×10^{-8}	64
P25	2.3×10^{-7}	5.2	1.0×10^{-8}	25
V0.80-CT	1.3×10^{-7}	20	6.8×10^{-8}	40
V2.5-CT	3.3×10^{-7}	49	5.0×10^{-8}	40

References

1. C. Bauer, P. Jacques and A. Kalt: *Chem. Phys. Lett.*, 1999, **307**, 397–406.
2. J. Fernández, J. Kiwi, J. Baeza, J. Freer, C. Lizama and H. D. Mansilla: *Appl. Catal. B*, 2004, **48**, 205–211.
3. S. Liu, T. Xie, Z. Chen and J. Wu: *Appl. Surf. Sci.*, 2009, **255**, 8587–8592.
4. M. A. Rauf, M. A. Meetani and S. Hisaindee: *Desalination*, 2011, **276**, 13–27.
5. K. Rajeshwar, M. E. Osugi, W. Chanmanee, C. R. Chenthamarakshan, M. V. B. Zanoni, P. Kajitvichyanukul and R. Krishnan-Ayer: *J. Photochem. Photobiol. C*, 2008, **9**, 171–192.
6. R. Doong, P.-Y. Chang and C.-H. Huang: *J. Non-Cryst. Solids*, 2009, **335**, 2302–2308.
7. M. A. Barakat, H. Schaeffer, G. Hayes and S. I. Shah: *Appl. Catal. B*, 2004, **57**, 23–30.
8. X. Chen and S. S. Mao: *Chem. Rev.*, 2007, **107**, 2891–2959.
9. M. Yan, F. Chen, J. Zhang and M. Anpo: *J. Phys. Chem. B*, 2005, **109**, 8673–8678.
10. Y. Cong, J. Zhang, F. Chen, M. Anpo and D. He: *J. Phys. Chem. C*, 2007, **111**, 10618–10623.
11. M. Hussain, N. Russo and G. Saracco: *Chem. Eng. J.*, 2011, **166**, 138–149.
12. J. Zhu, J. Zhang, F. Chen and M. Anpo: *Mater. Lett.*, 2005, **59**, 3378–3381.
13. F. S. Freyria, B. Bonelli, R. Sethi, M. Armandi, E. Belluso and E. Garrone: *J. Phys. Chem. C*, 2011, **115**, 24143–24152.
14. D. C. Hurum, A. G. Agrios, K. A. Gray, T. Rajh and M. C. Thurnauer: *J. Phys. Chem. B*, 2003, **107**, 4545–4549.
15. U. Stafford, K. A. Gray, P. V. Kamat and A. Varma: *Chem. Phys. Lett.*, 1993, **205**, 55–61.
16. J. C.-S. Wu and C.-H. Chen: *J. Photochem. Photobiol. A*, 2004, **163**, 509–515.
17. L. Pauling and E. Cameron: 'Cancer and vitamin C'; 1993, Philadelphia, PA, Camino Books.
18. G. R. Buettner and B. A. Jurkiewicz: *Radiat. Res.*, 1996, **145**, 532–541.
19. K. Kustin and D. L. Toppen: *Inorg. Chem.*, 1973, **12**, 1404–1407.
20. S. Fukuchi, R. Nishimoto, M. Fukushima and Q. Zhu: *Appl. Catal. B*, 2014, **147**, 411–419.
21. G. B. Shul'pin and E. R. Lachter: *J. Mol. Catal. A*, 2003, **197**, 65–71.
22. M. Piumetti, F. S. Freyria, M. Armandi, F. Geobaldo, E. Garrone and B. Bonelli: *Catal. Today*, 2014, **227**, 71–79.
23. B. Schwenzer, L. Wang, J. S. Swensen, A. B. Padmaperuma, G. Silverman, R. Korotkov and D. J. Gaspar: *Langmuir*, 2012, **28**, 10072–10081.
24. M. Piumetti, B. Bonelli, P. Massiani, S. Dzwigaj, I. Rossetti, S. Casale, M. Armandi, C. Thomas and E. Garrone: *Catal. Today*, 2012, **179**, 140–148.
25. M. Piumetti, B. Bonelli, P. Massiani, Y. Millot, S. Dzwigaj, L. Gaberova, M. Armandi and E. Garrone: *Microporous Mesoporous Mater.*, 2011, **142**, 45–54.
26. C. Bauer, P. Jacques and A. Kalt: *J. Photochem. Photobiol. A*, 2001, **140**, 87–92.
27. C. Von Sonntag and H. P. Schuchmann: 'Peroxyl radicals', (ed. Z. B. Alfassi), 173–234; 1997, New York, NY, John Wiley and Sons.
28. T. Ohno, K. Sarukawa, K. Tokieda and M. Matsumura: *J. Catal.*, 2001, **203**, 82–86.
29. A. A. Frimer: 'Oxygen radicals in biology and medicine', (ed. M. G. Simic, et al.), 29–39; 1988, New York, NY, Plenum Press.
30. W. Lin and H. Frei: *J. Am. Chem. Soc.*, 2002, **124**, 9292–9298.
31. M. G. Clerici and O. A. Khooldeva: *'Liquid phase oxidation via heterogeneous catalysis'*; 2013, NJ, Wiley.
32. J. Zou, J. Gao and Y. Wang: *J. Photochem. Photobiol. A*, 2009, **202**, 128–135.
33. G. V. Nizova, Y. N. Kozlov and G. B. Shul'pin: *Russ. Chem. Bull. Int. Ed.*, 2004, **53**, 2330–2333.
34. G. Süss-Fink, L. Gonzalez and G. B. Shul'pin: *Appl. Catal. A*, 2001, **217**, 111–117.
35. C. B. Rodella, P. A. P. Nascente, R. W. A. Franco, C. J. Magon, V. R. Mastelaro and A. O. Florentino: *Phys. Status Solidi A*, 2001, **187**, 161–169.
36. C. Gannoun, R. Delaigle, P. Eloy, D. B. Debecker, A. Ghorbel and E. M. Gaigneaux: *Catal. Commun.*, 2011, **15**, 1–5.
37. T.-B. Nguyen, M.-J. Hwang and K. S. Ryu: *Appl. Surf. Sci.*, 2012, **258**, 7299–7305.
38. S. Youn, S. Jeong and D. H. Kim: *Catal. Today*, 2014, **232**, 185–191.
39. E. Kleszczewska: *Polish J. Environ. Stud.*, 1999, **8**, 313–318.
40. E. I. Morosanova, M. V. Belyakov and Y. A. Zolotov: *J. Anal. Chem.*, 2012, **67**, 14–20.
41. H. J. H., Fenton: *J. Chem. Soc. Trans.*, 1894, **65**, 899–910.
42. J. J. Pignatello, E. Oliveros and A. MacKay: *Crit. Rev. Environ. Sci. Technol.*, 2006, **36**, 1–84.

5

Re-dispersion of gold supported on a 'mixed' oxide support

Kevin Morgan*[1], Robbie Burch[1], Muhammad Daous[2], Juan-José Delgado[3], Alexandre Goguet[1], Christopher Hardacre**[1], Lachezar A. Petrov[2] and David W. Rooney[1]

[1]CenTACat, School of Chemistry and Chemical Engineering, David Keir Building, Queen's University Belfast, Stranmillis Road, Belfast, BT9 5AG, Northern Ireland, UK
[2]Chemical and Materials Engineering Department, King Abdulaziz University, Jeddah, Saudi Arabia
[3]Departamento de Ciencia de los Materiales e Ingeniería Metalúrgica y Química Inorgánica, Facultad de Ciencias, Universidad de Cádiz. E-11510 Puerto Real, Cádiz, Spain

Abstract The ability to reactivate, stabilize and increase the lifetime of gold catalysts by dispersing large, inactive gold nanoparticles to smaller nanoparticles provides an opportunity to make gold catalysts more practical for industrial applications. Previously it has been demonstrated that mild treatment with iodomethane (*J. Am. Chem. Soc.*, 2009, 131, 6973; *Angew. Chem. Int. Ed.*, 2011, 50, 8912) was able to re-disperse gold on carbon and metal oxide supports. In the current work, we show that this technique can be applied to re-disperse gold on a 'mixed' metal oxide, namely a mechanical mixture of ceria, zirconia and titania. Characterization was conducted to guage the impact of the iodomethane (CH_3I) treatment on a previously sintered catalyst.

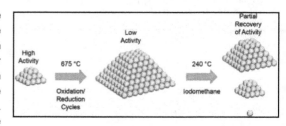

Keywords Gold catalysts, Catalyst regeneration, Gold re-dispersion, CO oxidation

Introduction

Since the 1980s when it was first demonstrated that gold was catalytically active,[1,2] there has been an ever increasing number of reports on gold catalyzed reactions.[3–5] The utilization of gold catalysts has been spurred on by advances in the ability to produce small nanoparticles, which have been found to be catalytically active in a range of reactions in sectors as varied as environmental remediation and bulk/pharmaceutical chemical production.[3,6]

It has been reported that the size of the deposited gold nanoparticles as well as the nature of the metal-support interaction are strongly linked to the catalytic performance.[3,7–9] However, while these systems often show excellent initial activity and selectivity, it is known that gold catalysts typically have low stability under thermal treatment and/or reaction conditions.[6] This deactivation has been proposed to be associated with surface poisoning, loss of interaction between gold and the support, as well as sintering of the gold nanoparticles.[6] Hence, if

gold catalysts are to have a long-term future in industrial applications, methods of stabilization and/or reactivation need to be developed in order to make this feasible.[3]

It has been reported previously that Au/C catalysts could be activated during the carbonylation of methanol to methyl acetate in the presence of iodomethane (CH_3I).[10,11] This activation has been attributed to increased gold nanoparticle dispersion from clusters of > 10 nm to gold dimers/trimers, which are stabilized by iodine. Owing to the harsh conditions (240°C and 16 bar), this process may not be applicable for all gold catalytic systems; however, it has recently been demonstrated that atmospheric pressure treatments using iodomethane for gold dispersion provide a treatment to atomically disperse these large particles of gold on both carbon[6,12] and metal oxide supports.[3]

While gold catalysts supported on both carbon and metal oxides are common, there are a range of mixed metal oxides which have been found to provide additional desirable properties over pure metal oxides. For example, mixed ceria–zirconia supports possess the ability to switch between Ce^{3+} and Ce^{4+} due to elevated oxygen storage and oxygen transport capacity.[13] It has been reported that catalysts supported on mixed metal oxides can be more active

*Corresponding author, email kmorgan08@qub.ac.uk
**Corresponding author, email c.hardacre@qub.ac.uk

than those supported on individual pure metal oxides.[13,14] Some of the reactions for which gold catalysts supported on mixed metal oxides have been found to be particularly active include CO oxidation,[14,15] water gas shift (WGS)[16,17] and preferential oxidation of CO (PROX).[18,19]

The catalytic oxidation of carbon monoxide is an important reaction in applications including cold start exhaust emission control for automobiles, respirators and clean hydrogen production. Mixed oxide catalysts have been reported for CO oxidation,[20,21] the activity of which have been found to improve upon doping with gold.[14,15] Au/CeZrTiO$_x$ is a mixed oxide catalyst which has been designed for applications in hydrocarbon oxidation as well as methanol, methane and carbon monoxide fuel cells.[22]

To date, there has been no reported application of the iodomethane treatment to gold catalysts supported on mixed metal oxide supports in order to tailor the properties of the catalyst. Herein, we report the effect of iodomethane treatment on the activity of a previously sintered 1.0 wt% Au/CeZrTiO$_x$ catalyst for CO oxidation.

Experimental

Catalyst preparation

To deposit gold on the catalysts, a LabMax®(Mettler-Toledo, USA) automated laboratory reactor system with a 500 cm^3 glass reactor was used. This system provides constant reactor temperature, constant pH of the reaction media, uniform constant intensive mixing and simultaneous controlled dosing of liquid components. The support was a mechanical mixture with composition CeO$_2$:ZrO$_2$:TiO$_2$ = 5.5 : 2.5 : 2.0.

Before gold deposition, the support was subjected to ultrasonic treatment for 30 min and dried at 120°C for 4 h. This was to ensure sufficient mixing of the support in order to facilitate a homogeneous gold dispersion. Gold was deposited from an aqueous solution of HAuCl$_4$ (Acros Organics, Belgium) at 80°C, with intensive mixing. An aqueous solution of magnesium citrate was added to maintain the pH of the liquid phase at 8.0 ± 0.1. At the end of the precipitation process, the suspension was aged for 1 h at 60°C. The suspension was then filtered and the catalyst (~ 10 g) was washed with 1 L of warm distilled water (50°C) to remove Cl$^-$ ions. The powder was then dried for 4 h at 120°C and then calcined by ramping the temperature at 10°C min^{-1} to 550°C before being held at this temperature for 6 h in air. As gold clusters are known to sinter with prolonged heating at temperatures greater than 395°C,[23] the sintered catalyst was prepared by cycling between reducing and oxidizing environments at 675°C.

Treatment with iodomethane

Iodomethane (Aldrich, USA) treatments were performed in a quartz reactor (O.D. 6 mm) using ca. 100 mg of catalyst. The quartz reactor was heated in flowing argon (40 cm^3 min^{-1}) with a tubular furnace and the temperature was increased to the set temperature (240°C) using a ramp rate of 10°C min^{-1}. Once at temperature, the iodomethane/argon was admitted to the reactor with a steady flow of 40 cm^3 min^{-1} iodomethane/argon with the vapor pressure of the iodomethane determined by a saturator held at 25°C, equating to 0.14 cm^3 min^{-1} of iodomethane.

Characterization

Inductively coupled plasma – optical emission spectrometry (ICP-OES) was used to quantify the concentration of gold on the fresh, sintered and iodomethane-treated catalysts. Sodium peroxide was used to remove the gold from the support, and a 4300 DV ICP-OES (Perkin Elmer, USA) was used to quantify the amount of gold for all the catalysts. The Brunauer, Emmett and Teller (BET) surface area characterisation was obtained using a Tristar 3000 (Micromeritics, USA) gas adsorption analyzer using N$_2$ to determine average pore volume. The sample was subjected to vacuum to remove any impurities, such as water, followed by flushing with helium gas for 2 min before the vacuum was then reintroduced.

Raman analysis of the samples was carried out in order to determine any effect of the treatments on the support. A Ramanstation fibre optic system (Avalon, UK) was used employing a 785 nm laser. Spectra were accumulated for 90 s and the reported data was averaged over five scans.

High Resolution Transmission Electron Micrographs were collected on a 200 kV JEM-2010F(JOEL, USA) instrument with a structural resolution of 0.19 nm at Scherzer defocus conditions. Scanning Transmission Electron Microscopy (STEM) images were recorded in the same instrument using a high-angle annular dark-field (HAADF) detector and an electron beam probe of 0.5 nm.

Powder X-Ray diffraction (XRD) measurements were carried out using an X'Pert Pro X-ray diffractometer (PANalytical, USA). The X-ray source used was copper K$_\alpha$ with a wavelength of 1.5405Å. Diffractograms were collected from 20° to 85° with a step size of 0.04°.

Diffuse reflectance ultraviolet–visible (DRUV) spectroscopy measurements were performed using a Lambda 650S UV–vis spectrophotometer (Perkin Elmer, USA) equipped with a Harrick stage chamber and an integrating sphere detector Magnesium oxide was used as the 'auto zero' reference for all sample spectra.

CO oxidation

In order to probe the ability of iodomethane to improve the activity of the sintered catalyst, CO oxidation was performed as this is known to require small gold nanoparticles for good activity.[24–26] Activity tests were performed in a quartz tubular reactor (O.D. 10 mm). One hundred milligrams of catalyst was held in place between two plugs of quartz wool and a thermocouple was placed in the center of the catalyst bed. The temperature of the reactor was controlled using a tubular furnace with a proportional-integral-derivative (PID) controller and the reaction was investigated by ramping the temperature to 500°C. The reactants and products were analyzed at the outlet of the reactor by gas chromatography (Clarus 500, Perkin Elmer) with a FID and TCD. The gas feed consisted of 2% CO, 2% O$_2$, 0.5% Kr and 95.5% Ar with a total flowrate of 100 cm^3 min^{-1}.

Results and discussion

Fig. 1 shows the comparison of the CO oxidation activity for the fresh 1.0 wt% Au/CeZrTiO$_x$ catalyst, sintered catalyst, iodomethane-treated sintered catalyst and support. The support showed 50% conversion at 400°C. As expected, on addition of gold, significantly higher activity was observed

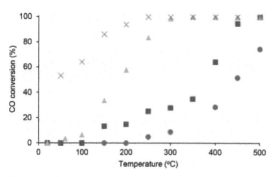

Figure 1 CO conversions as a function of temperature for the Au/CeZrTiO$_x$ catalyst samples; support (●), fresh (x), sintered (■) and sintered/CH$_3$I (▲)

with the fresh catalyst showing 50% conversion at 50°C. On sintering, the activity of the catalyst significantly decreased only achieving 50% conversion at ~375°C, i.e. close to that found for the support only. Importantly, after the sintered catalyst was treated with iodomethane, there was a marked increase in the activity with 50% conversion observed at ~175°C. In order to probe the reason for the enhancement following iodomethane treatment, the catalyst structure and textural characteristics were examined.

The nature of the catalyst (mechanical mixture of three metal oxide supports) has made material characterization difficult, and as such it has been a case of using a process of elimination in order to gain an understanding of the effect of the iodomethane treatment. Additionally, for purposes of clarity, while results from each of the techniques are discussed below, only results from techniques where discernible changes were observed are reported in the table and figures.

Gold elemental analysis for the catalysts are summarized in Table 1. These results show that the iodomethane treatment did not significantly change the gold loading. This is consistent with the previously reported data on carbon supported and metal oxide supported catalysts, where little leaching/removal of gold was observed following treatment with iodomethane even at 240°C.[3] Therefore, any changes observed with other characterization methods can be associated with structural changes in the gold particles rather than due to a significant loss of gold content.

Table 1 summarizes the BET surface areas of the studied systems. Comparing the BET surface area of the fresh catalyst to the support alone shows a small increase in surface area. However, on sintering the sample the surface area decreases from 116 to ~8 m^2 g^{-1} coupled with a significant decrease in the pore volume of the catalyst. This is reversed to some degree

on treatment with iodomethane where the surface area is found to increase to ~25 m^2 g^{-1}; however, this is still lower than that of the original fresh catalyst.

The XRD patterns for each catalyst system (Fig. 2) shows reflections which can be attributed to the individual components of the support; i.e. titania (at 25.2, 47.4 and 69.4°), ceria (at 28.5, 59.0, 76.6 and 79.0°) and zirconia (33.1 and 56.3°). On deposition of gold onto the support a change in the mixed oxide particle size is observed. This is shown by examining the dominant reflection for ceria at 28.5°, titania at 47.4° and zirconia at 56.3°. In each case, a small shift in the maximum, of ~0.3°, was observed associated with a decrease in the lattice spacing. The more striking effect is the broadening of the oxide features. For example, the full width at half maximum (FWHM) increased from ~0.3° to 1.1°, 1.6° and 1.8° for the ceria, titania and zirconia features, respectively. This would indicate a decrease in particle size of the oxide crystallites. This may be responsible for the small increase in BET surface area from the support to the fresh catalyst sample. It is also worth noting that, after the gold sintering procedure, the FWHM of these three reflections decreases to 0.7°, 1.1° and 0.8°. This is indicative of a growth in the oxide particle size which shows that the catalyst support also underwent sintering. The increase in the particle size may be contributing to the sharp decrease in the BET surface area; however, it is unlikely to be the predominant cause due to the large change observed. The latter is more likely to be due to the loss of pores as a result of the collapse of the pore structure.

It should be noted that the XRD patterns for all the gold based samples had weak diffraction features at 38.4° and 44.5° which can be attributed to gold. However, due to the low signal to noise ratio and interference from a weak support feature, in the case of the peak at ~38°, this did not allow accurate measurement of the gold particle size. It is interesting to note that, unlike in the previously reported systems where the iodomethane treatment resulted in an almost complete absence of gold diffraction features,[3,6,12,27] the gold features remained. This may be associated with some of the gold not dispersing following the treatment as the diffractograms are always dominated by the largest particle size.

In order to further probe the changes in the support and the gold following the treatment, Raman and TEM studies

Table 1 Gold elemental analysis and textural characteristics of the Au/CeZrTiO$_x$ catalyst samples

Catalyst Sample	Au content (wt-%)	BET surface area (m^2 g^{-1})	Pore volume (cm^3 g^{-1})	Pore diameter (nm)
Support	–	111.0	0.20	6.7
Fresh	0.98	116.0	0.18	6.6
Sintered	1.02	7.6	0.02	16.1
Sintered/CH$_3$I	1.00	24.6	0.08	14.2

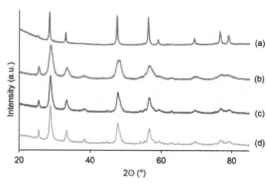

Figure 2 X-ray diffractograms for the Au/CeZrTiO$_x$ catalyst samples; support (a), fresh (b), sintered (c) and sintered/CH$_3$I (d)

were undertaken. There were no discernible changes in the band frequencies or the FWHM observed in the Raman spectra, showing little influence of the iodomethane treatment on the phase of the support, in agreement with the XRD patterns. Similarly, the TEM did not show any significant changes to the structure of the support and only dramatic changes of the particle size could be deciphered. The TEM did show large gold particles, the smallest average size of which were of the order of 25–50 nm; however, due to the complex chemical composition of the support (three cations) and the poor Z contrast between the gold and the ceria element of the support, the presence of small gold particles or clusters (<0.5 nm) cannot be dismissed. From energy-dispersive X-ray (EDX) spectroscopy, the gold is evenly distributed across the support and it does not appear that the gold has preferentially been located on one elemental phase over the others. However, the TEM was able to determine that no particles in the size range of 1–5 nm are likely to be present. Therefore, it was not possible to determine accurate particle size distribution.

In order to probe the smaller gold particles, diffuse reflectance UV–vis spectroscopy (DRUV) was employed as this is sensitive to the small rather than the large particles, in contrast with the XRD technique. Diffuse reflectance ultraviolet–visible height normalized spectra for each of the gold based materials are shown in Fig. 3. A broad characteristic gold adsorption band for the fresh and sintered catalyst, with the maximum at ~600 nm was observed; however, on treatment of the sintered catalyst with iodomethane, this feature disappears and there is an increase in the UV–vis absorption at lower wavelengths. The peak ~600 nm has been attributed to the plasmon resonance of gold metal nanoparticles[3,28,29] whereas absorption at lower wavelengths is owing to ionic gold ($\lambda \leq 250$ nm) and small clusters of gold ($280 \leq \lambda \leq 380$ nm).[30,31] The disappearance of the peak at 600 nm and the increase in absorption at around below 500 nm indicate dispersion of some of the gold nanoparticles into small clusters following treatment of the sintered sample with iodomethane.[3] Interestingly, these nanoparticles are present in both the fresh and sintered catalysts; however, as shown by the high activity of the fresh catalyst for CO oxidation, it is likely that this catalyst also has well dispersed gold. It is not possible to probe the presence of the very smallest particles/ionic gold

owing to the presence of the titania and its strong absorption band below 330 nm.[3]

The high CO oxidation activity of the fresh catalyst is likely to be owing to the presence of highly dispersed gold, as reported previously,[24–26] and high surface area of the support compared with the sintered catalyst. From the DRUV, it is clear that the iodomethane treatment of the sintered catalyst resulted in a reduction of gold particle size and this is reflected in the increased activity of the iodomethane-treated sintered sample for CO oxidation compared to the sintered catalyst, as well as an increase in the surface area of the sample. An increase in support surface area alone does not promote the catalysis since the support has a much larger surface area than the sintered and iodomethane-treated sintered samples, yet the support has a far inferior activity for the CO oxidation reaction than any of the catalyst samples. From previous studies it is known that the iodomethane treatment can result in almost complete dispersion of gold into atoms, dimers and trimers and this is likely to be the case in these samples too. However, while significantly higher activity is observed for the treated sintered catalyst compared with the system before treatment, the activity of the original catalyst is not recovered. This may be due to residual halide being present on the surface of the catalyst which is known to poison gold based CO oxidation catalysts.[32–34] In addition, the surface area and pore volume of the sintered catalyst following treatment is much lower than that of the fresh catalyst resulting in far fewer accessible active sites.

Conclusion

The reported results showed the applicability of iodomethane treatments for the reactivation of gold, probably owing to re-dispersion of gold particles in a mechanically mixed oxide support, namely ceria–zirconia–titania. The TEM analysis suggests that there is little difference in the samples as a consequence of the treatment with iodomethane. The DRUV does indicate that iodomethane-treated sintered sample did result in gold redispersion. Importantly, iodomethane treatment of the sintered sample was found to reactivate the deactivated sintered catalyst for CO oxidation and significantly reduced the temperature for 50% conversion by 200°C. Although the catalyst was not as active as the fresh catalyst treatment either owing to the presence of surface halide or the smaller surface area, this is the first time that an effectively completely deactivated gold catalyst has been reactivated using a technique which does not reduce the amount of gold on the surface of the catalyst. This treatment provides a significant opportunity to reactivate gold based catalysts without the need to remove the catalyst from the reactor system.

Notes

Supporting data are openly available on Queen's University Research Portal http://pure.qub.ac.uk./portal/en/datasets.

Acknowledgements

Authors gratefully acknowledge funding for this work from King Abdulaziz University (grant no. D-005/431), and the CASTech grant (EP/G012156/1) from the EPSRC. Authors

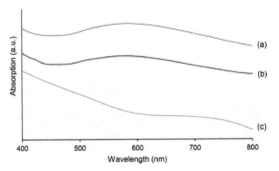

Figure 3 Diffuse reflectance ultraviolet–visible (DRUV) spectra for the Au/CeZrTiO$_x$ catalyst samples; fresh (*a*), sintered (*b*) and sintered/CH$_3$I (*c*)

would like to acknowledge the contributions of Yahia Alhamed, Abdulrahim Al-Zahrani and Ahmed Arafat (all of King Abulaziz University). Juan-José Delgado is grateful to Ramon y Cajal program and the Ce-NanoSurPhases project grant form MINECOx.

References

1. M. Haruta, T. Kobayachi, H. Sano and N. Yamada: *Chem. Lett.*, 1987, **2**, 405–408.
2. G. J. Hutchings: *J. Catal.*, 1985, **96**, 292–295.
3. J. Sá, R. Taylor, H. Daly, A. Goguet, R. Tiruvalam, Q. He, C. J. Kiely, G. J. Hutchings and C. Hardacre: *ACS Catal.*, 2012, **2**, 552–560.
4. A. S. K. Hashmi and G. J. Hutchings: *Angew. Chem. Int. Ed.*, 2006, **45**, 7896–7936.
5. G. C. Bond, C. Louis and D. T. Thompson: 'Catalysis by gold'; 2006, London, Imperial College Press.
6. J. Sá, S. F. R. Taylor, C. Paun, A. Goguet, R. Tiruvalam, C. Kiely, M. Nachtegaal, G. Hutchings and C. Hardacre: *Angew. Chem. Int. Ed.*, 2011, **50**, 8912–8916.
7. A. Abad, P. Concepcion, A. Corma and H. Garcia: *Angew. Chem. Int. Ed.*, 2005, **44**, 4066–4069.
8. S. Biella and M. Rossi: *Chem. Commun*, 2003, 378–379.
9. J. Guzman and B. C. Gates: *J. Am. Chem. Soc.*, 2004, **126**, 2672–2673.
10. J. R. Zoeller, A. H. Singleton, G. C. Tustin and D. L. Carver: 'Vapor phase carbonylation process using gold catalysts', Eastman Chemical Company, US Patent 6506933 2003.
11. J. R. Zoeller, A. H. Singleton, G. C. Tustin and D. L. Carver: 'Gold based heterogeneous carbonylation catalysts', Eastman Chemical Company, US Patent 6509293 2003.
12. A. Goguet, C. Hardacre, I. Harvey, K. Narasimharao, Y. Saih and J. Sá: *J. Am. Chem. Soc.*, 2009, **131**, 6973–6975.
13. L. Matějová, P. Topka, L. Kaluža, S. Pitkäaho, S. Ojala, J. Gaálová and R. L. Keiski: *Appl. Catal. B*, 2013, **142-143**, 54–64.
14. M. Haruta, N. Yamada, T. Kobayashi and S. Iijima: *J. Catal.*, 1989, **115**, 301–309.
15. K. Morgan, K. J. Cole, A. Goguet, C. Hardacre, G. J. Hutchings, N. Maguire, S. O. Shekhtman and S. H. Taylor: *J. Catal.*, 2010, **276**, 38–48.
16. H. Daly, A. Goguet, C. Hardacre, F. C. Meunier, R. Pilasombat ar D. Thompsett: *J. Catal.*, 2010, **273**, 257–265.
17. R. Pilasombat, H. Daly, A. Goguet, J. P. Breen, R. Burch, C. Hardac and D. Thompsett: *Catal. Today*, 2012, **180**, 131–138.
18. D. Cameron, R. Holliday and D. Thompson: *J. Power Sources*, 200 **118**, 298–303.
19. O. H. Laguna, M. A. Centeno, G. Arzamendi, L. M. Gandía, F. Romer Sarria and J. A. Odriozola: *Catal. Today*, 2010, **157**, 155–159.
20. T. H. Rogers, C. S. Piggot, W. H. Bahlke and J. M. Jennings: *J. Ar Chem. Soc.*, 1921, **43**, 1973–1982.
21. B. Solsano, G. J. Hutchings, T. Garcia, S. H. Taylor: New: *J. Chem* 2004, **28**, 708–711.
22. L. A. Petrov and V. M. Tatchev: 'Gold catalyst for fu cells', Laman Consultancy Limited, European Patent EP1027152-A 2000.
23. G. M. Veith, A. R. Lupini, S. Rashkeev, S. J. Pennycook, D. R. Mullin V. Schwartz, C. A. Bridges and N. J. Dudney: *J. Catal.*, 2009, **262**, 92–10
24. G. C. Bond and D. T. Thompson: *Catal. Rev. Sci. Eng.*, 1999, **4** 319–388.
25. M. Valden, X. Lai and D. W. Goodman: *Science*, 1998, **281**, 1647–165
26. M. Okumura, K. Tanaka, A. Ueda and M. Haruta: *Solid State Ionic* 1997, **95**, 143–149.
27. K. Morgan, R. Burch, M. Daous, J. J. Delgado, A. Gogue C. Hardacre, L. A. Petrov and D. W. Rooney: *Catal. Sci. Technol.*, 201 **4**, 729–737.
28. Y. -M. Kang and B. -Z. Wan: *Catal. Today*, 1995, **26**, 59–69.
29. N. Bogdanchikova, A. Pestryakov, I. Tuzovskaya, T. A. Zeped M. H. Farias, H. Tiznado and O. Martynyuk: *Fuel*, 2013, **110**, 40–47.
30. E. Smolentseva, N. Bogdanchikova, A. Simakov, A. Pestryako I. Tusovskaya, M. Avalos, M. H. Farías, J. A. Díaz and V. V. Gurin: *Sur Sci.*, 2006, **600**, 4256–4259.
31. I. V. Tuzovskaya, A. V. Simakov, A. N. Pestryakov, N. E. Bogdanchikova V. V. Gurin, M. H. Farías, H. J. Tiznado and M. Avalos: *Catal. Commun* 2007, **8**, 977–980.
32. C. K. Costello, M. C. Kung, H. -S. Oh, Y. Wang and H. H. Kung: *App Catal. A*, 2002, **232**, 159–168.
33. H. -S. Oh, J. H. Yang, C. K. Costello, Y. M. Wang, S. R. Bare, H. H. Kun and M. C. Kung: *J. Catal.*, 2002, **210**, 375–386.
34. P. Broqvist, L. M. Molina, H. Grönbeck and B. Hammer: *J. Catal.*, 200 **227**, 217–226.

6

Production of butylamine in the gas phase hydrogenation of butyronitrile over Pd/SiO$_2$ and Ba-Pd/SiO$_2$

Y. Hao, M. Li, F. Cárdenas-Lizana and M. A. Keane*

Chemical Engineering, School of Engineering and Physical Sciences, Heriot-Watt University, Edinburgh EH14 4AS, Scotland

Abstract The gas phase (1 atm, 473–563 K) hydrogenation of butyronitrile has been studied over Pd/SiO$_2$ and Ba–Pd/SiO$_2$. Catalysts characterization involved temperature-programmed reduction (TPR), H$_2$/NH$_3$ chemisorption/temperature programmed desorption (TPD), X-ray diffraction (XRD), and TEM

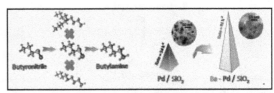

measurements. The incorporation of Ba with Pd resulted in the formation of smaller metal nano-particles (7 vs 28 nm) with a resultant (seven-fold) higher H$_2$ chemisorption and decreased total surface acidity (from NH$_3$ chemisorption/TPD). Temperature-related activity maxima were observed for both catalysts and are associated with thermal desorption of the nitrile reactant. Exclusivity to the target butylamine was achieved at $T \geq 543$ K where Ba–Pd/SiO$_2$ delivered higher selective hydrogenation rate (91 mol h^{-1} mol$_{Pd}^{-1}$) than Pd/SiO$_2$ (54 mol h^{-1} mol$_{Pd}^{-1}$), attributed to greater availability of surface-reactive hydrogen. Lower surface acidity served to minimize condensation to higher amines. The rate and selectivity to butylamine exceed those previously reported for gas phase operation.

Keywords Selective hydrogenation, Butyronitrile, Butylamine, Pd/SiO$_2$, Ba–Pd/SiO$_2$

Introduction

Primary aliphatic amines are widely used in the textile, pharmaceutical, fine chemical, and agrochemical sectors.[1] Industrial amine production is based on nitrile hydrogenation where undesired formation of secondary and tertiary amines is difficult to avoid.[1,2] Taking the hydrogenation of butyronitrile (Fig. 1), partial reduction (step (I)) generates an imine (butylidenimine) as reactive intermediate that is converted (step (II)) to the target butylamine. The latter can participate in a condensation (step (III)) with the imine and reduction (step (IV)) to produce the secondary amine (dibutylamine) with NH$_3$ elimination. Additional condensation of dibutylamine and imine (step (V)) and subsequent hydrogenation (step (VI)) generate the tertiary amine (tributylamine).[3] Nitrile hydrogenation has been predominantly conducted in batch liquid phase at elevated H$_2$ pressures (20–45 bar).[4–8] A range of supported transition metal (Ru,[5,9] Ni,[4,10–12] Co,[8,13,14] Rh,[5] Cu,[6] Pd,[5,15] and Pt[5,15]) catalysts has been used, where Ru, Ni, and Co favored primary amine formation, Rh and Cu generated the secondary

amine, and Pd and Pt promoted amine mixtures.[2] There is evidence that reaction rate is dependent on the catalytic metal where the following sequence of decreasing activity has been reported in batch liquid phase hydrogenation of butyronitrile over SiO$_2$ supported catalysts:[16] Ni > Co > Pt > Ru > Cu > Pd. A move from batch to continuous processes has been highlighted by the fine chemical industry as a priority to reduce downtime and increase throughput.[17] Supported Ni catalysts have been applied in gas phase operation but low selectivity to the target primary amine and loss of activity with time on-stream are decided drawbacks.[10,11,18–21] Use of Pd/ZnO has delivered high selectivity (99%) to ethylamine but low activity (conversions <6%) in acetonitrile conversion where (PdZn) alloy formation served to inhibit condensation to higher amines.[22]

Selectivity in nitrile reduction is affected by the acid–base properties of the support[4] and electronic character of the metal site[23] which, in turn, is influenced by the use of additives and/or promoters.[24] An increase in carrier basicity (by nitrogen doping[10] or LiOH treatment[13]) inhibits condensation in the hydrogenation of butyronitrile (Fig. 1, steps (III) and (V)). Moreover, use of ammonia provides a basic reaction medium that can enhance primary amine selectivity.[2] Modifications to the electron density

*Corresponding author, email m.a.keane@hw.ac.uk

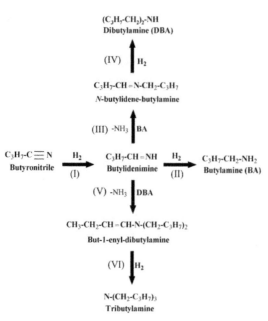

Figure 1 Reaction scheme for the hydrogenation of butyronitrile

of supported metal sites can also impact on product distribution where the formation of negatively charged (Cu) nanoparticles was deemed essential for the selective synthesis of *n*-propylamine.[24] Although supported nano-scale Pd has been widely adopted in catalytic hydrogenation,[25,26] application in nitrile conversion is limited. This is most likely due to the observed low activity and selectivity to primary amines.[2,7,15,16] In previous work, we reported the formation of a supported $Pd^{\delta-}$ phase on Ba–Pd/SiO$_2$ with greater H$_2$ uptake capacity relative to Pd/SiO$_2$.[27] Authors explore here the promoting effect of Ba in Pd/SiO$_2$ as an approach to enhanced butylamine production in gas phase butyronitrile hydrogenation.

Experimental

Catalyst preparation, activation, and characterization

A 5% w/w Pd/SiO$_2$ was prepared by impregnation of fumed SiO$_2$ (Aldrich) with Pd(OAc)$_2$ in dimethylformamide (DMF). The DMF was removed from the impregnated sample under vacuum over 12 h at ambient temperature. The Ba–Pd complex ((DMF)$_x$BaPd(CN)$_4$)$_\infty$ bimetallic precursor was prepared as described elsewhere.[28] The SiO$_2$ support was added to a solution of the precursor in DMF to deliver a 5% w/w Pd loading (Pd/Ba = 1 mol mol^{-1}). Samples were sieved (ATM fine test sieves) into batches of 75 μm mean particle diameter and activated by reduction in flowing (60 cm^3 min^{-1}) dry H$_2$ (BOC, 99.99%) at 10 K min^{-1} to 573 K, flushed in N$_2$/He, cooled to ambient temperature, and passivated in 1% v/v O$_2$/He for *ex situ* analysis. Metal loading was determined by inductively coupled plasma-optical emission spectrometry (ICP-OES, Vista-PRO, Varian Inc. (Palo Alto, USA)) from the diluted extract of aqua regia (25% v/v HNO$_3$/HCl). Nitrogen adsorption/desorption, temperature programmed reduction (TPR), H$_2$ (and NH$_3$) chemisorption, and temperature-

programmed desorption (TPD) were performed using the commercial CHEM-BET 3000 (Quantachrome) unit equipped with an on-line thermal conductivity detector (TCD) and data acquisition using the TPR Win™ software. Samples (0.05–0.1 g) were loaded in a U-shaped Quartz cell (3.76 mm i.d.), outgassed for 30 min, and the total specific surface area (SSA) recorded in a 30% v/v N$_2$/He flow with undiluted N$_2$ (BOC, 99.9%) as internal standard. Two cycles of N$_2$ adsorption–desorption were employed using the standard single-point BET method. Temperature-programmed reduction analysis was conducted in 17 cm^3 min^{-1} (Brooks mass flow controlled) 5% v/v H$_2$/N$_2$ at 10 K min^{-1} to 573 K. The reduced sample was maintained at 573 K in H$_2$/N$_2$ until the signal returned to baseline, swept with 65 cm^3 min^{-1} N$_2$/He for 1.5 h, cooled to ambient temperature, and subjected to an undiluted H$_2$ (BOC, 99.99%) or NH$_3$ (BOC, 99.98%) pulse (50–1000 μL) titration procedure. A contribution due to Pd hydride formation can be discounted as H$_2$ partial pressure (<2 torr) in the sample cell was well below that (>11 torr) required to generate the hydride.[29] Samples were thoroughly flushed in N$_2$/He (65 cm^3 min^{-1}) to remove weakly bound H$_2$ (or NH$_3$) and subjected to TPD (at 10–50 K min^{-1}) to 1000–1273 K with a final isothermal hold until the signal returned to baseline. The resultant profile was corrected using the TPD recorded in parallel directly following TPR to explicitly determine H$_2$ (or NH$_3$) release. Powder X-ray diffraction (XRD) analyses were conducted on a Bruker/Siemens D500 incident X-ray diffractometer using Cu Kα radiation; samples were scanned at 0.02° step^{-1} over the range 20° ≤ 2θ ≤ 90°. Diffractograms were identified using JCPDS-ICDD reference standards (Pd (05-0681) and BaSiO$_3$ (04-0504)). Palladium particle size was obtained from the Scherrer equation,[30]

$$d_c = \frac{K\lambda}{\beta \cos \theta} \tag{1}$$

where d_c is the mean size of the ordered (crystalline) domains, K is a dimensionless shape factor (0.9), λ the X-ray wavelength (1.5056 Å), β line broadening at half the maximum intensity, and θ the Bragg angle (2θ = 40.1°). Palladium particle morphology (size and shape) was determined by transmission electron microscopy analysis, conducted on a JEOL-2000 TEM/STEM microscope equipped with a UTW energy dispersive X-ray (EDX) detector (Oxford Instruments), and operated at an accelerating voltage of 200 kV. Samples were prepared by ultrasonic dispersion in 2-butanol, evaporating a drop of the resultant suspension onto a holey carbon/Cu grid (300 mesh). Up to 800 individual particles were counted for each catalyst and the mean metal diameter (d_{TEM}) calculated from

$$d_{TEM} = \frac{\sum_i n_i d_i}{\sum_i n_i} \tag{2}$$

where n_i is the number of particles of diameter d_i.

Hydrogenation of butyronitrile

Catalytic system

The hydrogenation of butyronitrile (Sigma-Aldrich, ≥99%) was conducted *in situ* following catalyst activation at 1 atm and 473–563 K in a fixed bed vertical glass reactor (i.d. = 15 mm). Reactions were conducted under operating conditions that ensured negligible internal or external mass and heat transfer limitations. The nitrile reactant was delivered at a fixed calibrated flowrate (0.6 cm^3 h^{-1}) via a glass/teflon air-tight syringe and teflon line using a microprocessor controlled infusion pump (Model 100 kDa Scientific). A layer of borosilicate glass beads served as preheating zone, ensuring that the reactant was vaporized and reached reaction temperature before contacting the catalyst. Isothermal conditions ($\pm$1 K) were maintained by diluting the catalyst bed with ground glass (75 µm); the ground glass was mixed thoroughly with the catalyst before insertion in the reactor. Reaction temperature was continuously monitored using a thermocouple inserted in a thermowell within the catalyst bed. A co-current flow of butyronitrile and H$_2$ (<1% volume by volume nitrile in H$_2$) was maintained at GHSV = 1.0 × 10^4 h^{-1}. The inlet nitrile flow (F) was constant (6.9 mmol h^{-1}) where the H$_2$ content was in excess (by a factor of 12 relative to the stoichiometric requirement for butylamine formation) and monitored using a Humonics (Model 520) digital flowmeter. Molar palladium (n) to F ratio spanned the range 3.4 × 10^{-4}–3.4 × 10^{-3} h. The reactor effluent was frozen in a liquid N$_2$ trap for subsequent analysis, which was made using a Perkin-Elmer Auto System XL chromatograph equipped with a programed split/splitless injector and a flame ionization detector, employing a DB-1 capillary column (i. d. = 0.33 mm, length = 50 m, film thickness = 0.20 µm). Data acquisition and manipulation were performed using the TurboChrom Workstation Version 6.1.2 (for Windows) chromatography data system. Fractional butyronitrile conversion (X) was obtained from

$$X = \frac{[\text{Butyronitrile}]_{in} - [\text{Butyronitrile}]_{out}}{[\text{Butyronitrile}]_{in}} \qquad (3)$$

and nitrile consumption rate from

$$\text{Rate (h}^{-1}) = \frac{X \times F}{n} \qquad (4)$$

where selectivity to product "i" (S$_i$) is given by

$$S_i(\%) = \frac{N_i x_i}{\sum N_i x_i} \times 100 \qquad (5)$$

[Butyronitrile]$_{in}$ and [Butyronitrile]$_{out}$ represent inlet and outlet concentration, respectively, and N$_i$ is the stoichiometric coefficient for product ^{a}i. In blank tests, passage of butyronitrile in a stream of H$_2$ through the empty reactor or over the SiO$_2$ support alone did not result in any detectable conversion. Repeated reactions with different samples of catalyst from the same batch delivered raw data reproducibility and carbon mass balance better than ±6%.

Thermodynamic analysis

Application of thermodynamics provides an important guide to the maximum conversion/selectivity possible under a given set of reaction conditions. Butyronitrile, hydrogen, and all products (butylamine, dibutylamine, tributylamine, and NH$_3$) were considered. Setting the inlet nitrile at 1 mol, product distribution at equilibrium was determined over 473–563 K at a total pressure of 1 atm, where the H$_2$/butyronitrile molar ratio was kept constant (= 24) to mimic catalytic reaction conditions. Equilibrium calculations

were made using CHEMCAD (Version 6) where the Gibbs reactor facility was applied to obtain product composition under conditions of minimized Gibbs free energy. The equation of state for fugacity employed the Soave–Redlich–Kwong approach.[31] The total Gibbs function is given by

$$G^t = \sum_{i=1}^{N} n_i \bar{G}_i = \sum_{i=1}^{N} n_i \bar{\mu}_i = \sum n_i G_i^0 + RT \sum n_i \ln \frac{\hat{f}_i}{f_i^0} \qquad (6)$$

For gas phase reaction equilibrium, $\hat{f}_i = \hat{\phi}_i y_i P$, $f_i^0 = P^0$ and $\Delta G^0 = \Delta G_{f_i}^0$ and the minimum Gibbs free energy of each gaseous species and total for the system can be expressed by

$$\Delta G_{f_i}^0 + RT \ln \frac{\hat{\phi}_i y_i P}{p^0} + \sum_k \lambda_k a_{ik} = 0 \qquad (7)$$

$$\sum_{i=1}^{N} n_i \left(\Delta G_{f_i}^0 + RT \ln \frac{\hat{\phi} y_i P}{p^0} + \sum_k \lambda_k a_{ik} \right) = 0 \qquad (8)$$

according to the Lagrange undetermined multiplier method with the elemental balance constraint

$$\sum_{i=1}^{N} n_i a_{ik} = A_k \qquad (9)$$

Results and discussion

Catalyst characterization

Critical physico-chemical characteristics of Pd/SiO$_2$ and Ba–Pd/SiO$_2$ are given in Table 1. The SSA of Pd/SiO$_2$ was close to that of the SiO$_2$ support (200 m^2 g^{-1}). In contrast, Ba–Pd/SiO$_2$ exhibited a significantly lower SSA, which can be ascribed to partial pore blockage as observed for silica supported Pd/lanthanide bimetallics prepared from analogous precursors.[32] The TPR profile for Pd/SiO$_2$ (Fig. 2(I)) exhibited a negative peak (H$_2$ release) at 368 K that can be attributed to Pd hydride decomposition.[33] The profile for Ba–Pd/SiO$_2$ (Fig. 2(II)) is also characterized by H$_2$ release with T_{max} at 354 K and an associated Pd hydride composition (H/Pd ratio = 0.19) that differed from Pd/SiO$_2$ (0.34). The lower H/Pd suggests inhibited hydride formation due to the incorporation of Ba with the possible formation of smaller Pd particles as size governs hydride composition with an upper H/Pd = 0.76 for bulk Pd.[34] A shift in hydride decomposition temperature to a lower value also correlates with a decrease

Table 1 Physico-chemical properties of SiO$_2$-supported Pd and Ba–Pd catalysts

	Pd/SiO$_2$	Ba–Pd/SiO$_2$
SSA (m^2 g^{-1})	191	154
TPR T_{max} (K)	368^a	354^a, 573^b
Pd hydride (H/Pd, mol mol^{-1})	0.34	0.19
H$_2$ chemisorption (µmol g^{-1})	7	46
H$_2$ TPD (µmol g^{-1})	19	91
NH$_3$ chemisorption (mmol g^{-1})	0.49	0.34
NH$_3$ TPD (mmol g^{-1})	0.48	0.32
d_{TEM} (nm)	28	7
d_c (nm)	33	9

^{a} H$_2$ release.
^{b} H$_2$ consumption.
TPD: temperature programmed desorption; TPR: temperature programmed reduction; SSA: specific surface area.

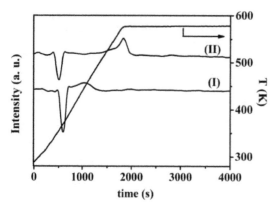

Figure 2 Temperature-programmed reduction (TPR) profiles for (I) Pd/SiO$_2$ and (II) Ba–Pd/SiO$_2$

in Pd particle size.[35] Hydrogen consumption at 573 K for Ba–Pd/SiO$_2$ suggests a temperature-induced reduction step, which may result from stabilization of surface Pd oxide with the addition of Ba.[27] Supported Pd morphology was determined by TEM analysis, and representative images (with associated size distributions) are presented in Fig. 3. Both catalysts display quasi-spherical Pd nano-particles with a narrower distribution of smaller particles for Ba–Pd/SiO$_2$ (mean size = 7 nm) compared with Pd/SiO$_2$ (28 nm). This result is in good agreement with Pd size obtained from application of the Scherrer equation to XRD line broadening (Table 1) and is inline with the work of Liu et al.[36] who observed an increase in metal dispersion for zeolite supported Pt following Ba exchange. The incorporation of Ba has been shown to minimize Pd agglomeration with the formation of smaller Pd particles.[27,37] Piacentini et al.[38] reported increased Pt dispersion at higher Ba loadings in Ba–Pt/Al$_2$O$_3$. Enhanced Pd dispersion can account for the greater H$_2$ chemisorption on Ba–Pd/SiO$_2$ relative to Pd/SiO$_2$ (Table 1), where smaller Pd particles facilitate dissociative H$_2$ adsorption.[39] Hydrogen release from both catalysts during TPD (Table 1) exceeded that adsorbed in ambient temperature pulse titration, diagnostic of spillover hydrogen during TPR.[40] Hydrogen TPD from Pd/SiO$_2$ generated the profile given in Fig. 4(I), characterized by a two-stage release with T_{max} = 775 and 1273 K (final isothermal hold). Drawing on available literature, the lower temperature peak can be ascribed to loss of chemisorbed hydrogen from Pd.[41] This is consistent with the equivalence of H$_2$ desorbed over 720–980 K (6 μmol g^{-1}) and that chemisorbed (Table 1). Hydrogen release at higher (1050–1273 K) temperatures has been attributed to desorption from the support and metal/support interface.[27,42] The TPD profile for Ba–Pd/SiO$_2$ (Fig. 4(II)) also showed two desorption peaks with a greater (by a factor of 5) amount of H$_2$ desorbed compared with Pd/SiO$_2$ (Table 1). This can be linked to the presence of smaller Pd particles that facilitate higher H$_2$ uptake and diffusion/spillover to the support.[43] The shift of the second H$_2$ desorption peak for Ba–Pd/SiO$_2$ to a lower temperature suggests that the incorporation of Ba impacts on spillover release.[27] Panagiotopoulou and Kondarides[44] recorded a displacement of H$_2$-TPD to a lower temperature (by 50 K) for

Ca-promoted Pt/TiO$_2$ that they attributed to a modification c the metal-support interface with the formation of Pt–Ca–Ti^{3-} sites that exhibit weaker Pt–H bond strength.

Surface acidity was probed by NH$_3$ chemisorption/desorption where TPD (Fig. 5a) from the silica support (I) exhibited NH$_3$ release with T_{max} = 343 K that can be attributed to weak acid sites.[45,46] Both Brønsted (hydroxyl groups acting as proton donors)[46] and weak Lewis acid sites[45,46] have been detected on silica surfaces by FTIR spectroscopy. Temperature programmed desorption from Pd/SiO$_2$ (II) and Ba–Pd/SiO (III) exhibited a common T_{max} (= 357 K) with NH$_3$ desorptio (0.15 mmol g^{-1}) close to that observed for SiO (0.19 mmol g^{-1}). Integration of NH$_3$ desorption signals gave а total release that correlated well with the chemisorptior measurements (Table 1). A secondary higher temperature (600–1000 K) NH$_3$ release was in evidence that must be linkee to surface acidity generated during metal incorporation ane sample treatment. Gao et al.[47] recorded an additional NH desorption peak at 475–775 K for Ni/SiO$_2$ that was no observed for SiO$_2$. Moreover, Jiang et al.[48] reported an increase in NH$_3$ release at ca. 773 K during TPD of Pd/SBA with increasing Pd content (0.005 → 0.01 wt%) Ammonia chemisorption/desorption was lower for Ba–Pd/SiO$_2$ compared with Pd/SiO$_2$ (Table 1). X-ray diffraction analysis of Ba–Pd/SiO$_2$ (Fig. 5b) has revealed peaks characteristic of metallic Pd and a BaSiO$_3$ phase The formation of BaSiO$_3$ can result in a consumption of surface acid sites. Labalme and co-workers[49] also reported a decrease ir total surface acidity as a result of Ba addition to Pt/Al$_2$O$_3$ based on NH$_3$ TPD, which was attributed to neutralization of surface hydroxyl groups. X-ray photoelectron spectroscopy (XPS analysis has established the formation of an electron-rich Pc phase in Ba–Pd/SiO$_2$.[27] This is consistent with Labalme's conclusion of electron donation from electro positive Ba to Pt o Al$_2$O$_3$[50,51] and is inline with reports, which have concluded that addition of Ba increases the electron density of Pd sites.[52,5]

Butyronitrile hydrogenation: thermodynamic considerations

A thermodynamic analysis of butyronitrile hydrogenatior was performed to determine system behavior at equilibrium. Under thermodynamic control, the nitrile reactant was fully converted under the reaction conditions employed in this study. The calculated equilibrium selectivity as a function of reaction temperature is presented in Fig. 6 where it can be seen that the tertiary amine is the predominant product (S = 83–90%). Production of equivalent amounts of butylamine and dibutylamine as by-products is favored by increasing temperature. Preferential tertiary amine production indicates that there is no thermodynamic barrier for the coupled reduction and condensation steps shown in Fig. 1.

Butyronitrile hydrogenation: catalytic activity/selectivity (at 473 K)

Taking 473 K as a benchmark temperature, fractional butyronitrile conversion (X) was time invariant over Pd/SiO$_2$ (I) and Ba–Pd/SiO$_2$ (II) where the latter exhibited higher nitrile conversion (Fig. 7a). This is significant given the temporal decline in activity reported in liquid[4–7,9,54–56] and gas phase[10,11,18,20] nitrile hydrogenation over supported

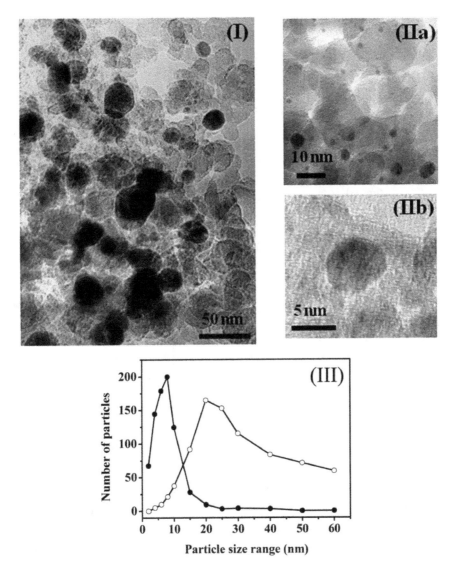

Figure 3 Representative transmission electron microscopy (TEM) images of (I) Pd/SiO$_2$ (○) and (II) Ba–Pd/SiO$_2$ (•) with (III) associated particle size distributions

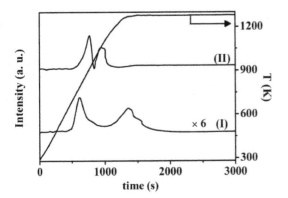

Figure 4 Hydrogen temperature-programmed desorption (TPD) profiles for (I) Pd/SiO$_2$ and (II) Ba–Pd/SiO$_2$

Pd,[5−7,9,55,56] Pt,[7,57] Ru,[7] and Ni.[11,19,20] Catalyst deactivation has been ascribed to metal particle agglomeration,[7,10] active site occlusion by amine product(s),[7,10,15,57,58] and catalyst coking associated with the formation of dehydrogenated surface species and carbides.[11,15,19,21] The greater levels of H$_2$ uptake/release exhibited by Ba–Pd/SiO$_2$ (Table 1) can account for the observed higher nitrile hydrogenation activity. Product distribution was invariant with conversion (Fig. 7b and c) where butylamine was the major product. This differs from predominant tertiary amine formation under thermodynamic equilibrium (Fig. 6) and demonstrates catalytic control. Nitrile transformation to amines via the pathways shown in Fig. 1 requires catalyst bifunctionality[59] where the metal phase serves to promote hydrogenation (steps (I), (II), (IV), and (VI)) and condensation of the imine intermediates with butyl- (step (III)) and dibutyl-amine

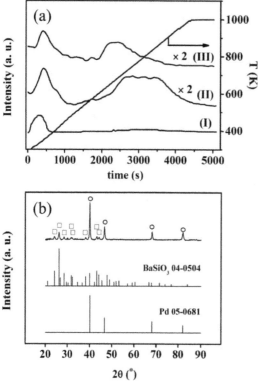

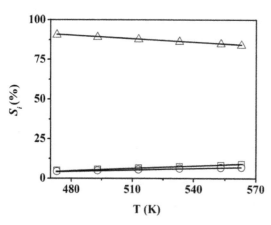

Figure 5 *a* Ammonia temperature-programmed desorption (TPD) profiles for (I) SiO_2, (II) Pd/SiO_2, and (III) Ba–Pd/SiO_2; *b* X-ray diffraction (XRD) pattern for Ba–Pd/SiO_2 with peak assignments for Pd (○) and $BaSiO_3$ (□) phases. JCPDS-ICDD reference diffractograms for Pd (05-0681) and $BaSiO_3$ (04-0504) are also included

Figure 6 Product selectivity (S_i) as a function of reaction temperature at the thermodynamic equilibrium: butylamine (□); dibutylamine (○); tributylamine (△)

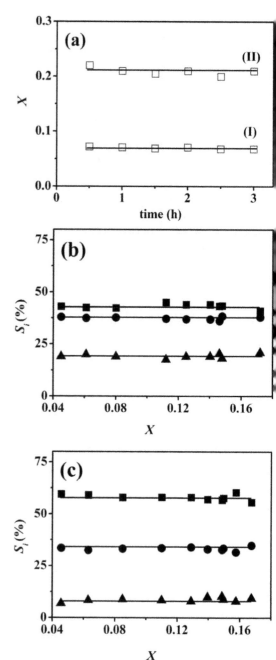

Figure 7 *a* Time on-stream butyronitrile fractional conversion (X) for reaction over (I) Pd/SiO_2 and (II) Ba–Pd/SiO_2 and selectivity (S_i) to butylamine (■), dibutylamine (•), and tributylmine (▲) as a function of fractional conversion over *b* Pd/SiO_2 and *c* Ba–Pd/SiO_2; *Reaction conditions: P* = 1 atm, *T* = 473 K, *n/F* = 3.4×10^{-4}–3.4×10^{-3} h

(step (V)) proceeds on surface acid sites.[2,4,18,19,59] Selectivity to the target butylamine was higher over Ba–Pd/SiO_2 (Fig. 7c) than Pd/SiO_2 (Fig. 7b). This can be partly attributed to the lower surface acidity of the bimetallic catalyst that served to suppress condensation to secondary and tertiary amines. Moreover, weaker butylamine interaction with electron-rich Pd sites on Ba–Pd/SiO_2 resulting from repulsion with the –NH_2 function can favor desorption of the primary amine without further reaction. Branco *et al.*[24] reached a similar conclusion for the conversion of propionitrile over lanthanide-promoted Cu where electron enriched Cu sites exhibited weaker adsorption of primary amine, limiting subsequent condensation.

Butyronitrile hydrogenation: temperature effects

Reaction temperature is a critical variable that impacts on reactant/intermediate activation and desorption dynamics, which in turn influence hydrogenation rate and product distribution.[60] The nitrile consumption rate delivered by Pd/SiO$_2$ (I) and Ba–Pd/SiO$_2$ (II) passed through maxima at 523 and 493 K, respectively, as shown in Fig. 8. Nieto-Márquez et al.[11] reported a maximum rate of butyronitrile hydrogenation over (carbon nanosphere) supported Ni at comparable temperatures (463 K $< T <$ 583 K) that they linked to reactant thermal desorption that decreased surface coverage. Reaction over Ba–Pd/SiO$_2$ delivered higher nitrile consumption rates with the maximum at a lower temperature (493 K). The latter can be linked to weaker butyronitrile interaction via the nitrogen electron lone pair at electron rich Pd sites, resulting in more facile desorption. There is evidence in the literature[11] for a higher T_{max} in butyronitrile hydrogenation rate over Ni particles with lower electron density that can be attributed to stronger $-C \equiv N$ adsorption and the requirement for higher desorption temperatures. The higher selectivity to primary amine over Ba–Pd/SiO$_2$ (Fig. 8) can be ascribed to the lower surface acidity that limited condensation. Butylamine selectivity over both catalysts increased with increasing temperature to 100% at $T \geq$ 543 K. This is quite distinct from the thermodynamic equilibrium composition

(Fig. 6) where tri-butylamine was the major product over the entire temperature range. An increase in temperature must induce desorption of the butylamine product, circumventing condensation. Cristiani and co-workers[61] have recorded an increase in primary amine formation (S = 16–41%) at higher temperatures (453–563 K) in stearonitrile conversion over CuO–Cr$_2$O$_3$. In contrast, Braos-García and co-workers[21] noted decreasing primary amine selectivity and preferential secondary amine formation with increasing temperature (378–418 K) for acetonitrile hydrogenation over mixed alumina/gallium oxide (16% w/w Ga$_2$O$_3$) supported Ni. Nieto-Márquez et al.[11] have proposed that selectivity maxima are the result of contributions due to mass transfer, thermodynamic limitations, and thermal poisoning. The higher butyronitrile consumption rate delivered by Ba–Pd/SiO$_2$ coupled with reaction exclusivity translates into higher butylamine productivity (91 mol h^{-1} mol$_{Pd}^{-1}$) than that achieved over Pd/SiO$_2$ (54 mol h^{-1} mol$_{Pd}^{-1}$) at 543 K. Authors attribute this to greater availability of surface reactive hydrogen. It is important to note that in previous reports, reaction over Ni delivered higher activity and primary amine selectivity than Pd in liquid phase nitrile hydrogenation.[6,16] Authors have achieved an order of magnitude higher rate over Ba–Pd/SiO$_2$ with full selectivity to the target butylamine when compared with reported gas phase continuous reaction over supported Ni under similar reaction conditions (T = 493 K, 1 atm).[10] Our results demonstrate that the combination of an alkaline earth metal (Ba) with Pd facilitates enhanced cleaner primary amine production.

Conclusions

Authors have attained a higher selective butyronitrile hydrogenation rate (91 mol h^{-1} mol$_{Pd}^{-1}$) to the target butylamine in the gas phase hydrogenation of butyronitrile over Ba–Pd/SiO$_2$ relative to Pd/SiO$_2$ (54 mol h^{-1} mol$_{Pd}^{-1}$). The increased rate can be attributed to greater available surface reactive hydrogen (from H$_2$ chemisorption coupled with TPD) on Ba–Pd/SiO$_2$ bearing smaller metal nanoparticles (7 vs 28 nm). Lower surface acidity (from NH$_3$ adsorption/TPD) served to minimize side condensation and the formation of higher amines over Ba–Pd/SiO$_2$. Temperature-related activity maxima (T_{max}) are attributed to thermal desorption that limits surface coverage by reactant where T_{max} for Ba–Pd/SiO$_2$ (493 K) was lower than Pd/SiO$_2$ (523 K) reflecting weaker surface interaction for the former. Authors have provided the first reported evidence for (i) full selectivity to primary amine in nitrile hydrogenation over supported Pd; (ii) enhanced selective hydrogenation rate for a Pd-alkaline earth metal formulation that was 10 times higher than that reported previously for supported Ni.

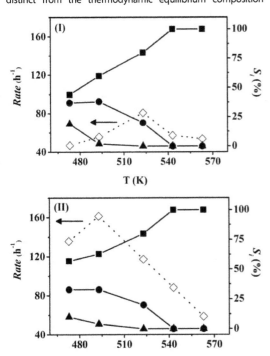

Figure 8 Reaction rate ($\diamond$, dashed lines) and selectivity (S_i, solid lines) to butylamine (■), dibutylamine (●), and tributylamine (▲) as a function of temperature for reaction over (I) Pd/SiO$_2$ and (II) Ba–Pd/SiO$_2$; *Reaction conditions: P* = 1 atm, *n/F* = 1.4 × 10^{-3} h

Acknowledgment

Authors acknowledge the contribution of Prof. S. G. Shore and Dr X. Wang to this work and financial support to Y. Hao and M. Li through the Overseas Research Students Award Scheme (ORSAS).

References

1. J. Krupka and J. Pasek: *Curr. Org. Chem.*, 2012, **16**, (8), 988–1004.
2. S. Gómez, J. A. Peters and T. Maschmeyer: *Adv. Synth. Catal.*, 2002, **344**, (10), 1037–1057.

3. S. Nishimura: 'Handbook of heterogeneous catalytic hydrogenation for organic synthesis', 265–265; 2001, New York, John Wiley.

4. H. Chen, M. Xue, S. Hu and J. Shen: *Chem. Eng. J.*, 2012, **181–182**, (1), 677–684.

5. Y. Huang and W. M. H. Sachtler: *Appl. Catal. A Gen.*, 1999, **182**, (2), 365–378.

6. Y. Huang and W. M. H. Sachtler: *J. Catal.*, 1999, **188**, (1), 215–225.

7. Y. Huang, V. Adeeva and W. M. H. Sachtler: *Appl. Catal. A Gen.*, 2000, **196**, (1), 73–85.

8. P. Schärringer, T. E. Müller and J. A. Lercher: *J. Catal.*, 2008, **253**, (1), 167–179.

9. Y. Y. Huang and W. M. H. Sachtler: *J. Catal.*, 1999, **184**, (1), 247–261.

10. A. Nieto-Márquez, D. Toledano, P. Sánchez, A. Romero and J. L. Valverde: *J. Catal.*, 2010, **269**, (1), 242–251.

11. A. Nieto-Márquez, D. Toledano, J. C. Lazo, A. Romero and J. L. Valverde: *Appl. Catal. A Gen.*, 2010, **373**, (1–2), 192–200.

12. H. Chen, M. W. Xue and J. Y. Shen: *Catal. Lett.*, 2010, **135**, (3–4), 246–255.

13. A. Chojecki, M. Veprek-Heijman, T. E. Müller, P. Schärringer, S. Veprek and J. A. Lercher: *J. Catal.*, 2007, **245**, (1), 237–248.

14. D. J. Segobia, A. F. Trasarti and C. R. Apesteguía: *Appl. Catal. A Gen.*, 2015, **494**, (1), 41–47.

15. M. C. Carrión, B. R. Manzano, F. A. Jalón, I. Fuentes-Perujo, P. Maireles-Torres, E. Rodríguez-Castellón and A. Jiménez-López: *Appl. Catal. A Gen.*, 2005, **288**, (1–2), 34–42.

16. D. J. Segobia, A. F. Trasarti and C. R. Apesteguía: *Appl. Catal. A Gen.*, 2012, **445–446**, (1), 69–75.

17. C. Jiménez-González, P. Poechlauer, Q. B. Broxterman, B.-S. Yang, D. Am Ende, J. Baird, C. Bertsch, R. E. Hannah, P. Dell'Orco, H. Noorrnan, S. Yee, R. Reintjens, A. V. Massonneau and J. Manley: *Org. Process Res. Dev.*, 2011, **15**, (4), 900–911.

18. A. C. Gluhoi, P. Mărginean and U. Stănescu: *Appl. Catal. A Gen.*, 2005, **294**, (1), 208–214.

19. P. Braos-García, P. Maireles-Torres, E. Rodríguez-Castellón and A. Jiménez-López: *J. Mol. Catal. A Chem.*, 2003, **193**, (1–2), 185–196.

20. A. Nieto-Márquez, V. Jiménez, A. M. Raboso, S. Gil, A. Romero and J. L. Valverde: *Appl. Catal. A Gen.*, 2011, **393**, (1–2), 78–87.

21. P. Braos-García, P. Maireles-Torres, E. Rodríguez-Castellón and A. Jiménez-López: *J. Mol. Catal. A Chem.*, 2001, **168**, (1–2), 279–287.

22. N. Iwasa, M. Yoshikawa and M. Arai: *Phys. Chem. Chem. Phys.*, 2002, **4**, (21), 5414–5420.

23. M. Arai, T. Ebina and M. Shirai: *Appl. Surf. Sci.*, 1999, **148**, (3–4), 155–163.

24. J. B. Branco, D. Ballivet-Tkatchenko and A. P. de Matos: *J. Phys. Chem. C.*, 2007, **111**, (41), 15084–15088.

25. M. Armbrüster, M. Behrens, F. Cinquini, K. Föttinger, Y. Grin, A. Haghofer, B. Klötzer, A. Knop-Gericke, H. Lorenz, A. Ota, S. Penner, J. Prinz, C. Rameshan, Z. Révay, D. Rosenthal, N. Rupprechter, P. Sautet, R. Schlögl, L. Shao, L. Szentmiklósi, D. Teschner, D. Torres, R. Wagner, R. Widmer and G. Wowsnick: *ChemCatChem.*, 2012, **4**, (8), 1048–1063.

26. I. Favier, D. Madec, E. Teuma and M. Gómez: *Curr. Org. Chem.*, 2011, **15**, (18), 3127–3174.

27. E. Ding, S. Jujjuri, M. Sturgeon, S. G. Shore and M. A. Keane: *J. Mol. Catal. A Chem.*, 2008, **294**, (1–2), 51–60.

28. D. W. Knoeppel, J. P. Liu, E. A. Meyers and S. G. Shore: *Inorg. Chem.*, 1998, **37**, (19), 4828–4837.

29. A. L. Bugaev, A. A. Guda, K. A. Lomachenko, V. V. Srabionyan, L. A. Bugaev, A. V. Soldatov, C. Lamberti, V. P. Dmitriev and J. A. van Bokhoven: *J. Phys. Chem. C.*, 2014, **118**, (19), 10416–10423.

30. U. Holzwarth and N. Gibson: *Nat. Nanotechnol.*, 2011, **6**, (9), 534–534.

31. X. Wang, S. Li, H. Wang, B. Liu and X. Ma: *Energy Fuel.*, 2008, **22**, (6), 4285–4291.

32. S. Jujjuri, E. Ding, S. G. Shore and M. A. Keane: *J. Mol. Catal. A Chem.*, 2007, **272**, (1–2), 96–107.

33. F. Menegazzo, T. Fantinel, M. Signoretto and F. Pinna: *Catal. Commun.*, 2007, **8**, (6), 876–879.

34. F. Cárdenas-Lizana, Y. Hao, M. Crespo-Quesada, I. Yuranov, X. Wang, M. A. Keane and L. Kiwi-Minsker: *ACS Catal.*, 2013, **3**, (6), 1386–1396.

35. S. Gómez-Quero, F. C. árdenas-Lizana and M. A. Keane: *Ind. Eng. Chem. Res.*, 2008, **47**, (18), 6841–6853.

36. P. Liu, X. Zhang, Y. Yao and J. Wang: *React. Kinet. Mech. Catal.*, 2010, **100**, (1), 217–226.

37. F. Klingstedt, H. Karhu, A. K. Neyestanaki, L. E. Lindfors, T. Salmi and J. Väyrynen: *J. Catal.*, 2002, **206**, (2), 248–262.

38. M. Piacentini, R. Strobel, M. Maciejewski, S. E. Pratsinis and A. Baiker: *J. Catal.*, 2006, **243**, (1), 43–56.

39. A. M. Doyle, S. K. Shaikhutdinov, S. D. Jackson and H.-J. Freund: *Angew. Chem. Int. Ed. Engl.*, 2003, **42**, (42), 5240–5243.

40. C. Tu and S. Cheng: *ACS Sust. Chem. Eng.*, 2013, **2**, (4), 629–636.

41. C. Amorim, G. Yuan, P. M. Patterson and M. A. Keane: *J. Catal.*, 2005, **234**, (2), 268–281.

42. M. Chettibi, A.-G. Boudjahem and M. Bettahar: *Trans. Met. Chem.*, 2011, **36**, (2), 163–169.

43. S. D. Lin and M. A. Vannice: *J. Catal.*, 1993, **143**, (2), 563–572.

44. P. Panagiotopoulou and D. I. Kondarides: *Appl. Catal. B Environ.*, 2011, **101**, (3), 738–746.

45. A. V. Biradar, S. B. Umbarkar and M. K. Dongare: *Appl. Catal. A Gen.*, 2005, **285**, (1–2), 190–195.

46. L. Óvári and F. Solymosi: *J. Mol. Catal. A Chem.*, 2004, **207**, (1), 35–40.

47. C.-G. Gao, Y.-X. Zhao and D.-S. Liu: *Catal. Lett.*, 2007, **118**, (1–2), 50–54.

48. J. Jiang, C. Yang, J. Sun, T. Li and F. Cao: *Adv. Chem. Eng. Res.*, 2013, **2**, (4), 73–78.

49. V. Labalme, B. Béguin, F. Gaillard and M. Primet: *Appl. Catal. A Gen.*, 2000, **192**, (2), 307–316.

50. V. Labalme, E. Garbowski, N. Guilhaume and M. Primet: *Appl. Catal. A Gen.*, 1996, **138**, (1), 93–108.

51. V. Labalme, N. Benhamou, N. Guilhaume, E. Garbowski and M. Primet: *Appl. Catal. A Gen.*, 1995, **133**, (2), 351–366.

52. N. Mahata, K. V. Raghavan, V. Vishwanathan and M. A. Keane: *React. Kinet. Catal. Lett.*, 2001, **72**, (2), 297–302.

53. K. Tanikawa and C. Egawa: *Appl. Catal. A Gen.*, 2011, **403**, (1–2), 12–17.

54. P. Kukula and K. Koprivova: *J. Catal.*, 2005, **234**, (1), 161–171.

55. L. Hegedűs, T. Máthé and T. Kárpáti: *Appl. Catal. A Gen.*, 2008, **349**, (1–2), 40–45.

56. L. Hegedűs and T. Máthé: *Appl. Catal. A Gen.*, 2005, **296**, (2), 209–215.

57. M. Arai, Y. Takada and Y. Nishiyama: *J. Phys. Chem. B.*, 1998, **102**, (11), 1968–1973.

58. P. F. Yang, Z. X. Jiang, P. L. Ying and C. Li: *J. Catal.*, 2008, **253**, (1), 66–73.

59. M. J. F. M. Verhaak, A. J. van Dillen and J. W. Geus: *Catal. Lett.*, 1994, **26**, (1–2), 37–53.

60. M. P. González-Marcos, J. I. Gutiérrez-Ortiz, C. González-Ortiz De Elguea, J. I. Alvarez and J. R. González-Velasco: *Can. J. Chem. Eng.*, 1998, **76**, (5), 927–935.

61. C. Cristiani, G. Groppi, G. Airoldi and P. Forzatti: 'Catalytic hydrogenation', in 'Studies in surface science and catalysis', Eds. K. K. Unger, G. Kreysa and J. P. Baselt, 128–128; 1986, Amsterdam, Elesvier Science Publishers.

7

Modification of Pd for formic acid decomposition by support grafted functional groups

Simon Jones, Amy Kolpin and Shik Chi Edman Tsang*

Wolfson Catalysis Centre, Department of Chemistry, University of Oxford, Oxford OX1 3QR, UK

Abstract Formic acid is proposed as a storage material to supply hydrogen gas for small portable fuel cell devices. However catalysts for its decomposition must be highly active and selective to provide a high quantity of hydrogen and carbon dioxide at ambient conditions but prevent any CO formation that can poison the catalysts. In this paper we report the functionalization of high surface area metal oxides with amine groups, which are then utilized as catalyst support to host Pd nanoparticles. It is demonstrated that the electronic and geometric properties of Pd nanoparticles can be substantially modified by these functionalized supports, resulting in improved activity and selectivity performance for the formic acid dehydrogenation.

Keywords Formic acid decomposition, Hydrogen production, Nanoparticles, Palladium, Support functionalization

Introduction

The development of fuel cells is currently an area of great interest, as they are seen to be an important part of a green energy infrastructure. Hydrogen is widely expected to play an important role in the future energy landscape, with it proposed as a major carbon independent energy carrier.[1] As such, effective storage of hydrogen has become a vitally important issue for the widespread utilization of hydrogen fuel cells.[2] For application in portable devices (laptops, mobile phones, etc.) a high volumetric hydrogen density is vital to enable small, high power systems.

Formic acid has been proposed as a suitable hydrogen storage material for these small devices, due to a volumetric hydrogen density of $53 \ g \ L^{-1}$ coupled with its facile decomposition to only H_2 and CO_2 even at low temperatures.[3] In solution, formic acid is reported to decompose via two mechanisms, displayed below.[4] The dehydrogenation mechanism (equation (1)) is clearly the desired reaction to produce hydrogen for use in fuel cells, whilst the dehydration mechanism (equation (2) is highly unfavourable, producing CO that is strongly poisoning to typical fuel cell anode catalysts resulting in dramatically reduced cell performance.[5]

$$HCOOH \rightarrow H_2 + CO_2 \qquad (1)$$

*Corresponding author, email edman.tsang@chem.ox.ac.uk

$$HCOOH \rightarrow H_2O + CO \qquad (2)$$

A wide range of both homogeneous and heterogeneous catalysts have been developed to decompose formic acid.[3,6] In terms of heterogeneous catalysts, Pd is widely reported as the most active monometallic system at close to ambient conditions, and tailoring of nanoparticles' electronic properties through alloy or core–shell formation has shown dramatic promotional effects on catalytic activity.[4,7] Addition of Ag to Pd, either as the core of a core–shell particle or in an alloy has resulted in approximately a five- and tenfold increase over Pd particles, respectively.[4,7] Tedsree et al.[8] demonstrated a linear correlation between the activity of metal-core Pd-shell particles and the work function of the core metal, indicating the reaction is highly dependent on surface electron density.

The mechanisms of the two decomposition reactions have been observed to occur through different surface bound intermediates, which have been reported to be favored on different catalytic sites.[4] It is proposed that a bridging bound formate (each oxygen atom adsorbed to a metal atom) is the intermediate for the dehydrogenation pathway, whereas a linearly bound formate (a single oxygen bound to the metal surface) is the intermediate for the dehydration pathway.[4] It has been indicated that the dehydration reaction is favored on low coordinate sites, whereas the dehydrogenation reaction is preferred on terrace sites.[9,10] Thus, it is clear that to facilitate production of high quality hydrogen from formic acid, tailoring the electronic properties of a catalyst must also be coupled with optimizing the geometric properties.

Recently, we demonstrated that organic polymers can be utilized to modify the electronic and/or geometric properties of Pd nanoparticles, which greatly affected the activity and selectivity of formic acid decomposition.[11] Organic molecules have been termed 'unconventional' promoters, as they differ from the typical techniques for modifying catalysts and are being widely researched for a range of reactions.[12] Small organic molecules have been shown to modify metallic surfaces geometrically (selective blocking of specific catalytic sites) and/or electronically (electronic donation or withdrawal to/from the surface). The use of support bound organic groups has also recently been reported to modify heterogeneous catalysts. Yadav et al.[13] used the formation of an amine containing silica shell around small Au particles to produce a yolk-shell structure with their particles encapsulated in a 'nanoreactor'. The activity of these catalysts for formic acid decomposition was dramatically increased compared to amine free structures. This agrees with previous reports that amines are required to achieve high formic acid decomposition activity with Au.[14] Whilst surface bound modifiers are shown to enhance formic acid decomposition, Xu et al. found that total encapsulation of the Au particles in the silica shell was crucial for enhanced activity.

In this work, we report the functionalization of metal oxide supports with organic groups to produce supported palladium nanoparticles with tailored electronic and geometric properties. These surface tailored Pd nanoparticles have been tested for room temperature formic acid decomposition. We demonstrate that the support bound organic functional groups can dramatically alter the activity and/or selectivity of the Pd nanoparticles for formic acid decomposition.

Experimental

Functionalized metal oxides were synthesized by a similar method to Harlick and Sayari.[15] Briefly, the metal oxide was dried under vacuum at 150°C, placed in a three-neck round bottom flask under a nitrogen atmosphere and dispersed in dry toluene by stirring. This suspension was heated to 85°C and, if required, water was added to the mixture followed by the desired functionalized silane (1.27 moles per gram metal oxide). After 16 h, the reaction was filtered, washed extensively with hot toluene and dried in air at 100°C. Supported palladium nanoparticles were synthesized using a deposition–precipitation method combined with an in situ hydrogen reduction, adapted from Cho et al.[16] 300 mg of support was dispersed in 200 mL of water by extensive sonication and stirring, before the required amount of $Pd(NO_3)_2$ solution was added. After 30 min stirring, the reaction mixture was adjusted to the required pH (8 for alumina or 11 for silica) using NaOH. It was then purged with nitrogen, before the gas was then changed to hydrogen and the reaction was heated to the desired temperature. After 1 h, the reaction mixture was cooled, filtered, washed extensively with water until pH was neutral and dried for 12 h at 85°C.

Thermal gravimetric analysis was performed on a TA instruments Q50. Powdered samples were heated at 10°C min^{-1} from room temperature to 950°C under air or nitrogen. TEM was performed on a JOEL 2010 operating at 200 kV. The dry catalyst was dispersed in ethanol by sonication before a drop was placed on a holey carbon coated copper grid that was then dried under vacuum for at least 1 h. Before CO chemisorption measurements, the sample was pre-reduced by 5% hydrogen in argon at 40°C for 1 h and then purged with helium. CO chemisorption measurements were performed using a Quantachrome Instruments ChemBET Pulsar TPR/TPD equipped with a 75 µL pulse loop. Ninety-nine per cent CO was used as the pulse gas, helium was used as the flow gas, and at least five pulses were performed after complete CO adsorption had occurred. X-ray diffraction (XRD) patterns were collected using a Philips PANalytical X'Pert Pro PW 1710 diffractometer operating in Bragg–Brentano focusing geometry with a generator at 40 kV and 40 mA producing Cu K_α radiation with a wavelength of 1.5406 Å. The patterns were collected between 2θ of 20 and 90°. Before measurements samples were thoroughly ground and placed in a fixed-depth well on a glass slide.

Formic acid decomposition activity was measured by monitoring the H_2/CO_x gas evolution. Catalyst containing approximately 4.25 mg of Pd was placed in a reaction vessel under a nitrogen atmosphere with 10.575 mL of 1.4M formic acid. The reaction mixture was maintained at 22–25°C and stirred at 1000 rev min^{-1}, and the volume of gas produced was recorded. For analysis of the gas composition, the gas produced was collected directly in a syringe before being injected into a GC equipped with both a TCD and FID, and into a GC-methanator FID. This provided accurate detection of H_2, CO_2, CO and other gases, with a detection limit below 10 ppm for CO.

Results and discussion

It has previously been reported that adsorbed amine groups electronically modify metal nanoparticles, and can show a preference for blockage of specific surface sites.[11,17] As such, we utilized the widely studied silane coupling reaction to functionalize oxide supports with an amine functional group. The majority of previous studies have focused on the functionalization of SiO_2 surfaces, so initially commercial high surface area silica (Davisil® 643, 300 m^2 g^{-1}) was functionalized. A primary amine silane, (3-aminopropyl)trimethoxysilane, was used to functionalize the support. The high hydroxyl surface coverage of silica facilitates a high degree of functionalization; however, it is reported that the addition of water during silane coupling can significantly increase the number of attached functional groups by promoting the formation of cross-linking bonds between the individual silane molecules on the surface.[15] The functionalized supports, synthesized with and without the addition of water are designated as $MO + NH_3$ and $MO + NH_3 + H_2O$, respectively. These supports were analyzed by thermal gravimetric analysis (TGA), details in Table 1. From the TGA profiles (SI) two phases were identified; phase 1 is attributed to the loss of water from the support surface and from pores, phase 2 is attributed to loss of surface hydroxyls, organic impurities and functional groups. The weight lost during phase 2 for the functionalized and unfunctionalized supports allows the degree of functionalization to be determined.

From Table 1, in agreement with literature reports, the addition of water clearly results in a significant increase in the number of functional groups on the silica surface.[15,18] Additionally, the quantities of functional groups per m^2 support are close to the values reported for similar methods in the literature.[15,18] To ensure the structure of the propylamine chain is maintained after attachment to the support surface, the samples were analyzed by FTIR, shown in Fig. 1.

In the FTIR spectra of silica, a number of peaks are observed; the broad peak around 3400 cm^{-1} is attributed to the OH stretch of surface hydroxyl groups and adsorbed water molecules, the peak at 3750 cm^{-1} is assigned to isolated surface hydroxyl species and the peak at 1630 cm^{-1} is assigned to deformation vibrations of adsorbed water.[19,20] For the functionalized supports additional peaks at 2900 and 3300 cm^{-1} are clearly present, assigned to C–H and N–H vibrations respectively.[20,21] Also for the functionalized supports the intensity of the broad hydroxyl region is decreased, indicating the conversion of surface Si–OH to Si–O–Si bonds through attachment of the silane molecules. These results demonstrate the silane molecules have coupled to the silica surface, and that the aminopropyl group has not decomposed under the functionalization or drying conditions. The increased intensity of the C–H and N–H peaks for $SiO_2 + NH_3 + H_2O$ over $SiO_2 + NH_3$ again indicates this support is more highly functionalized.

Palladium nanoparticles were synthesized on these supports using a deposition–precipitation method using $Pd(NO_3)_2$ combined with an *in situ* hydrogen reduction, adapted from that of Cho *et al.*, resulting in samples with 5 wt-%Pd.[16] X-ray diffraction patterns and TEM micrographs (SI) clearly demonstrate the formation of Pd nanoparticles. The actual Pd content of the catalysts has not been measured directly however the mass of Pd in typical catalysts (i.e. Pd/C) was determined by TGA analysis and was found to match with expected contents using the same method of synthesis. Additionally, the filtrate from catalyst synthesizes was analyzed using UV-visible spectroscopy to identify any un-reduced metal salts but in no cases was any identified. Details of the Pd nanoparticles on different supports are displayed in Table 2.

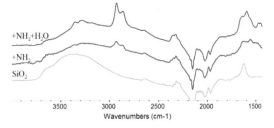

Figure 1 FTIR of SiO_2 and functionalized SiO_2

The size of the silica supported Pd particles decreased with increasing functional group content, indicating the amine group's ability to stabilize the formation of the metal nanoparticles. However, the surface areas of all the silica supported Pd particles, measured by CO chemisorption, are significantly less than expected from their particle size and from comparison with a Pd/C catalyst with similar Pd particle size (Table 2). These low metal surface areas can be partly attributed to the formation of Pd oxide species, which are reported to be stabilized on oxide surfaces.[22] These oxide particles are not reduced by H_2 until higher temperatures, but are not detectable by XRD due to their small size. To explore this proposal a sample of 5% Pd on $SiO_2 + NH_2 + H_2O$ was pre-treated under hydrogen at 350°C (just below the amine decomposition temperature) before measuring CO chemisorption. The resulting surface area was greatly increased, to 163.8 m^2 $g(Pd)^{-1}$, confirming the Pd oxide species reduce metal surface area and as such cause a reduction in CO adsorption.

Furthermore, the surface area results of the Pd/SiO_2 based samples do not display the increase in surface area expected with decreasing Pd particle size. This effect is then attributed to the blockage of the metal surface by the amine groups tethered to the support, similar to effects reported when a Pd/C catalyst was modified with varying amounts of an amine polymer.[11]

The overall formic acid decomposition activity of all the silica based Pd catalysts (L $g(Pd)^{-1}$) is significantly lower than that of a commercial Pd/C catalyst, which is attributed to their much lower surface areas from the presence of small unreduced PdO species. Once this much lower surface area is taken into account, the TOF clearly shows that the Pd surface sites of the unfunctionalized silica supported catalyst is comparable to Pd/C. Additionally, the TOFs of the different

Table 1 Details of SiO_2 and γ-Al_2O_3 functionalization

	Weight loss/%			Amine content	
	Phase 1*	Phase 2†	FG loss‡	mmol(N) g^{-1}§	μmol(N) m^2¶
SiO_2	1.77	2.14	N/A	N/A	N/A
+NH_2	1.28	6.90	4.76	0.821	2.74
+$NH_2 + H_2O$	1.53	14.00	11.86	2.045	6.82
γ-Al_2O_3	12.87	15.93	N/A	N/A	N/A
+$NH_2 + H_2O$	11.83	21.60	5.67	0.978	5.43

*Phase 1: 0–200°C for SiO_2, 0–300°C γ-Al_2O_3.
†Phase 2: 200/300°C to 900°C for SiO_2/γ-Al_2O_3.
‡wt- % difference in phase 2 of functionalized and unfunctionalized supports.
§Moles of propylamine, M_w=58.
¶Calculated using a surface area of 300 m^2 g^{-1} and 180 m^2 g^{-1} for SiO_2 and γ-Al_2O_3, respectively.

silica based catalysts highlight the effect of amine functionalization. Pd/SiO$_2$+NH$_2$ has double the TOF of the Pd/SiO$_2$ catalyst, with SiO$_2$+NH$_2$+H$_2$O (that has over double the number of functional groups on the support) showing almost double the TOF again. The electronic promotion of the Pd by the amine groups is also apparent from the initial activities of the silica based catalysts, with SiO$_2$+NH$_2$+H$_2$O being over 3 times more active than that unfunctionalized SiO$_2$. The electronically modified Pd on functionalized supports overcome the lower surface areas to produce more gas than Pd/SiO$_2$, indicating that whilst the amine groups are blocking the particle surface they are sufficiently modifying the remaining surface to produce more active catalysts.

However, the SiO$_2$ based catalysts still show a low mass activity compared to other Pd samples. In order to increase the mass activity an alternate metal oxide was investigated. Due to its high degree of surface hydroxyl groups and reasonable surface area, a commercial γ-Al$_2$O$_3$ was examined.[22,23] Following the synthesis procedure for SiO$_2$+NH$_2$+H$_2$O, γ-Al$_2$O$_3$ was functionalized with (3-aminopropyl)trimethoxysilane using the addition of water to maximize functionalization. As can be seen in Table 1, although the absolute quantity of functional groups is lower than for silica, the surface density of functional groups is similar suggesting that at this level of functionalization interactions between the groups limit further functionalization. Pd nanoparticles were synthesized on both the unfunctionalized and functionalized γ-Al$_2$O$_3$ using the same method as for silica supports; however, the pH was adjusted to 8 to prevent the formation of aluminium hydroxide which can fill the pores of the γ-Al$_2$O$_3$ structure.

From XRD, the formation of Pd nanoparticles is apparent however size determination is problematic due to significant overlap between the alumina and Pd peaks (SI). As such, TEM and CO chemisorption were used to analyze the Pd particles (Table 2). The γ-Al$_2$O$_3$ supported particles were found to have a much higher Pd surface area than those on SiO$_2$; however, the measured areas are still lower than expected from the particle size, again attributed to stabilized Pd oxide species. As with the SiO$_2$ based supported Pd particles, the presence of the amine functional groups stabilized the formation of the Pd resulting in greatly reduced particle size.

However, this reduction in particle size again did not result in the expected increase in Pd surface area, indicating a substantial interaction between the amine groups and the metal surface that overcomes increasing surface area of the small metal particles.

The promotional effect of the surface bound amine groups is again dramatic, resulting in a 3.5 times increase in gas production over two hours compared to unfunctionalized γ-Al$_2$O$_3$. The Pd/γ-Al$_2$O$_3$+NH$_2$+H$_2$O catalyst overcomes the surface area being almost one third of Pd/C, to result in a catalyst with double the activity. Additionally, the initial activity is almost 3 times that of Pd/C, demonstrating that the amine groups have produced a very highly active Pd surface. Thus, the formation of Pd supported on amine functionalized γ-Al$_2$O$_3$ displays one of the highest reported TOFs for the room temperature formic acid decomposition.

The promotional role of amine functional groups has previously been reported to be possible by two methods; firstly an electronic effect, through electron donation, and secondly a geometric modification, by blocking low coordinate sites on the Pd surface preventing formation of poisoning CO.[11] By comparing the initial and final activities, a measure of the catalyst's poisoning/deactivation can be examined and the mechanisms of promotion can be identified. To probe the site preference for the adsorption of short chain alkyl amines on Pd nanoparticles, a model system was examined. PVP stabilized Pd nanoparticles were synthesized, and the effect of pre-adsorbing a free form of amine on the particles was probed by monitoring the binding modes of subsequently adsorbed CO (details in SI). Typically on Pd, CO will bind in a linear mode (around 2040 cm^{-1}) or a bridging mode (around 1930 cm^{-1}) on low coordinate and terrace sites, respectively.[24]

The FTIR results in Fig. 2 demonstrate that the free form of ethylamine adsorbed more on the terrace sites. Marshall et al. reported similar results with alkanethiols that overcame a preference for binding on low coordinate Pd sites by forming stable assemblies on the terrace sites.[25,26] From their results the sterically hindered propylamine appeared to prevent the formation of these stable assemblies, and was found to favor blockage of low coordinated sites. These FTIR results also indicate that, by using the sterically hindered

Table 2 Details and activity of 5 wt-%Pd nanoparticles on various supports

Support	Size/nm	Surface area/ m^2 g(Pd)$^{-1}$	Initial rate*/ L g(Pd)$^{-1}$ min^{-1}	Final rate†/ L g(Pd)$^{-1}$ min^{-1}	Deactivation‡	Total gas§/ L g(Pd)$^{-1}$	TOF¶
SiO$_2$	5.8**	25.4	0.083	0.014	5.9	5.13	107
+NH$_2$	5.4**	15.3	0.126	0.039	3.2	6.94	240
+NH$_2$+H$_2$O	5.2**	14.6	0.282	0.054	5.2	11.93	432
Pd/C	5.3**/4.5††	85.1	0.534	0.122	4.4	27.12	169
γ-Al$_2$O$_3$	4.1††	58.0	0.522	0.040	13.1	14.59	133
+NH$_2$+H$_2$O	2.4††	33.1	1.459	0.209	7.0	53.22	851
+PEI	4.1††	62.0	0.526	0.258	2.0	42.91	366

*Rate of gas production in the first 10 min.
†Rate of gas production in the final 20 min.
‡Calculated from initial rate divided by final rate.
§Total gas produced in 2 h at 25°C.
¶TOF calculated from total moles of H2 divided by moles of Pd per hour.
**Particle size from XRD.
††Particle size from TEM.

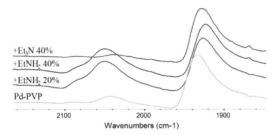

Figure 2 FTIR of CO adsorbed on Pd-PVP with ethylamine or triethylamine. Intensity normalized to bridging CO peak

amines on the surface of the support, selective blockage of low coordinate sites could be promoted.

However, from Table 2 it is also evident that significant deactivation is still taking place, with only one of the aminopropyl functionalized catalysts resulting in a lower deactivation than Pd/C. This indicates that this surface attached amine is unable to offer the selectively blocking for the high proportion of the low coordinate sites at long duration presumably due to the rigidity of the short terminal pendant amine groups from the support surface. Thus, the short length of the propyl chains from the support surface may limit the chance for the total encapsulation of the Pd particles and also reduce the site specificity for the low coordinated sites. To overcome both potential issues a support was functionalized with a highly-branched poly-ethyleneimine (PEI) polymer, which contains a high number of longer tertiary amines. A two-step process was utilized to attach this polymer to the γ-Al$_2$O$_3$ support. Firstly, an epoxide containing silane ((3-glycidoxypropyl)trimethoxysilane) is attached to the support surface, then this surface bound species is reacted with PEI through an epoxide coupling reaction to chemically bind the polymer to the support. Pd nanoparticles were then formed on this support as with the Pd/γ-Al$_2$O$_3$ sample.

From Table 2 it is clear the Pd particles on γ-Al$_2$O$_3$ + PEI are very similar to those supported on unmodified γ-Al$_2$O$_3$, both in terms of size and surface area. Additionally, the initial activities of these two catalysts are very similar and much lower than Pd/γ-Al$_2$O$_3$ + NH$_3$ + H$_2$O, indicating a very low interaction between the PEI and the Pd surface. However, over the two hours of reaction the PEI catalyst experiences much less deactivation, resulting in a substantially more active catalyst than Pd/γ-Al$_2$O$_3$. This agrees with our FTIR results that this sterically hindered with longer and branched amines will favor selectively blocking low coordinate sites. Further, from the FTIR in Fig. 2, it is clear that the electronic modification from this sterically hindered amine to the bridging CO species on terrace sites is much less than for primary amines. So for the sterically hindered amines in PEI that are more favorably bound to low coordinate sites, the electronic modification of the terrace sites is greatly reduced preventing the promotion in initial activity observed with propylamine modified supports (Fig. 3). This limited electronic influence of low coordinate site decoration over terrace sites has previously been reported by Tsang et al., who found that for decoration of Pt particles with Co only terrace areas near corners had increased activity.[27]

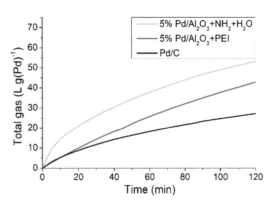

Figure 3 Formic acid decomposition activity of supported Pd catalysts at 25°C

Conclusion

Two methods for modifying Pd particles with support bound amines have been demonstrated. Figure 3 clearly shows the promotion methods result in two distinct activity profiles; primary amine functionalization gives a curved profile, with a high initial rate but suffering substantial deactivation, indicating a strong electronic promotion but minimal blockage of low coordinate sites. However, PEI functionalization gives a much straighter profile, attributed to a more selective blocking of low coordinate sites preventing the formation of poisoning CO. Thus, we have shown the feasibility of support functionalization to tailor both the electronic and geometric properties of a supported catalysts' surface and used it to produce two types of catalyst that exceed the activity and TOF of a typical Pd/C catalyst.

Conflicts of interest

The authors declare no conflicts of interest.

Acknowledgements

We thank the EPSRC for the financial support of this work and also for providing a DPhil studentship to SJ.

References

1. K. Bennaceur and E. Stout: Oilf. Rev., 2005, 17, (1), 30–41.
2. S. Enthaler: ChemSusChem, 2008, 1, 801–804.
3. S. Enthaler, J. von Langermann and T. Schmidt: Energy Environ. Sci., 2010, 3, 1207–1217.
4. K. Tedsree, T. Li, S. Jones, C. W. A. Chan, K. M. K. Yu, P. A. J. Bagot, E. A. Marquis, G. D. W. Smith and S. C. E. Tsang: Nat. Nanotechnol., 2011, 6, 302–307.
5. X. Chenga, Z. Shia, N. Glassa, L. Zhang, J. Zhang, D. Song, Z.-S. Liu, H. Wang and J. Shen: J. Power Sources, 2007, 165, 739–756.
6. M. Grasemann and G. Laurenczy: Energy Environ. Sci., 2012, 5, 8171–8181.
7. S. Zhang, Ö. Metin, D. Su and S. Sun: Angew. Chemie Int. Ed., 2013, 52, 3681–3684.
8. K. Tedsree, C. W. A. Chan, S. Jones, Q. Cuan, W.-K. Li, X.-Q. Gong, and S. C. E. Tsang: Science, 2011, 332, 224–228.
9. G. Samjeské, A. Miki, S. Ye and M. Osawa: J. Phys. Chem. B, 2006, 110B, 16559–16566.
10. K. Tedsree, A. T. S. Kong and S. C. Tsang: Angew. Chem. Int. Ed., 2009, 48, 1443–1446.
11. S. Jones, J. Qu, K. Tedsree, X.-Q. Gong and S. C. E. Tsang: Angew. Chem. Int. Ed., 2012, 51, 11275–11278.
12. Y. J. Tong: Chem. Soc. Rev., 2012, 41, 8195–8209.

13. M. Yadav, T. Akita, N. Tsumori and Q. Xu: *J. Mater. Chem.*, 2012, **22**, 12582–12586.

14. Q.-Y. Bi, X.-L. Du, Y.-M. Liu, Y. Cao, H.-Y. He and K.-N. Fan: *J. Am. Chem. Soc.*, 2012, **134**, 8926–8933.

15. P. J. E. Harlick and A. Sayari: *Ind. Eng. Chem. Res.*, 2007, **46**, 446–458.

16. H. Cho, J. Park, B. Hong and Y. Park: *Bull. Korean Chem. Soc.*, 2008, **29**, 328–334.

17. J. H. Ryu, S. S. Han, D. H. Kim, G. Henkelman and H. M. Lee: *ACS Nano*, 2011, **5**, 8515–8522.

18. K. M. K. Yu, I. Curcic, J. Gabriel, H. Morganstewart and S. C. Tsang: *J. Phys. Chem. A*, 2010, **114A**, 3863–3872.

19. M. Hair: *J. Non-Cryst. Solids*, 1975, **19**, 299–309.

20. R. Al-Oweini and H. El-Rassy: *J. Mol. Struct.*, 2009, **919**, 140–145.

21. H.-S. Jung, D.-S. Moon and J.-K. Lee: *J. Nanomater.*, 2012, **2012**, 593471.

22. N. Acerbi, S. Golunski, S. C. E. Tsang, H. Daly, C. Hardacre, R. Smith and P. Collier: *J. Phys. Chem. C*, 2012, **116C**, 13569–13583.

23. T. Chang, J. Chen and C. Yeh: *J. Catal.*, 1985, **96**, 51–57.

24. F. Hoffmann: *Surf. Sci. Rep.*, 1983, **3**, 107–192.

25. S. T. Marshall, M. O'Brien, B. Oetter, A. Corpuz, R. M. Richards, D. K. Schwartz and J. W. Medlin: *Nat. Mater.*, 2010, **9**, 853–858.

26. I. Makkonen, P. Salo, M. Alatalo and T. S. Rahman: *Surf. Sci.*, 2003, **532–535**, 154–159.

27. S. C. E. Tsang, N. Cailuo, W. Oduro, A. T. S. Kong, L. Clifton, K. M. K. Yu, B. Thiebaut, J. Cookson and P. Bishop: *ACS Nano*, 2008, **2**, 2547–2553.

8

Investigation of support effect on CO adsorption and CO + O_2 reaction over $Ce_{1-x-y}M_xCu_yO_{2-\delta}$ (M = Zr, Hf and Th) catalysts by in situ DRIFTS

Tinku Baidya[1] and Parthasarathi Bera*[2]

[1]Solid State and Structural Chemistry Unit, Indian Institute of Science, Bangalore 560012, India
[2]Surface Engineering Division, CSIR−National Aerospace Laboratories, Bangalore 560017, India

Abstract Adsorption of CO as well as CO + O_2 reaction over Cu^{2+} ion substituted $Ce_{1-x}M_xO_2$ (M = Zr, Hf and Th) supports have been studied by DRIFTS. Linear Cu^+—CO bands are observed over all catalysts upon introduction of CO. But, Cu^+—CO band positions are shifted to little higher frequencies in $Ce_{0.68}M_{0.25}Cu_{0.07}O_{2-\delta}$ compared to $Ce_{0.93}Cu_{0.07}O_{2-\delta}$. However, Cu^+—CO bands are in same positions when CO and O_2 are adsorbed simultaneously over all the catalysts. Ramping the temperature in the DRIFTS cell after simultaneous CO and O_2 adsorption shows the formation of CO_2 as well as decrease of CO. Comparison of intensities of CO_2 bands of different catalysts as a function of temperature indicates that $Ce_{0.68}Th_{0.25}Cu_{0.07}O_{2-\delta}$ shows lowest temperature CO oxidation among all the catalysts that is because of its more electron withdrawing power.

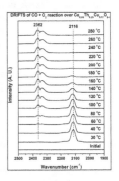

Keywords Adsorption, CO, Oxidation, DRIFTS, $Ce_{0.93}Cu_{0.07}O_{2-\delta}$, $Ce_{0.68}M_{0.25}Cu_{0.07}O_{2-\delta}$

Introduction

CeO_2 is well known as an oxygen storage material and redox support for various noble and transition metals.[1] It finds extensive applications in NO_x reduction, three-way catalysis and many other oxidation reactions.[2−5] The unique oxygen storage property of CeO_2 based oxides act as a buffer during oxidation/reduction cycle in a reaction. The oxygen storage capacity (OSC) has been enhanced by doping with tetravalent cations like Zr^{4+}, Hf^{4+}, Sn^{4+} and Ti^{4+} into CeO_2 matrix, but their effect is observed to be different on the extent of reducibility of the mixed oxides.[6−9] However, active metal dispersed on these doped cerium oxide catalysts acts as an adsorption site for reactants that remains in close proximity of lattice oxygen. Therefore, reactant can easily utilize the lattice oxygen to form oxidized product. Thus, created oxygen vacancies act as adsorption sites for the dissociation of feed oxygen. This was reported as bi-functional mechanism in the literature.[4,10−12]

In recent years, CeO_2 and TiO_2 based noble metal ionic catalysts have been found to show significant enhancement of CO oxidation at a lower temperature.[3−5] It has been demonstrated that substituted noble metal ions on CeO_2 matrix are catalytically more active than fine metal particles dispersed on Al_2O_3. Cu^{2+}, Ag^+, Au^{3+}, Pd^{2+} and Pt^{2+} are the active sites for the catalytic reactions. Structural studies show that metal ions are incorporated into CeO_2 substrate to a certain limit in the solid solution form of $Ce_{1-x}M_xO_{2-\delta}$ (x ≤ 0.05). Among the transition metals, Cu based oxide catalysts have been found to be very active for CO oxidation, NO reduction and CO-PROX reaction.[13−17] However, in our previous study, it has been found that Pd^{2+} ion substituted in different CeO_2 based supports show different activities.[18] It has been observed that CO oxidation activation energy decreases with increasing ionic character of Pd^{2+} in CeO_2 based oxides. Based on this finding, it would be worthwhile to extend this idea of ionic character in other metal ions such as Cu^{2+} that are very much active for several catalytic reactions when they are supported on different matrices. This can provide a more general outlook into how active metal ion can be more activated by choosing suitable support. In this regard, it would be interesting to study CO adsorption and variation of catalytic activities of Cu^{2+} ion in different Cu substituted CeO_2 supports toward CO oxidation.

Diffuse reflectance infrared Fourier transform spectroscopy (DRIFTS) is a versatile tool to get information about the nature of adsorbed species on the surface of catalyst during reaction.[19-21] It can also provide the trends of reaction with respect to the temperature. The present work focus on DRIFTS study to understand CO adsorption at room temperature as well as CO oxidation by O_2 at different temperatures over different Cu substituted $Ce_{1-x}M_xO_2$ (M = Zr, Hf and Th) oxides. The effect of support on the substituted Cu^{2+} ions toward CO adsorption/oxidation behavior has been investigated.

Experimental methods

Preparation

Cu substituted $Ce_{1-x}M_xO_2$ (M = Zr, Hf and Th) were prepared by solution combustion method. The preparation of the above catalysts was slightly different because of preferring suitable fuel for the combustion reaction. For the preparation of $Ce_{0.93}Cu_{0.07}O_{2-\delta}$ 30 mL aqueous solution of 5 g $(NH_4)_2Ce(NO_3)_6.6H_2O$, 1.255 g $Cu(NO_3)_2.5H_2O$ and 2.33 g of $C_2H_6N_4O_2$ (oxalyl dihydrazide) was taken in a 300 mL borosilicate dish. The dish was kept in the furnace preheated at 400 °C. Initially, the solution boiled with frothing and foaming and underwent dehydration. At the point of complete dehydration, the combustion started with a burning flame on the surface leading to a voluminous solid product within 5 min. $Ce_{0.68}Zr_{0.25}Cu_{0.07}O_{2-\delta}$ was prepared from $(NH_4)_2Ce(NO_3)_6.6H_2O$, $Zr(NO_3)_4$, $Cu(NO_3)_2.5H_2O$ and $C_2H_5NO_2$ (glycene) in 0.73:0.25:0.02:2.5 molar ratio. Similarly, $Ce_{0.68}Hf_{0.25}Cu_{0.07}O_{2-\delta}$ was prepared from $(NH_4)_2Ce(NO_3)_6.6H_2O$, $Hf(NO_3)_4$, $Cu(NO_3)_2.5H_2O$ and $C_2H_5NO_2$ (glycene) in 0.73: 0.25: 0.02:2.5 molar ratio. However, $Hf(NO_3)_4$ was prepared in situ using $HfCl_4$ as a precursor. Initially, the required amount of $HfCl_4$ was hydrolyzed in water giving a white precipitate of $Hf(OH)_4$ which was dissolved in a minimum volume of concentrated HNO_3. $Ce_{0.68}Th_{0.25}Cu_{0.07}O_{2-\delta}$ was prepared from the solution of 5 g of $(NH_4)_2Ce(NO_3)_6.6H_2O$, 1.61 g of $Th(NO_3)_4$, 0.272 g of $Cu(NO_3)_2.5H_2O$ and 2.46 g of $C_2H_5NO_2$ (glycene). The solution of ~30 mL taken in a 300 mL borosilicate dish was placed in the furnace at 400 °C to carry out the combustion and similar procedures were followed as above to obtain the solid product. All as-prepared catalysts were heated at 500 °C for 1 h to remove all organic residues present in the samples.

Characterization

BET surface area of Cu substituted $Ce_{1-x}M_xO_2$ (M = Zr, Hf and Th) catalysts were measured using Quantachrome Autosorb Automated Gas Sorption System (Quantachrome Instruments). X-ray diffraction (XRD) patterns of all catalysts were recorded on a Philips PANalytical X'Pert PRO diffractometer at a scan rate of 0.12° min[-1] with 0.02° step size in the 2θ range between 20° and 65°. X-ray photoelectron spectra (XPS) of Cu/CeO_2 based materials were recorded in a Thermo Fisher Scientific Multilab 2000 with AlKα radiation (1486.6 eV) operated at 15 kV and 10 mA (150 W). Binding energies reported here are with reference to graphite at 284.5 eV and they are accurate within ± 0.1 eV. For XPS analysis, the powder samples were pressed to 0.5 mm thick pellets and placed into a preparation chamber with ultrahigh vacuum (UHV) at 10[-9] Torr for 5 h in order desorb any adsorbed species present on the sample surface

and then it was transferred to analyzer chamber with UHV at 10[-9] Torr. All the spectra were obtained with a pass energy of 30 eV and step increment of 0.05 eV.

DRIFTS measurements

CO. adsorption at room temperature and in situ DRIFT spectra of the CO. + O_2 reaction as a function of temperature over Cu substituted $Ce_{1-x}M_xO_2$ (M = Zr, Hf and Th) catalysts were recorded with an accumulation of 20 scans at a resolution of 4 cm[-1] using a FTIR spectrometer from Thermo Scientific Nicolet 380 FTIR with a liquid N_2 cooled high sensitivity MCT detector and a DRIFTS cell. Aliquots of ca. 100 mg catalyst powder were placed inside the DRIFTS cell with ZnSe windows and a heating cartridge that allowed samples to be heated up to 300 °C. DRIFTS cell was connected with a gas handling system in order to measure in situ spectra under controlled gas environments at atmospheric pressure. Catalysts were activated in the DRIFTS cell by calcination for 2 h in a diluted O_2/He stream at 300 °C before introducing the reaction mixture. Subsequently, the system was cooled down to room temperature and the background spectra were recorded. Adsorption studies over the catalysts were carried out by passing CO (0.5% in He). The reaction mixture (0.5% CO. +0.5% O_2 in He) was passed over the catalysts at a total flow rate of 100 cm[3] min[-1] at atmospheric pressure for in situ studies. DRIFTS cell was heated to different temperatures and the spectra were collected after getting 10 min of steady state. Temperature of the DRIFTS cell was controlled with a thermocouple in direct contact with the sample. Circulating water was used to cool the body of the reaction chamber. All spectra obtained were transformed into absorption spectra by using Kubelka-Munk (K-M) function which is linearly related to the absorber concentration in the DRIFT spectra. CO, O_2 and He gases used in this study were supplied by Bhoruka Gas (purity higher than 99.95%).

Results and discussion

Characterization studies

BET surface areas of 7 at.% Cu substituted CeO_2, $Ce_{0.75}Zr_{0.25}O_2$, $Ce_{0.75}Hf_{0.25}O_2$ and $Ce_{0.75}Th_{0.25}O_2$ catalysts are 15, 22, 38 and 26 m[2] g[-1], respectively. At $P/P_o = 0.3$, pore volumes of these catalysts are 4.68, 7.22, 9.25 and 8.37 cm[3] g[-1], respectively and pore diameter is 5 Å.

XRD patterns of the Cu substituted $Ce_{0.75}M_{0.25}O_2$ (M = Zr, Hf and Th) catalysts are presented in Fig. 1. Observed diffraction lines in the patterns are indexed to fluorite structure. The diffraction peak shifts indicate the substitution of Zr, Hf and Th into CeO_2 lattice. As Zr^{4+}(0.84 Å) and Hf^{4+}(0.83 Å) ions are smaller than Ce^{4+}(0.97 Å),[22] the peak shift toward higher angle compared to CeO_2 is observed that is an indication of $Ce_{0.75}M_{0.25}O_2$ solid solution formation. Similarly, peak is shifted toward lower angle in Th doped CeO_2 as Th^{4+}(1.05 Å) ion is larger than Ce^{4+}. In Cu substituted $Ce_{0.75}M_{0.25}O_2$ oxides, Cu metal or CuO related peaks are not detected even after magnification, indicating incorporation of ionic Cu^{2+} into CeO_2 or $Ce_{0.75}M_{0.25}O_2$. However, finely dispersed CuO species can also be present over CeO_2 or $Ce_{0.75}M_{0.25}O_2$ supports which is likely to be XRD silent

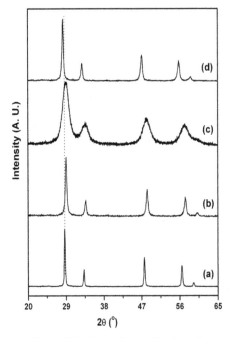

Figure 1 X-ray diffraction patterns of *a* Ce$_{0.93}$Cu$_{0.07}$O$_{2-\delta}$; *b* Ce$_{0.68}$Zr$_{0.25}$Cu$_{0.07}$O$_{2-\delta}$; *c* Ce$_{0.68}$Hf$_{0.25}$Cu$_{0.07}$O$_{2-\delta}$ and *d* Ce$_{0.68}$Th$_{0.25}$Cu$_{0.07}$O$_{2-\delta}$ catalysts

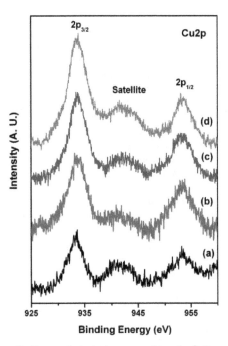

Figure 2 X-ray photoelectron spectra of Cu2p core levels of *a* Ce$_{0.93}$Cu$_{0.07}$O$_{2-\delta}$; *b* Ce$_{0.68}$Zr$_{0.25}$Cu$_{0.07}$O$_{2-\delta}$; *c* Ce$_{0.68}$Hf$_{0.25}$Cu$_{0.07}$O$_{2-\delta}$ and *d* Ce$_{0.68}$Th$_{0.25}$Cu$_{0.07}$O$_{2-\delta}$ catalysts

because of their low crystal size or amorphous nature. Crystallite size of Cu/Ce$_{0.75}$Zr$_{0.25}$O$_2$ and Cu/Ce$_{0.75}$Th$_{0.25}$O$_2$ catalysts obtained by using Scherrer formula is ~25 nm. However, the average crystallite size of Cu/Ce$_{0.75}$Hf$_{0.25}$O$_2$ is 10 nm.

XPS has been carried out to know the oxidation states of elements present in these catalysts. Cu2p core level spectra in various Cu substituted Ce$_{0.75}$M$_{0.25}$O$_2$ (M = Zr, Hf and Th) catalysts are shown in Fig. 2. Cu2p spectrum of Ce$_{0.93}$Cu$_{0.07}$O$_{2-\delta}$ is also included in the figure for comparison. Cu2p$_{3/2,1/2}$ peaks are resolved into spin-orbit doublets. Accordingly, Cu2p$_{3/2,1/2}$ core level peaks around 933.9 and 953.6 eV with corresponding satellite peaks and spin-orbit separation of 19.7 eV are assigned to Cu^{2+} in these type of catalysts.[13,14,23] Satellite peaks are characteristics of oxidized transition metals especially, Fe, Co, Ni and Cu.[24] It is well documented that an additional excitation of a second electron occurs during emission of a photoelectron of a core level creating a hole in it. Sudden creation of a hole in Cu2p^6 filled orbital from Cu^{2+} ion present in the catalyst makes Cu^{3+} ion and it becomes unstable. Therefore, an electron transfer from O2p level to Cu3d level occurs that leads to satellite peaks in the Cu2p core level spectrum as seen in the figure. It is to be noted that satellite peak (S) to main Cu2p$_{3/2}$ core level peak (M) intensity ratio (I$_S$/I$_M$) in CuO is found to be 0.55.[25-27] The intensity ratio varies from 0.39 to 0.48 in the case of Cu^{2+} ion in the square pyramidal position in Bi$_2$-Ca$_{1-x}$R$_x$Sr$_2$Cu$_2$O$_{8+\delta}$ (R = Y, Yb).[28] Further, I$_S$/I$_M$ values are 0.55 and 0.85 for Cu^{2+} in the octahedral and tetrahedral sites of a cubic spinel.[29] In the present study, intensity ratios obtained from the areas under the satellite and main peaks after background subtraction are calculated to be in the

range of 0.35 to 0.5 that is lower than the value of CuO. Therefore, Cu^{2+} ions in these catalysts are not in tetrahedral coordination and amount of finely dispersed CuO species is less in these catalysts. Figure 3 displays Ce3d core level spectra of Ce$_{0.93}$Cu$_{0.07}$O$_{2-\delta}$ and Ce$_{0.68}$M$_{0.25}$Cu$_{0.07}$O$_{2-\delta}$ (M = Zr, Hf and Th) catalysts. Ce3d$_{5/2,3/2}$ peaks at 882.5 and 900.7 eV with characteristic satellites are attributed to CeO$_2$ where Ce is in +4 oxidation state.[13,14,23,30] Satellite peak at 916.5 eV is relatively well separated from the rest of the spectrum and is the characteristic of the presence of tetravalent Ce (Ce^{4+}) in Ce compounds. However, possibility of the presence of

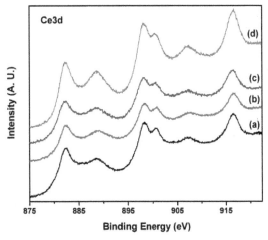

Figure 3 X-ray photoelectron spectra of Ce3d core levels of *a* Ce$_{0.93}$Cu$_{0.07}$O$_{2-\delta}$; *b* Ce$_{0.68}$Zr$_{0.25}$Cu$_{0.07}$O$_{2-\delta}$; *c* Ce$_{0.68}$Hf$_{0.25}$Cu$_{0.07}$O$_{2-\delta}$ and *d* Ce$_{0.68}$Th$_{0.25}$Cu$_{0.07}$O$_{2-\delta}$ catalysts

a small amount of Ce³⁺ on the surface level cannot be ruled out. In this regard, Ce3d spectra of all catalysts can be curve-fitted into several Ce3d$_{5/2,3/2}$ spin-orbit doublet peaks related to Ce⁴⁺ and Ce³⁺ species along with satellites. Peak areas of Ce⁴⁺ and Ce³⁺ components are commonly used to estimate their relative concentrations in the catalysts.[31] Concentrations of Ce³⁺ in all catalysts are evaluated to be in the range of 20–23 at.% with respect to the total amount of Ce species. Zr3d, Hf4f and Th4f core level spectra of respective catalysts are displayed in Fig. 4. Zr3d$_{5/3,3/2}$ peaks observed at 182.6 and 184.9 eV in $Ce_{0.68}Zr_{0.25}Cu_{0.07}O_{2-\delta}$ are associated with Zr⁴⁺.[32] In $Ce_{0.68}Hf_{0.25}Cu_{0.07}O_{2-\delta}$ Hf4f$_{7/2,5/2}$ core level peaks at 16.9 and 18.2 eV correspond to Hf⁴⁺.[33] Th4f$_{7/2,5/2}$ peaks at 334.4 and 343.6 eV in $Ce_{0.68}Th_{0.25}Cu_{0.07}O_{2-\delta}$ are related to Th⁴⁺.[32,34] Thus, XPS studies demonstrate that Cu, Ce, Zr, Hf and Th are present in +2, +4, +4, +4 and +4 oxidation states, respectively. From XPS studies, relative surface concentrations of Cu, Ce, Zr, Hf and Th in respective catalysts have been estimated by the relation given below[35]

$$C_M = \frac{\frac{I_M}{\sigma_M \lambda_M D_M}}{\Sigma \left(\frac{I_M}{\sigma_M \lambda_M D_M} \right)} \qquad (1)$$

where C_M, I_M, σ_M, λ_M and D_M are relative surface concentration of a particular metal (M), integrated intensity of the metal core level peak, photoionization cross-section of metal core level photoelectron, mean escape depth of the respective metal core level photoelectron and geometric factor, respectively. Integrated intensities of Cu2p, Ce3d, Zr3d, Hf4f and Th4f core level peaks have been taken into account to estimate their concentrations, whereas photoionization cross-sections and mean escape depths have been obtained from the literature.[36,37] The geometric factor has been taken as one, since the maximum intensity in this spectrometer is obtained at 90°. Accordingly, surface concentrations of Cu are evaluated to be in the range of 20 to 26 at.% in these catalysts which is around three times more than what is taken in the preparation indicating surface segregation of Cu²⁺ ions on CeO₂ and $Ce_{1-x}M_xO_2$ (M = Zr, Hf and Th) surfaces. Therefore, some of the Cu²⁺ species present in the catalysts can be related to finely dispersed CuO entities typically observed over these types of catalysts.

DRIFTS studies

Adsorption of CO has been carried out over $Ce_{0.93}Cu_{0.07}O_{2-\delta}$ and $Ce_{0.68}M_{0.25}Cu_{0.07}O_{2-\delta}$ (M = Zr, Hf and Th) catalysts to know the differences in the nature of their surfaces. Detailed *in situ* DRIFTS studies of CO + O₂ reaction have also been carried out over these catalysts.

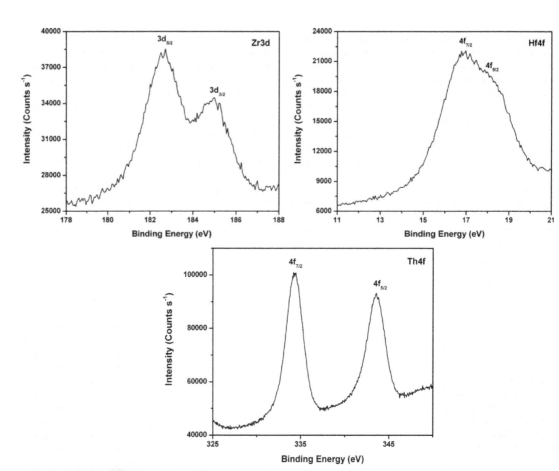

Figure 4   X-ray photoelectron spectra of Zr3d, Hf4f and Th4f core levels of $Ce_{0.68}Zr_{0.25}Cu_{0.07}O_{2-\delta}$, $Ce_{0.68}Hf_{0.25}Cu_{0.07}O_{2-\delta}$ and $Ce_{0.68}Th_{0.25}Cu_{0.07}O_{2-\delta}$ catalysts

CO adsorption over $Ce_{0.93}Cu_{0.07}O_{2-\delta}$ and $Ce_{0.68}M_{0.25\text{-}}$ $Cu_{0.07}O_{2-\delta}$ catalysts

The interaction of CO over $Ce_{0.93}Cu_{0.07}O_{2-\delta}$ catalyst is shown in Fig. 5. In general, CO interaction with catalyst is associated with various chemical transformations of CO molecule. Usual CO derived products during adsorption are carbonates (CO_3^{2-}), bicarbonates (HCO_3^-), formates ($HCOO^-$), carboxylates (COO^-) and carbonites (CO_2^{2-}). Carboxylates and bicarbonates are formed when cations can be easily reduced such as Cu^{2+} or when non-stoichiometric oxygen exists on the surface. On the other hand, carbonites are observed on strongly basic surfaces like MgO, CaO. Surface OH^- species are necessary to form bicarbonates and formates. However, all these species are characterized in the stretching frequencies below $1800\,cm^{-1}$. In the present study, several peaks are occurred in the carbonyl ($1900 - 2300\,cm^{-1}$) and carbonate ($900 - 1600\,cm^{-1}$) regions when CO is introduced over $Ce_{0.93}Cu_{0.07}O_{2-\delta}$ catalyst. Carbonyl and carbonate type species are formed upon the first contact of the catalyst with CO at room temperature. A band at $2100\,cm^{-1}$ which is gradually developed on the catalyst surface is because of CO adsorption over Cu^+ sites[19,38,39] and it is shifted to $2102\,cm^{-1}$ over time with higher intensity. In general, bands in the region of $2123-2114\,cm^{-1}$ are attributed to Cu^+ carbonyl species, where as bands in $2110-2080\,cm^{-1}$ regime correspond to Cu metal carbonyl.[39] In this sense, observed band at $2102\,cm^{-1}$ over $Ce_{0.93}Cu_{0.07}O_{2-\delta}$ may create confusion about its nature as it is in the borderline. Its low frequency suggests that they may be related to metallic copper carbonyls. No significant change in the intensities of this band is observed even after purging the DRIFTS cell with He for 5 min. Again, Cu^0—CO species is unstable at room temperature.[40] Therefore, in this study, adsorption center can correspond to

partially oxidized copper species and observed peak at $2102\,cm^{-1}$ is related to Cu^+—CO species. Peaks related to mono, bi and polydentate carbonates, bicarbonate and formate species are found in the $900 - 1600\,cm^{-1}$ region. It is important to note that carbonate and hydroxyl species are difficult to remove by oxidizing thermal treatment requiring temperatures above 750 °C which can induce catalyst sintering and loss of surface area.[19,38] Carbonate related species are formed over CeO_2 region of the catalysts. According to the literature, peaks observed at 1575, 1300, 1014 and $940\,cm^{-1}$ are attributed to bidentate carbonate species, whereas the peaks at 1470 and $1040\,cm^{-1}$ correspond to mono or polydentate carbonates formed on the surface.[19,20,38,39,41] Carbonyl peaks at 1396 and $1190\,cm^{-1}$ are assigned to hydrogen carbonate species. Observed peak at $1550\,cm^{-1}$ is related to formate species generated by interacting with hydroxyl species present on the surface (CO + $OH^- \rightarrow HCOO^-$).[38] In XPS studies, it has been demonstrated that Cu is in the +2 oxidation state. When CO is introduced over the catalyst, surface Cu species in +2 oxidation state gets partially reduced by CO and consequently, CO derived species such as carbonates are formed over the CeO_2 support ($2Cu^{2+} + 2O^{2-} + CO \rightarrow 2Cu^+ + CO_3^{2-}$) that is reflected in the DRIFT spectra. Complete reduction of Cu^{2+} to Cu^0 cannot be possible as CeO_2 is a reducible support. Therefore, Cu may remain as Cu^+ form in the catalyst. In this sense, frequencies related to Cu and CO interaction in $Ce_{0.93}Cu_{0.07}O_{2-\delta}$ catalyst are ascribed for Cu^+—CO species. A similar kind of reduction process at room temperature and formation of Cu^+—CO species has also been observed over Cu/CeO_2 based catalysts in earlier studies.[19,38,39]

Interactions of CO with $Ce_{0.68}Zr_{0.25}Cu_{0.07}O_{2-\delta}$ and $Ce_{0.68\text{-}}$ $Hf_{0.25}Cu_{0.07}O_{2-\delta}$ are displayed in Fig. 6. Similar to the

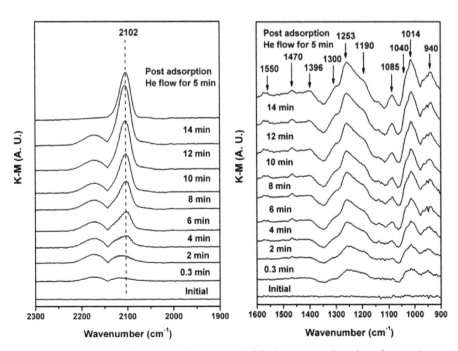

Figure 5 Diffuse reflectance infrared Fourier transform spectra of CO adsorption on $Ce_{0.93}Cu_{0.07}O_{2-\delta}$ catalyst

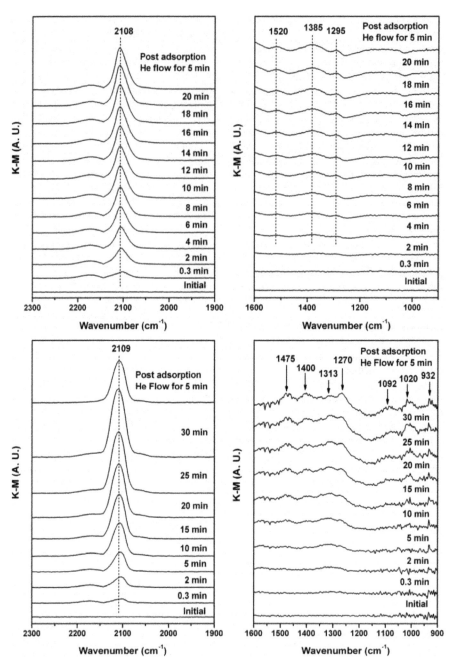

Figure 6 Diffuse reflectance infrared Fourier transform spectra of CO adsorption on (top) $Ce_{0.68}Zr_{0.25}Cu_{0.07}O_{2-\delta}$ and (bottom) $Ce_{0.68}Hf_{0.25}Cu_{0.07}O_{2-\delta}$ catalysts

$Ce_{0.93}Cu_{0.07}O_{2-\delta}$ catalyst, an intense peak at 2108 cm^{-1} observed over $Ce_{0.68}Zr_{0.25}Cu_{0.07}O_{2-\delta}$ in carbonyl region is ascribed to CO adsorption over reduced Cu (Cu^{+}—CO) species. Cu^{+}—CO peak positions are found to be at 2110 and 2109 cm^{-1} in cases of $Ce_{0.68}Hf_{0.25}Cu_{0.07}O_{2-\delta}$ and $Ce_{0.68}Th_{0.25}Cu_{0.07}O_{2-\delta}$ catalysts. It is important to note that Cu^{+}—CO bands have been shifted to higher frequencies (blue shift) because of presence of different environment of Cu^{2+} in $Ce_{0.68}M_{0.25}Cu_{0.07}O_{2-\delta}$ (M = Zr, Hf and Th) catalysts compared to $Ce_{0.93}Cu_{0.07}O_{2-\delta}$ catalyst. This shift is related to

the electron withdrawing ability of the substituent Zr^{4+}, Hf^{4+} and Th^{4+} ions. With more electron withdrawing character of the metals around Cu^{2+} in Zr, Hf and Th substituted CeO_2 catalysts, C—O bond becomes stronger because of lower π back donation. Peaks associated with mono, bi and poly dentate carbonate, bicarbonate and formate species are discernible in the $900 - 1600\ cm^{-1}$ region. These species are already produced upon the first contact of the samples with CO at room temperature. Peaks observed at 1520, 1430 and 1290 cm^{-1} on the surface of $Ce_{0.68}Zr_{0.25}Cu_{0.07}O_{2-\delta}$

are assigned for formate, hydrogen carbonate and bidentate carbonate. In $Ce_{0.68}Hf_{0.25}Cu_{0.07}O_{2-\delta}$, peaks at 1475, 1310 and 1075 cm^{-1} correspond to monodentate carbonate species, whereas 1010 cm^{-1} is related to bidentate species.[41] Peaks at 1400, 1270 and 1200 cm^{-1} are attributed to hydrogen carbonate species. However, intensities of these carbonate species are less in $Ce_{0.68}M_{0.25}Cu_{0.07}O_{2-\delta}$ (M = Zr, Hf and Th) catalysts compared to $Ce_{0.93}Cu_{0.07}O_{2-\delta}$. In the present study, it has been confirmed with XPS technique that Cu is in +2 oxidation state in $Ce_{0.68}M_{0.25}Cu_{0.07}O_{2-\delta}$ catalysts. Similar to $Ce_{0.93}Cu_{0.07}O_{2-\delta}$, introduction of CO over $Ce_{0.68}M_{0.25}Cu_{0.07}O_{2-\delta}$ partially reduces the surface Cu species which is in +2 oxidation state and consequently, CO derived species such as carbonates are formed over the catalysts which is confirmed from DRIFT spectra. There is no significant change in the intensities of all bands even after purging the DRIFTS cell with He for 5 min after CO exposure.

CO + O$_2$ reaction over $Ce_{0.93}Cu_{0.07}O_{2-\delta}$ and $Ce_{0.68}M_{0.25}Cu_{0.07}O_{2-\delta}$ catalysts

DRIFTS studies of CO + O$_2$ reaction have been carried out over $Ce_{0.93}Cu_{0.07}O_{2-\delta}$ and $Ce_{0.68}M_{0.25}Cu_{0.07}O_{2-\delta}$ catalysts to monitor the reaction sequences during the reaction. After simultaneous CO and O$_2$ adsorption, catalysts have been heated up to 300 °C and every 20 °C DRIFT spectra are recorded. Spectra of carbonyl and carbonate regions of $Ce_{0.93}Cu_{0.07}O_{2-\delta}$ catalyst at different temperatures are shown in Fig. 7. Cu$^+$—CO peak in $Ce_{0.93}Cu_{0.07}O_{2-\delta}$ is shifted to higher frequency (2114 cm^{-1}) in presence of O$_2$. Intensity of carbonyl peak at 2114 cm^{-1} slowly comes down above 100 °C and peaks at 2362 and 2335 cm^{-1} associated with gas phase CO$_2$ are started to be observed. As temperature increases CO oxidation is observed to accelerate that can be

seen from the increase in the gas phase CO$_2$ and corresponding carbonate bands because of CO$_2$ adsorption on the catalyst. Cu$^+$—CO peak is observed to disappear at higher temperatures over the catalyst. However, strong CO$_2$ bands are found even after disappearance of CO bands at higher temperatures during reaction indicating CO conversion. DRIFT spectra of $Ce_{0.68}M_{0.25}Cu_{0.07}O_{2-\delta}$ catalysts are similar to those of $Ce_{0.93}Cu_{0.07}O_{2-\delta}$ except the differences in onset temperatures and changes in the nature of carbonate species in presence of O$_2$. Figure 8 displays DRIFT spectra of $Ce_{0.68}Zr_{0.25}Cu_{0.07}O_{2-\delta}$ and $Ce_{0.68}Hf_{0.25}Cu_{0.07}O_{2-\delta}$ catalysts during CO + O$_2$ reaction. It has been observed that CO oxidation starts at higher temperature over $Ce_{0.68}Zr_{0.25}Cu_{0.07}O_{2-\delta}$ in comparison with $Ce_{0.68}Hf_{0.25}Cu_{0.07}O_{2-\delta}$. During CO + O$_2$ reaction, intensities of bands at 1335 and 1080 cm^{-1} related to monodentate carbonate species and bands at 1290 and 990 cm^{-1} corresponding to bidentate carbonate species increase over $Ce_{0.93}Cu_{0.07}O_{2-\delta}$ catalyst. Increase in the intensities of bands at 1285 and 1038 cm^{-1} associated with bidentate carbonate species along with hydrogen carbonate species (1480 and 1395 cm^{-1}) is observed in $Ce_{0.68}Zr_{0.25}Cu_{0.07}O_{2-\delta}$ catalyst. Intensities of monodentate carbonate (1340 and 1070 cm^{-1}) and bidentate carbonate (1300 and 990 cm^{-1}) species are found to increase over $Ce_{0.68}Hf_{0.25}Cu_{0.07}O_{2-\delta}$ catalyst. Increase in intensities of monodentate and bidentate carbonate species and decrease in hydrogen carbonate species is noticed over $Ce_{0.68}Th_{0.25}Cu_{0.07}O_{2-\delta}$ catalyst.

Ionic substitution of Cu in CeO$_2$ based oxides forming $—Ce^{4+}—\square—Cu^{2+}—$ type of linkages on their surfaces is established in the literature where metal ions and oxide ion vacancies are two distinct sites for adsorption of reducing and oxidizing molecules and subsequent catalysis.[3,4,13,14]

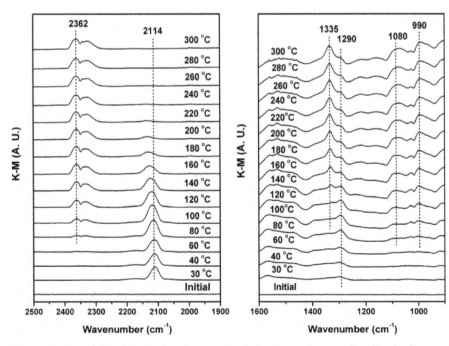

Figure 7　Diffuse reflectance infrared Fourier transform spectra during temperature ramping after simultaneous CO and O$_2$ adsorption on $Ce_{0.93}Cu_{0.07}O_{2-\delta}$ catalyst

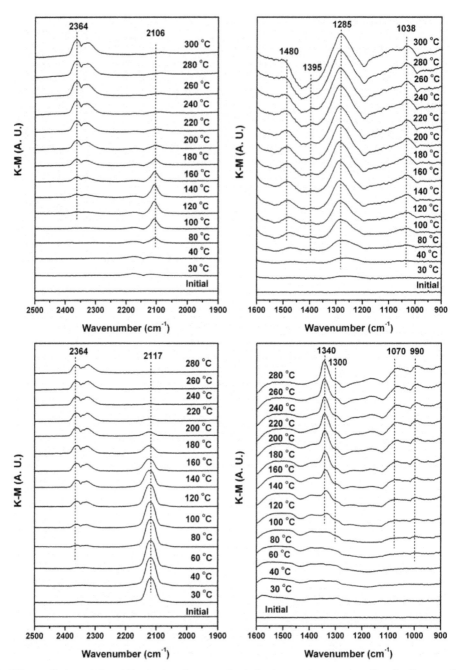

Figure 8 Diffuse reflectance infrared Fourier transform spectra during temperature ramping after simultaneous CO and O_2 adsorption on (top) $Ce_{0.68}Zr_{0.25}Cu_{0.07}O_{2-\delta}$ and (bottom) $Ce_{0.68}Hf_{0.25}Cu_{0.07}O_{2-\delta}$ catalysts

More than 10 at.% Cu substitution into CeO_2 matrix has been reported in the literature.[42] In this study, 7 at.% Cu has been substituted in the CeO_2 matrix so that effect of surrounding environment in different CeO_2 based matrices can be understood. Therefore, Zr, Hf and Th have been chosen because their oxides are non-reducible and having stable tetravalent oxidation state with similar structure of substituted CeO_2 based oxides. Surface areas of the catalysts are not significantly different in these catalysts except in $Ce_{0.68}Hf_{0.25}Cu_{0.07}O_{2-\delta}$. Preparation procedure of

$Ce_{0.68}Hf_{0.25}Cu_{0.07}O_{2-\delta}$ can be responsible for its higher surface area, because its preparation was started from hydroxide precipitation of $HfCl_4$, followed by washing of chloride and then dissolving in excess HNO_3.

In the present work, change of nature of ionic Cu species with different substituents in CeO_2 based oxides has been investigated as substitution of Cu in CeO_2 based oxides occur up to high concentration. The effect of support can be observed when CO adsorption shows variation in the intensity of Cu^+—CO in different supports in similar conditions. In this study, variation of Cu^+—CO peak intensities during CO oxidation have been accepted as a probe to find the differences in supports and it is shown in Fig. 9. The observed intensities of Cu^+—CO peaks increase with the following substitutions: Zr < Hf ~ Th. This trend can also be explained based on the higher effective nuclear charge of Hf^{4+} and Th^{4+} as compared to Zr^{4+}. As all the substituents are non-reducible, only electron-withdrawing power as determined by effective nuclear charge can influence the activity of these catalysts. Effective nuclear charge of Hf^{4+} or Th^{4+} is higher because of presence of f electrons as compared to Zr^{4+} indicating that lattice oxygen is more polarized toward Hf^{4+} and Th^{4+}. This means charge on Cu^{2+} ion is more localized and able to withdraw more electron toward itself or more reducing nature in the Hf/Th containing oxides. Therefore, transformation of more Cu^{2+} ion to Cu^+ occur significantly in the case of Hf and Th doped catalysts giving higher Cu^+—CO intensities. This also influences

the CO oxidation activities of these catalysts, because localized charge on Cu^{2+} induces more + ve charge on 'C' atom end of the adsorbed CO molecule while reacting with lattice oxygen. This explains lowest activity of Zr substituted catalyst that is shown in Fig. 10.

Conclusion

CO adsorption and oxidation are observed to be lowest over $Ce_{0.68}Zr_{0.25}Cu_{0.07}O_{2-d}$ among $Ce_{0.68}M_{0.25}Cu_{0.07}O_{2-d}$ (M≠Zr Hf and Th) catalysts. It has been found that Cu^+—CO vibrational intensity increases as Zr < Ce < Th ~ Hf. With increasing effective nuclear charge of the substituent metal (Zr, Hf and Th) in CeO_2 matrix, ionic charge on the Cu^{2+} is more localized as negative charge on oxygen gets depleted because of more polarization in the M—O bond around Cu^{2+} ion. This effect may be alleviated by converting more Cu^{2+} to Cu^+ and thereby, increasing the intensity of Cu^+—CO peak. Therefore, CO oxidation activity increases in the following substitutions: Zr < Ce < Hf ~ Th.

Acknowledgements

The authors are grateful to Prof. M. S. Hegde, Indian Institute of Science, Bangalore, for providing DRIFTS and XPS facilities.

References

1. A. Trovarelli: Catal. Rev. Sci. Eng., 1996, **38**, 439.
2. A. Trovarelli: 'Catalysis by ceria and related materials'; 2002, London, Imperial College Press.
3. M. S. Hegde, K. C. Patil and G. Madras: Acc. Chem. Res., 2009, **42**, 704.
4. P. Bera and M. S. Hegde: Catal. Surv. Asia, 2011, **15**, 181.
5. P. Bera and M. S. Hegde: J. Indian Inst. Sci., 2010, **90**, 299.
6. G. Dutta, U. V. Waghmare, T. Baidya, M. S. Hegde, K. R. Priolkar and P. R. Sarode: Catal. Lett., 2006, **108**, 165.
7. G. Dutta, U. V. Waghmare, T. Baidya, M. S. Hegde, K. R. Priolkar and P. R. Sarode: Chem. Mater., 2006, **18**, 3249.
8. T. Baidya, M. S. Hegde and J. Gopalakrishnan: J. Phys. Chem. B, 2007, **111**, 5149.
9. T. Baidya, A. Gupta, P. A. Deshpandey, G. Madras and M. S. Hegde: J. Phys. Chem. C, 2009, **113**, 4059.
10. T. Baidya, A. Marimuthu, M. S. Hegde, N. Ravishankar and G. Madras: J. Phys. Chem. C, 2007, **111**, 830.
11. S. Roy, A. Marimuthu, M. S. Hegde and G. Madras: Appl. Catal. B, 2007, **71**, 23.
12. S. Roy, A. Marimuthu, M. S. Hegde and G. Madras: Catal. Commun., 2008, **9**, 101.
13. P. Bera, S. T. Aruna, K. C. Patil and M. S. Hegde: J. Catal., 1999, **186**, 36.
14. P. Bera, K. R. Priolkar, P. R. Sarode, M. S. Hegde, S. Emura, R. Kumashiro and N. P. Lalla: Chem. Mater., 2002, **14**, 3591.
15. D. Gamarra, G. Munuera, A. B. Hungría, M. Fernández-García, J. C. Conesa, P. A. Midgley, X. Q. Wang, J. C. Hanson, J. A. Rodríguez and A. Martínez-Arias: J. Phys. Chem. C, 2007, **111**, 11026.
16. A. Martínez-Arias, D. Gamarra, M. Fernández-García, A. Hornés, P. Bera, Zs Koppány and Z. Schay: Catal. Today, 2009, **143**, 211.
17. A. Martínez-Arias, D. Gamarra, A. B. Hungría, M. Fernández-García, G. Munuera, A. Hornés, P. Bera, J. C. Conesa and A. López Cámara: Catalysts, 2013, **3**, 378.
18. T. Baidya, G. Dutta, M. S. Hegde and U. V. Waghmare: Dalton Trans., 2009, **38**, 455.
19. A. Hornés, P. Bera, A. López Cámara, D. Gamarra, G. Munuera and A. Martínez-Arias: J. Catal., 2009, **268**, 367.
20. T. Baidya, P. Bera, B. D. Mukri, S. K. Parida, O. Kröcher, M. Elsener and M. S. Hegde: J. Catal., 2013, **303**, 117.
21. D. D. Miller and S. S. C. Chuang: Catal. Commun., 2009, **10**, 1313.
22. R. D. Shanon: Acta Crystallogr. A, 1976, **32**, 751.
23. J. Huang, Y. Kang, T. Yang, Y. Wang and S. Wang: React. Kinet. Mech. Catal., 2011, **104**, 149.
24. S. Hüfner: 'Photoelectron spectroscopy: principles and applications'. 3rd edn; 2003, Berlin, Springer.

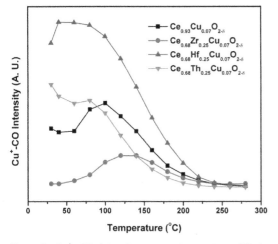

Figure 9 Cu^+—CO intensity versus temperature (°C) in $Ce_{0.93}Cu_{0.07}O_{2-δ}$ and $Ce_{0.68}M_{0.25}Cu_{0.07}O_{2-δ}$ (M = Zr, Hf, and Th) catalysts

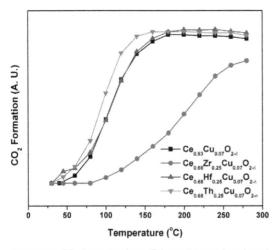

Figure 10 CO_2 formation from IR intensity as a function of temperature (°C) over $Ce_{0.93}Cu_{0.07}O_{2-δ}$ and $Ce_{0.68}M_{0.25}Cu_{0.07}O_{2-δ}$ (M = Zr, Hf and Th) catalysts

25. D. C. Frost, A. Ishitani and C. A. McDowell: *Mol. Phys.*, 1974, **24**, 861.
26. G. Moretti, G. Fierro, M. Lo Jacono and P. Porta: *Surf. Interface Anal.*, 1989, **14**, 325.
27. Y. Okamoto, K. Fukino, T. Imanaka and S. Teranishi: *J. Phys. Chem.*, 1983, **87**, 3740.
28. C. N. R. Rao, G. R. Rao, M. K. Rajumon and D. D. Sarma: *Phys. Rev. B*, 1990, **42**, 1026.
29. R. Bechara, A. Aboukaïs and J. -P. Bonnelle: *J. Chem. Soc. Faraday Trans.*, 1993, **89**, 1257.
30. D. D. Sarma, M. S. Hegde and C. N. R. Rao: *J. Chem. Soc., Faraday Trans.*, 1981, **77**, 1509.
31. C. Anandan and P. Bera: *Appl. Surf. Sci.*, 2013, **283**, 297.
32. J. F. Moulder, W. F. Stickle, P. E. Sobol and B. D. Bomben: 'Handbook of X-ray photoelectron spectroscopy', (ed. J. Chastain); 1992, Eden Prairie, PerkinElmer Corporation.
33. W. -E. Fu and Y. -O. Chang: *Appl. Surf. Sci.*, 2011, **257**, 7436.
34. Z. -W. Lin, Q. Kuang, W. Lian, Z. -Y. Jiang, Z. -X. Xie, R. -B. Huang and L -S. Zheng: *J. Phys. Chem. B*, 2006, **110**, 23007.
35. C. J. Powell and P. E. Larson: *Appl. Surf. Sci.*, 1978, **1**, 186.
36. J. H. Scofield: *J. Electron Spectrosc. Relat. Phenom.*, 1976, **8**, 129.
37. D. R. Penn: *J. Electron Spectrosc. Relat. Phenom.*, 1976, **9**, 29.
38. P. Bera, A. López Cámara, A. Hornés and A. Martínez-Arias: *J. Phys. Chem. C*, 2009, **113**, 10689.
39. P. Bera, A. López Cámara, A. Hornés and A. Martínez-Arias: *Catal. Today*, 2010, **155**, 184.
40. M. B. Padley, C. H. Rochester, G. J. Hutchings and F. King: *J. Catal.*, 1994, **148**, 438.
41. K. I. Hadjiivanov and G. N. Vayssilov: *Adv. Catal.*, 2002, **47**, 307.
42. A. Gayen, T. Baidya, A. S. Prakash, N. Ravishankar and M. S. Hegde: *Indian J. Chem.*, 2005, **44A**, 34.

9

Reduction properties of Ce in CeO$_x$/Pt/Al$_2$O$_3$ catalysts

P. P. Wells*[1,2], E. M. Crabb[3], C. R. King[1], S. Fiddy[4], A. Amieiro-Fonseca[5], D. Thompsett[5] and A. E. Russell[1]

[1]Department of Chemistry, University of Southampton, Highfield, Southampton SO17 1BJ, UK
[2]UK Catalysis Hub, Research Complex at Harwell, Rutherford Appleton Laboratory, Didcot OX11 0FA, UK
[3]Department of Life, Health and Chemical Sciences, The Open University, Milton Keynes MK7 6AA, UK
[4]CCLRC, Daresbury Laboratory, Warrington WA4 4AD, UK
[5]Johnson Matthey Technology Centre, Blounts Court, Sonning Common, Reading RG4 9NH, UK

Abstract A controlled surface reaction (CSR) technique has been successfully employed to prepare a series of CeO$_x$ modified Pt/Al$_2$O$_3$ catalysts, offering a unique system to specifically probe the relationship between Ce and Pt without any bulk CeO$_2$ present. Ce L$_3$ edge X-ray absorption near edge structure (XANES) analysis was used to ascertain the oxidation state of the Ce in the catalyst materials in atmospheres of air, H$_2$ (g), and CO (g) at room temperature. The XANES data showed that the Ce was present as both Ce^{3+} and Ce^{4+} oxidation states in an atmosphere of air, becoming predominantly present as Ce^{3+} in H$_2$ and CO. The results indicate the role of Pt in the process, and show that with the absence of bulk CeO$_2$, changes in Ce oxidation state can be observed at non-elevated

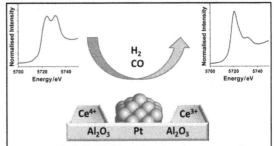

Note: Novel CeOx/Pt/Al2O3 catalysts have been prepared and characterised under different environments using XANES spectroscopy. The XANES spectra show that conversion to Ce3+ in an atmosphere of H2 is achieved, indicating that Ce is readily reduced at room temperature when well dispersed in the presence of Pt.

temperatures. The CeO$_x$/Pt/Al$_2$O$_3$ catalysts were tested for their performance toward the water gas shift (WGS) reaction and showed improved performance compared to the unmodified Pt/Al$_2$O$_3$, even at very low concentrations of Ce (~0.35 wt-%).

Keywords Water gas shift, Platinum, Ceria, Redox, Catalysis, XANES

Introduction

Decreasing the level of CO found in proton exchange membrane fuel cell (PEMFC) H$_2$ fuel streams derived from reformed hydrocarbons is a major challenge. Levels of CO as low as 10 ppm act as a poison toward Pt-based catalysts used within the PEMFC. One of the ways to reduce the amount of CO in the fuel stream is by a clean-up process utilizing the water gas shift (WGS) reaction

$$CO + H_2O \rightleftharpoons CO_2 + H_2 \, \Delta H^{o}_{298} = -41.1 \, kJmol^{-1}$$

At present, a multi-stage clean-up process involving high-temperature WGS (HTS), low-temperature WGS (LTS), and preferential oxidation (PROX) are employed to achieve levels of CO less than 10 ppm. Low-temperature WGS and HTS steps are used in conjunction as the exothermic nature of the WGS reaction leads to an 80-fold increase in the equilibrium constant when the temperature is reduced from 600 to 200°C.[1] A typical set-up exploits the kinetics of the process by employing an initial HTS step, in turn helping to decrease the required catalyst load. Subsequently, the gas stream is cooled before entering the LTS stage where the favorable thermodynamics are then exploited. The combination of HTS and LTS steps can reduce CO levels to less than 2000 ppm.[2]

Conventionally, Fe/Cr and Cu/Zn catalysts are used for the HTS and LTS processes, respectively.[1–3] These catalysts are not suitable for use in conjunction with fuel cell applications for several reasons. The catalysts need to be activated before

*Corresponding author, email peter.wells@rc-harwell.ac.uk

operation to control the reduction of surface oxides to produce the catalytic active states. The exothermic nature of the reduction means the activation in a reducing gas must be carried out with a slow heating rate. This is impractical in areas in which low-temperature fuel cells are expected to be employed, such as the automotive sector. For such an application, the sintering of the catalysts that would be encountered for a rapid activation step would lead to poor mechanical stability as a result of start-up/shut down cycles.[1,2,4] When in operation, the active reduced surface of the catalyst is pyrophoric. If the catalyst is exposed to air as a result of an accident, an exothermic reaction resulting in temperatures in excess of 650°C would be typical.[2]

Ceria-supported metal catalysts are also known to promote the WGS reaction. Early mechanistic work regarding the promotion was carried out by Shido and Iwasawa.[5,6] They proposed that bridging —OH groups on the surface of the ceria support react with CO to form bidentate formates, which then decompose in the presence of water to yield H_2 and CO_2. It was suggested that the role of the promoter metal was to aid the reduction of surface CeO_2 sites to form the active bridging —OH groups. The exact mechanism by which the promoter aids the formation of bridging —OH groups is unclear. One possibility is that the removal of surface oxygen by H_2 or CO is followed by dissociative adsorption of water. Another possibility is that the H_2 is dissociated directly over the promoter metal, with hydrogen spilling over onto the ceria surface leading to the formation of the bridging —OH groups.[7–9]

Another mechanism based on a surface redox process has also been put forward as a possible pathway for the WGS reaction on ceria-supported precious metal catalysts.[10,11] The redox mechanism proposes that CO adsorbs on the surface of reduced metal sites and reacts with oxygen from ceria to form CO_2. The relative merits and significance of the formate and redox pathways are still being debated.[12,13] One of the problems with the Pt/CeO_2 system is the deactivation of the catalyst over time. This deactivation has been linked to carbon deposits on the ceria surface and poor mechanical stability as a result of sintering.[10,14] Consequently, ceria-supported precious metal catalysts are doped with other metal oxides to help prevent sintering and also to promote the reduction of Ce(IV) to Ce(III).[4,15–17]

With much research carried out on conventional Pt/CeO_2 systems, it is desirable to prepare a novel system, maximizing the Pt/Ce interface[18] and incorporating a support more resistant to sintering under reducing conditions. Traditional methods for preparing bimetallic catalysts such as impregnation, electrochemical deposition, and precipitation lack the required specificity to deposit the secondary metal onto the surface sites of the primary metal in a controlled manner. One method that offers this specificity is the reaction between the reduced surface of the primary metal and an organometallic compound of the secondary metal. The work by Basset et al.[19–21] looked at the reactions between supported Ru, Rh, and Pt metals and alkyl complexes of Ge, Sn, Pb, Se, and Zn, the method was called surface organometallic chemistry (SOMC). The procedure exploited the interaction between the organometallic complex and the metal surface in the presence of a partial pressure of hydrogen.

The work by Crabb et al.[22–25] transferred this novel preparatory method into the design of fuel cell electrocatalysts. Carbon-supported Pt particles were modified with Ge, Cr, Ru,

and Pd using alkyl, allyl, and metallocene compounds for use as CO electrooxidation catalysts. As the method used organometallics other than alkyl complexes, these reactions are known as controlled surface reactions (CSRs). The reaction was assumed to be initiated by the physisorption of the organometallic species onto the reduced surface of the primary metal. Subsequent heating under hydrogen was believed to cleave the organic fragment, leaving behind adatoms on specific crystallographic sites. Thermal treatment of the resulting catalyst was used to obtain the surface alloy. In this study, the method developed by Crabb et al.[26–28] was used to prepare a series of $CeO_x/Pt/Al_2O_3$ catalysts using $Ce(acac)_3.H_2O$ as the organometallic precursor. X-ray absorption near edge structure (XANES) spectroscopy was used to probe the oxidation state of Ce in these materials. The co-locality of Ce and Pt allows for the relationship between the two metals and the effect this has on the oxidation state of Ce to be probed in more detail than previous studies have provided.[8,9,29,30] Studying conventional Pt/CeO_2 catalysts probes the entirety of Ce sites and fails to distinguish between surface and bulk interactions, which is detailed in this work.

Experimental

Catalyst preparation

Modifications were made to a 4 wt-% $Pt/\gamma-Al_2O_3$ catalyst supplied by Johnson Matthey. The preparation of $CeO_x/Pt/Al_2O_3$ catalysts was carried out in a manner similar to that employed by Crabb et al.[26–28] and involved reduction of the core catalyst and subsequent reaction with an organometallic precursor of the modifying metal. Briefly, the $Pt/\gamma-Al_2O_3$ substrate catalyst was first reduced in the reactor under flowing $H_2(g)$ (flow 60–100 $cm^3 min^{-1}$) at 350°C for 3 h and then cooled to room temperature under flowing $N_2(g)$. The required amount of $Ce(acac)_3.H_2O$ was dissolved in toluene, purged with $N_2(g)$ and added to the reactor. $H_2(g)$ was then passed through the reactor while stirring and heating at 90°C for 8 h. The reactor was then allowed to cool and flushed with $N_2(g)$ for 30 min. The contents of the reactor were discharged, filtered, and washed with toluene. The filtered catalyst was allowed to dry in air and then returned to the cleaned reactor, and the initial reduction step at 350°C was repeated. Catalysts were prepared with the following nominal Ce:Pt surface atomic ratios, 0.5, 1, 2, and 4 as well as a control reaction with the Al_2O_3 support (CeO_x/Al_2O_3) and Pt/ C, ($CeO_x/Pt/C$). Elemental analysis of the prepared catalyst was carried out using inductively coupled plasma–optical emission spectroscopy (ICP-OES).

Characterization

Ce L_3 edge X-ray absorption spectroscopy (XAS) spectra were acquired on station 7.1 at the SRS, Daresbury Laboratory. The ring operated with 2.0 GeV energy and 100–250 mA current. Station 7.1 utilizes a harmonic rejecting sagitally focusing double crystal Si(111) monochromator to acquire XAS data in the range of 4–10 keV. The sagital focusing allows for the X-ray beam to be maintained as a focused spot by dynamically bending the crystal while carrying out XAS experiments. This gives rise to an increase of X-ray flux and, as a consequence, improved signal-to-noise ratio.

The fluorescence detector uses a monolithic structure nine-channel array on a germanium wafer with a diameter of 21.8mm.[31]

All catalysts were prepared as BN pellets for XAS measurements, with the sample being placed in a gas treatment cell. The preparation of pellets involves grinding a set amount of catalyst and BN to form a homogeneous mixture and then compacting this in a purpose built press to form a self-supporting wafer. BN was used in preference to polyethylene because of the better gas flow through the pellet. The gas treatment cell contains a gas inlet and outlet ports so that different gases can be purged through and contained within the cell. The gas environments used within these studies were air, H_2, and CO. All gas treatments were performed away from the beamline, for safety, with a minimum purge time of 1 h. The cell consists of a series of Kapton windows to allow for the entry and exit of the X-ray beam and can be set up for both fluorescence and transmission acquisition, although only fluorescence experiments were performed. The raw data have been processed to give the normalized XANES spectra.

Powder samples for TEM EDX were crushed between two glass slides, and samples positioned onto a lacey carbon-coated copper 'finder' grid with the aid of a micromanipulator. The samples were examined in a Tecnai F20 transmission electron microscope. Both bright field and high resolution electron microscopy modes were used.

Catalytic testing

Fixed bed reactor testing was employed to assess the performance of the CeOx/Pt/Al2O3 samples for the gas phase WGS reaction. About 0.45g of catalyst was loaded into the reactor tube on a bed of glass wool, ensuring that the thermocouple was in the catalyst bed layer. Gas inlet and outlet lines were enclosed in an oven operating at a temperature of 110°C to prevent water condensation. The inlet gases were 5% CO, 30% steam, and 65% N_2 with a total flow of $300cm^3 min^{-1}$. The heating rate was $5°C min^{-1}$ and the maximum temperature set to 500°C. The analysis of the exit gases was performed using a Maihak s710 analyzer comprising a H_2 detector and IR CO and CO_2 sensors. Three thermocouples are placed inside the reactor chamber to monitor the temperature of the gas inlet, gas outlet and the catalyst bed.

Results

Catalyst preparation

Elemental analyses of the prepared catalysts were carried out using Inductively coupled plasma–atomic emission spectroscopy (ICP-AES). The experimental values, along with the theoretically calculated values, are shown in Table 1.

The catalysts are represented in the form n CeOx/Pt/Al2O3, where n is equal to the desired monolayer coverage (surface atomic ratio). In general, good agreement was found between the experimental and theoretical values, confirming that the desired loadings were achieved. The large amounts of CeOx deposited indicate that it is unlikely to be present at the Pt surface alone. A control reaction with Al2O3, in the absence of Pt, confirmed that the deposition of Ce was not dependent on the presence of Pt. However, this is not the case with carbon as the support. It is believed that this is caused by greater reducibility of the Al2O3 in comparison to carbon or the presence of other deposition sites on the support. The large amount of CeOx being deposited can be rationalized by the targeting of both Pt and Al2O3 sites or by the CeOx preferentially reacting with Pt sites before migrating onto the support.

TEM characterization

The TEM micrographs (Fig. 1) confirm the presence of small Pt particles supported on alumina. In some cases, the supported particles were found in isolation on the alumina support as is the case with Fig. 1a, b, and d; however, in other instances, dense agglomerations were found as illustrated in Fig. 1c. Figure 1a also shows that some of the particles found were faceted. Owing to the small size of the Pt particles (~1–2nm) and the dense agglomerations, particle size analyses were not performed. However, significantly larger particles were not observed for any of the prepared catalysts as evidenced by the TEM micrographs.

EDX analysis was used in an attempt to confirm the co-locality of Pt and Ce on the Al2O3 support. However, the method is not very sensitive to components present at concentrations below 1 wt-%; the 0.5 CeOx/Pt/Al2O3 sample consisted of 0.35wt-% Ce. Thus, the analysis is not expected to be very accurate and is only used as a rough screening method. In addition, the small size of the Pt particles (1–2nm particles) does not facilitate the accurate use of line scans across a single particle.

XANES characterisation

The 3+ and 4+ oxidation states of Ce give rise to different characteristic Ce L_3 edge XANES spectra. The XANES spectrum (Fig. 2) of CeO_2 (Ce IV) shows two distinct peaks b_2 and c, with two shoulders a and b_1 also visible. The weak shoulder a is attributed either to the dipole forbidden $2p_{3/2} \rightarrow 4f$ transitions[32] or to transitions to the conduction band, where unoccupied, delocalized final states with d character are present.[33] There is still some debate about the nature of peak b_1, although peak b_2 has been attributed to transitions to the final state Ce $[2p^5 4f^1 5d^1]$ O $[2p^5]$ and the final state configuration of peak c has been widely reported as transitions to Ce $[2p^5 4f^0 5d^1]$ O $[2p^6]$.[32–35]

Table 1 Elemental analysis of prepared CeOx/Pt/Al2O3 catalysts determined from ICP-OES measurements

Sample	Wt-% Pt Theoretical	Wt-% Pt Experimental	Wt-% Ce Theoretical	Wt-% Ce Experimental
0.5 CeOx/Pt/Al2O3	3.99	3.67	0.37	0.33
1 CeOx/Pt/Al2O3	3.97	3.67	0.74	0.70
2 CeOx/Pt/Al2O3	3.94	3.60	1.47	1.30
4 CeOx/Pt/Al2O3	3.88	3.56	2.90	2.33

ICP-OES: inductively coupled plasma–optical emission spectroscopy.

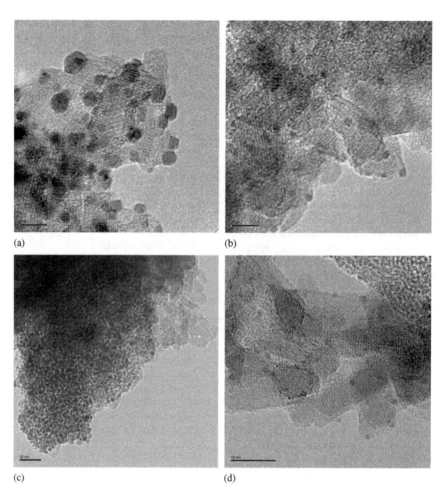

(a)   (b)

(c)   (d)

1   TEM micrographs of *a* 0.5, *b* 1, *c* 2, and *d* 4 CeO_x/Pt/Al_2O_3. The scale bars for figures, *a*, *b*, *c*, and *d* are 5, 5, 10, and 20 nm, respectively

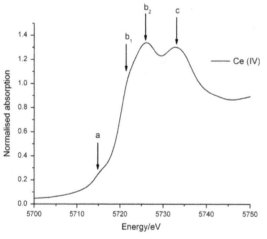

2   Ce L_3 edge X-ray absorption near edge structure (XANES) spectrum of CeO_2 in air

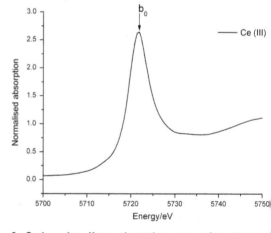

3   Ce L_3 edge X-ray absorption near edge structure (XANES) spectrum of $Ce(NO_3)_6$ in air

The XANES spectrum (Fig. 3) of the Ce L_3 edge of $Ce(NO_3)_6$ (Ce III) has starkly different characteristic from the spectrum of CeO_2. Ce (III) marks the occupation of the *4f* energy levels, and thus, certain transitions become less favored. There is one observable peak b_0, and this is because of transitions to the final state Ce $[2p^5 4f^1 5d^1]$ O $[2p^6]$. These differences in XANES characteristics for the (III) and (IV) oxidation states of Ce at the Ce L_3 edge make comparison of XANES spectra an appropriate technique for probing the reducibility of Ce under different conditions. In this study, only a qualitative assessment of the Ce oxidation state is employed, although there are procedures for a quantitative determination.[34]

Other *in situ* XANES studies of Pt/CeO$_2$ utilized a 23% H$_2$/He gas mix at elevated temperatures to confirm the formation of Ce (III) species.[8,30] Only a small change in overall Ce valency was observed, as the XANES technique provides the per-atom average oxidation state of the ceria at the surface and in the bulk. Further XANES studies, again carried out in a 23% H$_2$/He environment at elevated temperatures, showed a correlation between the Pt loading and the reducibility of the Pt/CeO$_2$ catalysts.[9] As the Pt loading was increased, the catalysts exhibited more Ce (III) character indicating that the metal promoter is linked to the formation of Ce (III) species. A maximum Ce(III) content of 25% was shown for a 5 wt-% Pt/CeO$_2$ catalyst at 300°C. This partial reduction of the ceria support appeared as a slight increase of peak b_2 with respect

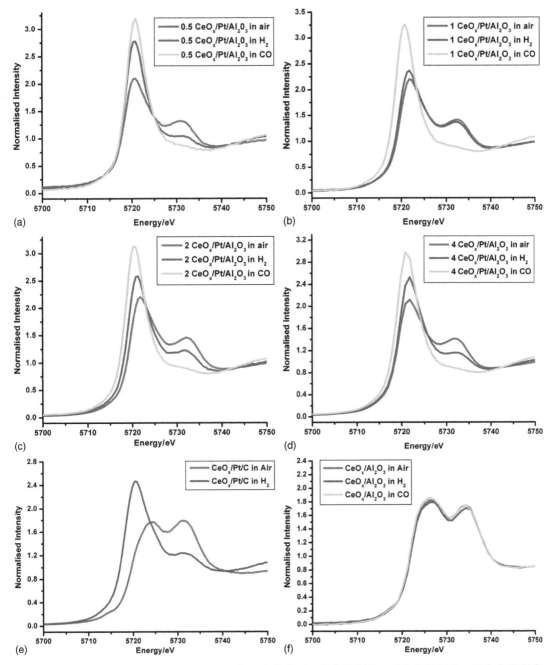

4   Normalized Ce L$_3$ edge XANES spectra of *a* 0.5 CeO$_x$/Pt/Al$_2$O$_3$, *b* 1 CeO$_x$/Pt/Al$_2$O$_3$, *c* 2 CeO$_x$/Pt/Al$_2$O$_3$, *d* 4 CeO$_x$/Pt/Al$_2$O$_3$, *e* CeO$_x$/Pt/C, and *f* CeO$_x$/Al$_2$O$_3$ in gas environments of air, H$_2$, and CO at room temperature

to peak c for the CeO₂ XANES spectrum. Only a small change in Ce valency is expected because of the large proportion of Ce in the bulk of the support in comparison to the surface.

The Ce L_3 edge XANES spectra in Fig. 4 of the CeO$_x$/Pt/Al$_2$O$_3$ catalysts show that at non-elevated temperatures under normal atmospheric conditions, a mixed valency of Ce(III) and Ce(IV) is achieved, which to the best of our knowledge has not been shown previously. This is evidenced by the characteristic shape of the Ce(IV) XANES spectrum to become distorted as a consequence of the contribution of the peak b_0 in Fig. 2. The XANES spectra show that the degree of Ce(III) character becomes more prominent in the gas environment of H$_2$ and that the materials are almost 100% Ce(III) when placed in atmospheres of CO. This is a confirmation that the CeO$_x$ is very finely dispersed as it is likely that subsurface Ce would remain as Ce(IV). This level of interaction between the Ce and Pt makes these systems ideal for XAS studies, which sample the whole population of Ce sites. The control reaction with Al$_2$O$_3$ provided the material CeO$_x$/Al$_2$O$_3$, which acted as a system to probe the changes in Ce oxidation state in the absence of Pt. In this system, identical XANES spectra were acquired for all of the gas environments, with the Ce being present in oxidation state (IV) in all instances. This proves that the presence of nanoparticulate Pt is required to facilitate the reduction of Ce sites. Anecdotally, this provides evidence of the co-locality of the Ce and Pt on the Al$_2$O$_3$ surface.

It is of particular interest that a partial reduction of Ce(IV) to Ce(III) is observed in air at room temperature for the CeO$_x$/Pt/Al$_2$O$_3$ catalysts, while no reduction was observed for the CeO$_x$/Al$_2$O$_3$ catalyst (as discussed above) and also the CeO$_x$/Pt/C catalyst. Thus, the partial reduction requires the presence of both Pt and Al$_2$O$_3$. This observation may reflect differences in the distribution of CeO$_x$ on the two supports. On carbon, the CeO$_x$ is less able to spread across the surface of the support, resulting in larger clusters of CeO$_x$ or encapsulation of the Pt particles, either of which would decrease the fraction of the Ce present in the +3 oxidation state.

In a gas environment of H$_2$, the reduction of Ce(IV) to Ce(III) species is more pronounced than that in air for all of the CeO$_x$/Pt/Al$_2$O$_3$ catalysts, with the exception of the 1 CeO$_x$/Pt/Al$_2$O$_3$ catalyst. As a further reduction is noted for all the other catalysts, the result for the 1 CeO$_x$/Pt/Al$_2$O$_3$ catalysts seems anomalous and may be the result of a leak in the gas cell, or insufficient purging with hydrogen. The additional reduction observed for the other CeO$_x$/Pt/Al$_2$O$_3$ catalysts may be caused by the dissociation of H$_2$ over surface Pt sites. The resultant hydrogen could then subsequently spill over onto neighboring Ce atoms, causing the formation of the bridging —OH groups,[36] thus creating the proposed active sites for the WGS reaction proceeding by the formate mechanism.[9]

Experiments performed in an environment of CO show signs of further reduction of Ce(IV) to Ce(III). Jacobs et al.[9] noted a similar effect after a reduction pre-treatment in CO for Pt/Ceria catalysts. This observation was attributed to the ability of CO to remove capping oxygen atoms at the ceria surface. As no partial reduction is observed for the CeO$_x$/Al$_2$O$_3$, this infers that the Pt is required for the CO/H$_2$ to reduce Ce(IV) or that previously formed Ce(III) species at the surface are required.

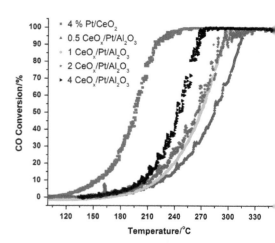

5 Water gas shift (WGS) testing results for different CeO$_x$/Pt/Al$_2$O$_3$ catalysts. Data for 4 wt-% Pt/CeO$_2$ are included for comparison. Data for Pt/Al$_2$O$_3$ (not shown) give a zero/baseline response over this temperature range.

Table 2 Temperature at 50% CO conversion for the water gas shift (WGS) testing results

Catalyst	T_{50}/°C
4 wt-% Pt/CeO₂	193.5
0.5 CeO$_x$/Pt/Al₂O₃	285.1
1 CeO$_x$/Pt/Al₂O₃	269.3
2 CeO$_x$/Pt/Al₂O₃	263.5
4 CeO$_x$/Pt/Al₂O₃	245.5

Catalytic testing

The WGS testing results (Fig. 5 and Table 2) show that the modification of Pt/Al$_2$O$_3$ with CeO$_x$ greatly enhances the activity of the catalyst. The response of an unmodified Pt/Al$_2$O$_3$ catalyst over the same temperature range gives a baseline response and exhibits no activity toward the WGS reaction. It is of great interest that the addition of ⊠0.35 wt-% Ce, as is the case with the 0.5 CeO$_x$/Pt/Al$_2$O$_3$, has such a pronounced effect on the activity of the catalyst. Further improvements in performance are attained as the number of monolayers of Ce are increased, with a 40°C reduction in T_{50} value for the 4 CeO$_x$/Pt/Al$_2$O$_3$ catalysts (245°C), compared to the 0.5 CeO$_x$/Pt/Al$_2$O$_3$ catalyst (285°C). However further improvements are required to reach the performance levels of a conventional Pt/CeO$_2$ catalyst (194°C).

Conclusion

A series of CeO$_x$/Pt/Al$_2$O$_3$ catalysts have been prepared by a controlled surface modification technique, which were found to contain Ce in a state where it is readily reduced to Ce(III) at room temperature. CeO$_2$ is widely known for its oxygen storage capacity, being able to store and release O$_2$ depending on the necessity at hand. This ability is widely reported as the property, which enables CeO$_2$ to promote various catalytic reactions. However, in most systems, CeO$_2$ is a support or part of a mixed oxide support in which any change in overall Ce valency is hard to observe using

XANES as the subsurface Ce(IV) is in huge excess compared to the surface Ce. The preparation method used within this study has produced a catalytic system with highly disperse CeO_x where the extent of reduction can be easily visualized using XANES spectroscopy. Previous studies on Pt/CeO_2[8,30] have had to reduce the catalysts in H_2 containing atmospheres at elevated temperatures (100–300°C) to witness only a partial reduction. The ability to be able to monitor the oxidation state more closely in conjunction with other techniques such as DRIFTS should enable a greater mechanistic insight into CeO_2 promoted catalytic reactions. The catalytic testing undertaken showed that the performance failed to match that offered by an optimized Pt/CeO_2 catalyst. However, it is noteworthy that only a small amount of Ce is required to see a large improvement in performance compared to Pt/Al_2O_3.

Acknowledgements

This work was supported by the EPSRC and Johnson Matthey. Beamtime at the SRS, Daresbury Laboratory, was provided by the CCLRC. The assistance of Chris Corrigan at Daresbury Laboratory is also acknowledged.

References

1. J. R. Ladebeck and J. P. Wagner: 'Chapter 16: catalyst development for water gas shift', in 'Handbook of fuel cells – Volume 3: fuel cell technology and applications'; ed. Wolf Vielstich, Arnold Lamm, Hubert Gasteiger, John Wiley & Sons, Chichester, 2003, 190–201.
2. R. Farrauto, S. Hwang, L. Shore and W. Ruettinger: *Annu. Rev. Mater. Res.*, 2003, **33**, 1–27.
3. A. F. Ghenciu: *Curr. Opin. Solid State Mater. Sci.*, 2002, **6**, 389–399.
4. S. Y. Choung, M. Ferrandon and T. Krause: *Catal. Today*, 2005, **99**, 257–262.
5. T. Shido and Y. Iwasawa: *J. Catal.*, 1993, **141**, 71–81.
6. T. Shido and Y. Iwasawa: *J. Catal.*, 1992, **136**, 493–503.
7. G. Jacobs and B. H. Davis: *Int. J. Hydrog. Energy*, 2010, **35**, 3522–3536.
8. G. Jacobs, E. Chenu, P. M. Patterson, L. Williams, D. Sparks, G. Thomas and B. H. Davis: *Appl. Catal. A*, 2004, **258**, 203–214.
9. G. Jacobs, U. M. Graham, E. Chenu, P. M. Patterson, A. Dozier and B. H. Davis: *J. Catal.*, 2005, **229**, 499–512.
10. A. Goguet, F. Meunier, J. P. Breen, R. Burch, M. I. Petch and A. F. Ghenciu: *J. Catal.*, 2004, **226**, 382–392.
11. A. Goguet, S. O. Shekhtman, R. Burch, C. Hardacre, E. Meunier and G. S. Yablonsky: *J. Catal.*, 2006, **237**, 102–110.
12. C. -H. Lin, C. -L. Chen and J. -H. Wang: *J. Phys. Chem. C*, 2011, **115**, 18582–18588.
13. C. M. Kalamaras, S. Americanou and A. M. Efstathiou: *J. Catal.*, 2011, **279**, 287–300.
14. L. Fu, N. Q. Wu, J. H. Yang, F. Qu, D. L. Johnson, M. C. Kung, H. H. Kung and V. P. Dravid: *J. Phys. Chem. B*, 2005, **109**, 3704–3706.
15. R. M. Navarro, M. C. Alvarez-Galvan, M. C. Sanchez-Sanchez, F. Rosa and J. L. G. Fierro: *Appl. Catal. B*, 2005, **55**, 229–241.
16. G. Kolb, H. Pennemann and R. Zapf: *Catal. Today*, 2005, **110**, 121–131.
17. G. Germani, P. Alphonse, M. Courty, Y. Schuurman and C. Mirodatos: *Catal. Today*, 2005, **110**, 114–120.
18. S. Aranifard, S. C. Ammal and A. Heyden: *J. Catal.*, 2014, **309**, 314–324.
19. J. P. Candy, B. Didillon, E. L. Smith, T. B. Shay and J. M. Basset: *J. Mol. Catal.*, 1994, **86**, 179–204.
20. B. Didillon, J. P. Candy, F. Lepeletier, O. A. Ferretti and J. M. Basset: 'Surface organometallic chemistry on metals - selective hydrogenation of citral on silica-supported rhodium modified by tetra-N-butyl germanium, tin and lead', in 'Heterogeneous catalysis and fine chemicals III', Edited by M. Guisnet, J. Barbier, J. Barrault, C. Bouchoule, D. Duprez, G. Pérot and C. Montassier, 147–154; 1993, Elsevier, ISBN: 978-0-444-89063-4.
21. F. D. Lefebvre, J. P. Candy, C. C. Santini and J. M. Basset: *Top. Catal.*, 1997, **4**, 211–216.
22. E. M. Crabb, R. Marshall and D. Thompsett: *J. Electrochem. Soc.*, 2000, **147**, 4440–4447.
23. E. M. Crabb and M. K. Ravikumar: *Electrochim. Acta*, 2001, **46**, 1033–1041.
24. E. M. Crabb, M. K. Ravikumar, Y. Qian, A. E. Russell, S. Maniguet, J. Yao, D. Thompsett, M. Hurford and S. C. Ball: *Electrochem. Solid-State Lett.*, 2002, **5**, A5–A9.
25. E. M. Crabb, M. K. Ravikumar, D. Thompsett, M. Hurford, A. Rose and A. E. Russell: *Phys. Chem. Chem. Phys.*, 2004, **6**, 1792–1798.
26. P. P. Wells, Y. D. Qian, C. R. King, R. J. K. Wiltshire, E. M. Crabb, L. E. Smart, D. Thompsett and A. E. Russell: *Faraday Discuss.*, 2008, **138**, 273–285.
27. P. P. Wells, E. M. Crabb, C. R. King, R. Wiltshire, B. Billsborrow, D. Thompsett and A. E. Russell: *Phys. Chem. Chem. Phys.*, 2009, **11**, 5773–5781.
28. E. M. Crabb and R. Marshall: *Appl. Catal. A*, 2001, **217**, 41–53.
29. I. D. Gonzalez, R. M. Navarro, W. Wen, N. Marinkovic, J. A. Rodriguez, F. Rosa and J. L. G. Fierro: *Catal. Today*, 2010, **149**, 372–379.
30. G. Jacobs, P. M. Patterson, L. Williams, E. Chenu, D. Sparks, G. Thomas and B. H. Davis: *Appl. Catal. A*, 2004, **262**, 177–187.
31. G. Derbyshire and K. C. Cheung: *J. Synchrotron Radiat.*, 1999, **6**, 62–63.
32. A. M. Shahin, F. Grandjean, G. J. Long and T. P. Schuman: *Chem. Mater.*, 2005, **17**, 315–321.
33. M. Balasubramanian, C. A. Melendres and A. N. Mansour: *Thin Solid Films*, 1999, **347**, 178–183.
34. A. M. Shahin, T. P. Schuman, F. Grandjean and G. J. Long: *Abstr. Pap. Am. Chem. Soc.*, 2004, **228**, 133.
35. J. Elfallah, S. Boujana, H. Dexpert, A. Kiennemann, J. Majerus, O. Touret, F. Villain and F. Lenormand: *J. Phys. Chem.*, 1994, **98**, 5522–5533.
36. M. Meng, Y. Q. Zha, J. Y. Luo, T. D. Hu, Y. N. Xie, T. Liu and J. Zhang: *Appl. Catal. A*, 2006, **301**, 145–151.

Promoting χ-Fe$_5$C$_2$(100)$_{0.25}$ with copper – a DFT study

Eric van Steen* and Michael Claeys

Centre for Catalysis Research, Department of Chemical Engineering, University of Cape Town, Private Bag X3, Rondebosch 7701, South Africa

Abstract The role of copper in iron based Fischer–Tropsch catalysts was investigated using DFT with χ-Fe$_5$C$_2$(100)$_{0.25}$ as a model surface. The presence of atomic copper on the iron-rich χ-Fe$_5$C$_2$(100)$_{0.25}$-surface is more favorable than its presence in surface. Nevertheless, the segregation of copper from the surface yielding fcc-Cu remains an exergonic process. Carbon monoxide at a coverage of 2.2 CO per nm^2 stabilizes atomic copper on this surface. The presence of copper results in the redshift in the stretching frequency of adsorbed CO. The mobility of copper atoms was investigated on χ-Fe$_5$C$_2$(100)$_{0.25}$ in the presence of CO. The hopping frequency is reduced due to the presence of CO, although never enough to avoid formation of fcc-Cu on a shorter time scale than typically required for the formation of hydrocarbons in the Fischer–Tropsch synthesis.

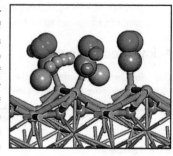

Keywords Iron, Copper, Fischer–Tropsch, Promoter, DFT

Introduction

CO-hydrogenation over transition metal catalysts is a useful step in the transformation of gaseous and solid carbon-containing feedstock to liquid fuels and/or chemicals. It has been argued that iron based catalysts may be useful, in particular, in the conversion of solid carbon-containing feedstock such as biomass or coal.[1] The nature of the catalytically active phase in iron based Fischer–Tropsch catalysts has been debated extensively in literature. Iron carbide, e.g. Hägg carbide, is generally thought to be a catalytically active phase.[2–5] Hence, DFT investigations to gain a better understanding of iron based Fischer–Tropsch catalysts have lately focused on Hägg iron carbide,[6–13] although other carbides have been considered as well.[14]

The low temperature iron based Fischer–Tropsch catalyst contains a number of promoters, such as alkali, silica/alumina, and copper.[15] These promoters have been classified according to their perceived role within the catalyst although these promoters may fulfill more than one role in the catalyst (e.g. silica is added as a binder/structural promoter, but it may also act as a chemical promoter).[16] Copper is typically regarded as a reduction promoter, i.e. facilitating the reduction of trivalent iron to divalent iron during the catalyst activation step.[15,17–24] A more facile reduction would allow a lower activation temperature, and

thus resulting in reduced extent of sintering during catalyst activation. A higher degree of transformation of iron into its active phase in conjunction with smaller crystallites would yield an increased activity.[20]

Copper may affect the reactivity of iron based Fischer–Tropsch catalysts in other ways than by altering the available surface area of the catalytically active phase. A kinetic study by O'Brien and Davis[23] showed that evaluating the reaction data using the Huff–Satterfield rate expression,[25] copper promotion resulted in an increase in the intrinsic rate constant and simultaneously in a decrease in the adsorption parameter, describing the ratio of the adsorption constant for water and the equilibrium constant for the dissociative adsorption of. This implies that copper would strengthen the dissociative adsorption of CO or weaken the adsorption of water. Furthermore, it has been reported that the incorporation of copper into the catalyst alters the product selectivity,[19,20,22] which has been attributed to a change in the surface chemistry during CO hydrogenation.[24] These effects would imply some role of copper as a chemical promoter for iron based Fischer–Tropsch catalysts.

Chemical promotion occurs by a change of the electronic state of the catalytically active site affecting the energetics of adsorption and transition of the various surface species. Various approaches can be taken to study the promotional effect of an adatom, namely the incorporation of the promoter in the sub-surface structure,[26] the incorporation of the promoter in the surface structure[27] or the adsorption

*Corresponding author, email eric.vansteen@uct.ac.za

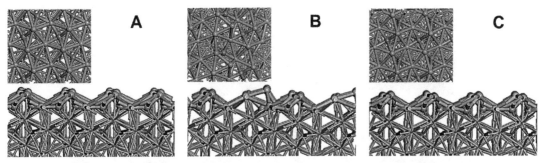

Figure 1   Top view and side view of copper in χ-Fe$_5$C$_2$(100)$_{0.25}$-surface (A) corrugated χ-Fe$_5$C$_2$(100)$_{0.25}$-surface; (B) copper in χ-Fe$_5$C$_2$(100)$_{0.25}$-surface by substituting (sub)surface carbon atom with copper; (C) copper in χ-Fe$_5$C$_2$(100)$_{0.25}$-surface by substituting iron atom on ridge with copper

of the promoter element on the surface.[29] The choice of the model system is crucial in understanding the role of the promoter element in working catalyst system. Hence, the stability of the promoter element in the various proposed structures has to be considered. Here, we report on the possible role of a single copper atom on a surface of Hägg iron carbide on the bonding of CO, with a particular emphasis on the stability of single copper atoms on a carbide surface.

Computational methods

The stability of well dispersed copper in or on an iron carbide surface in the presence of CO was investigated using spin polarized DFT calculations embedded in VASP.[30–33] The exchange–correlation energy was calculated using the RPBE functional[34] and the electron–ion interactions were described using the projector augmented wave method.[35,36] A cut-off energy of 350 eV was used throughout this study. The Brillioun zone was sampled using a Monkhorst–Pack[37] k-point grid with a spacing of less than 0.04 Å^{-1} and Gaussian smearing with σ=0.1 eV was used.

In preliminary studies (see Supplementary Material) the replacement of an iron or carbon atom the Hägg iron carbide structure was probed. It could be shown that the substitution of carbon with copper is much less favored than the substitution of iron by copper. The substitution of carbon for copper in Hägg iron carbide leads to a significant restructuring of the Hägg iron carbide structure with copper taking in a position, in which the nearest metal–metal distance resembles the nearest metal–metal distance in Hägg iron carbide rather than the metal–carbon distance. Hence, the stability of copper in or on the relatively stable, iron-rich χ-Fe$_5$C$_2$(100)$_{0.25}$-surface (following the nomenclature proposed by Steynberg et al.[8]) was investigated. The optimum Cu loading is typically reported as 0.01–0.04 mol Cu/mol Fe or more rigorously expressed as a surface concentration of about 1–2 Cu per nm^2.[2,20,21] Hence, copper as a chemical promoter was investigated by placing a single copper in or on a p(2 × 1) surface unit cell of the χ-Fe$_5$C$_2$(100)$_{0.25}$-surface (this would yield a surface concentration of 2.2 Cu nm^{-2}). The surface slabs were separated by a vacuum of at least 10 Å.

The effect of copper on co-adsorbed carbon monoxide was investigated. During the geometry optimizations the position of the adsorbates and the top four layers of the slab were optimized with forces on the unconstrained atoms less than 0.02 eV Å^{-1} while the bottom six layers of the slab were constrained. Dipole correction to the total energy has been included.[38]

The mobility of copper atoms on this Hägg carbide surface in the presence and absence of co-adsorbed carbon monoxide was investigated using the nudged elastic band[39] to locate transition state for hopping of Cu between two stable sites on the p(2 × 1) cell leaving all adsorbate species unconstrained. The transition state was further optimized using quasi-Newton algorithm to minimize the forces. All stationary points identified on the respective potential energy surfaces have been characterized based on a normal mode analysis within the frozen phonon approximation by perturbing the adsorbate atoms (Cu, C, and O) by 0.01 Å in the direction of each of the Cartesian coordinates, while keeping the substrate atoms fixed at their optimized positions. The change in the charge on the adsorbates was determined using the Bader analysis.[40,41]

Results and discussion

Hägg iron carbide contains three different types of iron atoms when classified according to the coordination to the nearest carbon atom. Our preliminary studies (see Supplementary Material) showed that substitution of either iron or carbon in the Hägg iron carbide structure is energetically highly unfavorable.

Direct chemical promotion requires the presence of the promoter near the 'catalytically active site', i.e. either in the surface of the carbide or as an adatom on the surface of the carbide. Copper can be positioned in the surface of χ-Fe$_5$C$_2$(100)$_{0.25}$ by either substituting a subsurface carbon atom or a surface iron atom with a copper atom. The substitution of a subsurface carbon atom with copper yields a structure in which the position of the copper atoms has shifted to a more exposed position with simultaneous displacement of the nearest surface iron atoms (see Fig. 1). The copper atom attempts to take in a position not too dissimilar to copper in the Cu(100) surface with the average distance to the seven neighboring iron atoms of 2.575±0.054 Å. The distance between copper and the nearest (two) carbon atoms in this configuration are 3.210 and 3.816 Å, respectively. Copper can be removed from this

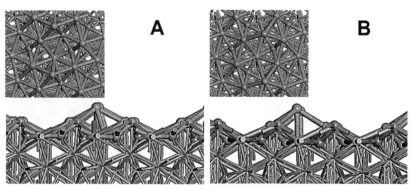

Figure 2 **Top and side view of copper adsorbed on χ-Fe$_5$C$_2$(100)$_{0.25}$-surface**

surface in an exergonic reaction with synthesis gas yielding fcc-Cu, χ-Fe$_5$C$_2$(100)$_{0.25}$ and water. This reaction is associated with an enthalpy change (at 0 K) of -1.96 eV (it should be noted that a large part of the enthalpy change, -1.46 eV, is due to formation of carbon from synthesis gas; the decomposition with graphene yielding Hägg iron carbide and fcc-Cu is less exergonic with a change in enthalpy of -0.50 eV). This implies that the substitution of a subsurface carbon atom with copper is not favored under Fischer–Tropsch conditions.

The substitution of a surface iron atom on the ridge of the surface of χ-Fe$_5$C$_2$(100)$_{0.25}$ by a copper atom does not alter the atom position much, with only a slight protrusion of the copper atom into the vacuum by 0.126 ± 0.005 Å. In this configuration copper is surrounded by five nearest neighbors iron atoms and two nearest neighbor carbon atoms. The distance between the copper in the surface and the nearest (two) carbon atoms is in this configuration only 2.078 and 2.214 Å, respectively, whereas the average distance to the nearest iron atoms is 2.678 ± 0.051 Å. The enthalpy change associated with the removal of a copper atom in this position yielding fcc-Cu, carbon (modeled as graphene) and χ-Fe$_5$C$_2$(100)$_{0.25}$ is then -0.97 eV.

Two stable positions were found for the adsorption of copper on the corrugated, iron-rich χ-Fe$_5$C$_2$(100)$_{0.25}$-surface. Copper is then present above the valley between the ridges formed by the protruding iron atoms (see Fig. 2). The more stable of the two configurations is closer to the surface, and is further distanced from the nearest subsurface carbon (see Table 1). The stability of these species was evaluated by comparing the state in which a single copper atom is adsorbed on χ-Fe$_5$C$_2$(100)$_{0.25}$ to the bare χ-Fe$_5$C$_2$(100)$_{0.25}$-surface and bulk fcc-Cu (copper is often found as fcc-Cu in Fischer–Tropsch catalysts and sometimes as relatively large crystallites[42,43]). The generation of atomically dispersed copper on this surface from fcc-Cu is endergonic at 0 K with the adsorbed state being at least 0.48 eV (including ZPE) higher than the state in which copper is separate from the carbide phase as fcc-Cu. Furthermore, it can be deduced that the diffusion of a copper atom along the valley between the protruding iron atoms forming the outer ridge of the χ-Fe$_5$C$_2$(100)$_{0.25}$-surface will have a barrier of at least 0.14 eV, somewhat higher than estimated for the diffusion of potassium on iron.[44] A Bader charge analysis of the adsorbed atomic copper on the surface of χ-Fe$_5$C$_2$(100)$_{0.25}$ shows

some charge transfer from the iron carbide structure to copper ($0.17e^-$ per Cu) from the surrounding iron atoms, but also the subsurface carbon atoms, due to the higher electron negativity of copper in comparison to iron.

The functioning of copper as a chemical promoter was investigated by co-adsorbing carbon monoxide and copper on the χ-Fe$_5$C$_2$(100)$_{0.25}$-surface. The presence of copper in its most stable position will restrict access of CO to the surface near the copper atom, and in particular access to the fourfold hollow site, the threefold hollow site, and the top position of iron atoms in the valley will be restricted. Only adsorption of CO in the most stable[5,8] top position on the iron atoms on the ridge was considered. The strength of CO adsorption is evaluated relative to CO in the gas phase, which was modeled in box with a size of $20 \times 20 \times 20$ Å using projector augmented wave–RPBE ($E_{cut-off}=1000$ eV; $\sigma=0.1$ eV; with dipole correction) resulting in a C=O bond length of 1.1427 Å and C–O stretching frequency 2113 cm^{-1} (in good agreement with the experimental values of 1.13 Å and 2170 cm^{-1} respectively).[9] The obtained adsorption energy for CO in the top position on the iron atoms at the ridge at a coverage of 2.2 CO per nm^2 is similar to that obtained for the on-top site on Fe$_5$C$_2$(110)$_{0.80}$, but stronger than on either Fe$_5$C$_2$(110)$_{0.00}$ and Fe$_5$C$_2$(010)$_{0.25}$[10] (see Table 2). The obtained bond lengths for C–O and Fe–CO are similar to those reported before for CO adsorbed on different iron carbide surfaces. The Fe–CO angle is almost linear with an angle of 179.6°, with CO almost parallel to the

Table 1 **Structural parameters for two stable positions of Cu on χ-Fe$_5$C$_2$(100)$_{0.25}$ (see Fig. 2 for optimized structural arrangement)**

	A	B
E_{ads}*/(eV/Cu)	0.46	0.60
ZPE/eV	0.02	0.02
h†/Å	1.139	1.285
$d_{Cu-Fe(ridge)}$/Å	2.582	2.582
	2.584	2.584
	2.692	2.592
$d_{Cu-Fe(valley)}$/Å	2.436	2.565
		2.587
d_{Cu-C}/Å	3.252	2.708

*Adsorption energy for Cu [incl. zero point energy (ZPE)] with reference to the bare χ-Fe$_5$C$_2$(100)$_{0.25}$-surface and fcc-Cu $E_{ads} = E_{slab+Cu} + ZPE - E_{slab} - E_{fcc-Cu}$
†Height above the plane through the outer iron atoms.

surface normal to the surface. Some slight structural rearrangements are observed in the carbide structure upon adsorption of CO on the on-top site, namely a shortening of the distance between iron and the carbidic carbon in the structure. The adsorption of CO on iron carbide is associated with a charge transfer from the carbide structure to CO. The Bader analysis shows that the charge on CO at a coverage of 2.2 CO per nm^2 is 0.46e$^-$. The increase in electron density on CO is associated with a decrease in the electron density around mainly the two iron atoms forming the ridge on the χ-Fe$_5$C$_2$(100)$_{0.25}$-surface, on which CO is adsorbed, with also the other surface and sub-surface layer iron and carbon atoms contributing.

Increasing the CO-coverage to 4.4 CO per nm^2 does not change the picture much (see also Cao et al.[6]). A further increase in the CO-coverage to 6.5 CO per nm^2 does affect the heat of adsorption; CO occupying two adjacent iron atoms on the ridge will then adopt a geometry which minimizes the interaction with co-adsorbed CO resulting in an angle between adsorbed CO and the surface normal of ~20°. The obtained adsorption energy as a function of surface coverage is a composite of the adsorption energy of CO in this particular geometry (1.35 eV/CO) and the adsorption energy of CO almost perpendicular to the surface (1.66 eV/CO). Carbon monoxide adsorbed in the angled position has a shorter C=O bond length and an elongated Fe–C-bond in comparison to CO adsorbed almost perpendicularly to the surface plane. This is associated with a blue-shift of the

C=O stretching frequency. Furthermore, the Bader analysis shows that the net charge transfer to adsorbed CO is reduced by 30% to 0.32e$^-$ per CO.

The presence of copper enhances the bonding of CO to the surface at a low coverage of CO (see Table 3). The adsorption of CO becomes 0.2 eV more stable due to the presence of copper at a low coverage of 2.2 CO per nm^2. Alternatively, it can be stated that the presence of CO stabilizes the presence of atomically dispersed copper on the carbide surface, although the adsorption of atomic copper remains endergonic at 0 K even at a low coverage of 2.2 CO per nm^2. The Bader analysis showed that co-adsorbing CO and copper on the χ-Fe$_5$C$_2$(100)$_{0.25}$-surface results in electron withdrawal from the copper atom (a change in the Bader charge of 0.5e$^-$ noted in comparison to the surface without any CO adsorbed). This resulted in an increase in the negative charge on CO and in particular CO in close proximity to the copper atom. The increased strength of adsorption is similar to that what is typically observed upon promotion of the surface with potassium,[44,45] albeit more moderately than observed with potassium at similar loading. Increasing the coverage to 6.5 CO per nm^2 or higher negates the effect of copper on the heat of adsorption, but the charge transfer from copper to adsorbed CO remains pronounced.

The increased charge transfer to CO adsorbed in the top position on the ridge of the χ-Fe$_5$C$_2$(100)$_{0.25}$-surface in the presence of co-adsorbed copper results in a red-shift of the

Table 2 Structural parameters describing top adsorption of CO on χ-Fe$_5$C$_2$(100)$_{0.25}$ as function of coverage with CO on protruding iron atoms

Coverage in p(2 × 1)-unit cell	1	2	3	4
Θ_{CO}/nm^{-2}	2.2	4.4	6.5	8.7
E_{ads}*/(eV/CO)	−1.66	−1.65	−1.46	−1.35
Demagnetization†/(μ_B/CO)	−1.43	−1.43	−1.03	−0.80
$v_{C=O(stretch)}$/cm^{-1}	1902	1897	1912	1924
		1941	1932	1951
			2000	1957
				2035
d_{C-O}/Å	1.175	1.172	1.170	1.161
		1.172	1.163	1.162
			1.163	1.161
				1.162
Angle of CO from surface normal/° Å	0.8	0.8	0.6	18.1
		0.5	22.3	17.0
			23.2	18.0
				16.6
$d_{Fe-C\ in\ CO}$/Å	1.784	1.786	1.790	1.828
		1.788	1.825	1.829
			1.826	1.828
				1.829
∠$_{Fe-C-O}$/°	179.6	179.5	179.9	177.7
		179.8	179.2	177.8
			179.1	177.6
				177.8

*Adsorption energy for CO including zero point energy correction with respect to gas phase CO and the bare surface defined as

$$E_{ads} = \frac{E_{slab+nCO} + ZPE_{slab+nCO} - E_{slab} - nE_{CO_g} + ZPE_{CO_g}}{n}$$

†Change in magnetization of the slab upon adsorption of CO.

C=O stretching frequency, an increase in the C=O bond length, and a (slight) decrease in the Fe–C bond length.

Copper may act as a chemical promoter for CO-conversion over Hägg iron carbide as visualized by the electron donation of copper resulting in an increased charge density on adsorbed CO. However, the endergonic nature of the dispersion of copper in its atomic form on one of the more favorable surfaces of Hägg iron carbide would imply that copper cannot act as a chemical promoter over a prolonged time. It was recently pointed out that Hägg iron carbide may be one of the phases through which the system cycles, i.e. it is an intermediate phase[1,42,43] not unlike an intermediate in a catalytic cycle. Thus, the stability of atomic copper is not the only point of consideration, but also its mobility over the carbide phase, which may be impeded by the presence of co-adsorbed CO. Copper may diffuse along the valley or over the ridge on the χ-Fe$_5$C$_2$(100)$_{0.25}$-surface with the latter expected to be energetically more demanding. The energetically more demanding pathway for diffusion of copper over the ridge on the χ-Fe$_5$C$_2$(100)$_{0.25}$-surface was investigated to see whether co-adsorbed CO could reduce the mobility of copper on these surfaces.

Figure 3 shows the minimum energy pathway for the diffusion of a copper atom over the ridge on the χ-Fe$_5$C$_2$(100)$_{0.25}$-surface. The diffusion pathway of a copper atom over the ridge of the bare χ-Fe$_5$C$_2$(100)$_{0.25}$-surface proceeds via the bridge between the two protruding iron atoms forming the ridge. The iron–copper distance in the transition state (2.44 Å) is only slightly shorter than in the most stable configuration (2.58 Å). The activation barrier for the movement of a copper atom over the bare Hägg iron carbide surface was determined to be 0.62 eV, and a hopping frequency at a typical temperature for the Fischer–Tropsch synthesis of 525 K is then determined as 8×10^6 s^{-1} (alternatively, the average time for a copper atom to travel 10 nm is then 3×10^{-5} s; see supplementary material for the determination of the hopping frequency) indicating a rather facile migration of atomic copper over this surface.

Co-adsorption of copper with carbon monoxide on the χ-Fe$_5$C$_2$(100)$_{0.25}$-surface not only stabilizes copper, but also impedes the mobility of atomic copper. The diffusion pathway over the ridge of this surface formed by the iron atoms opposite the ridge on which CO is adsorbed results in an activation barrier which is at a coverage of 2.2 CO molecules per nm^2 0.16 eV higher than on the bare surface. The transition state is again located on the midpoint between the protruding iron atoms on the ridge void of adsorbed CO. However, the movement of the copper atom is

Table 3 Structural parameters describing top adsorption of CO on χ-Fe$_5$C$_2$(100)$_{0.25}$ as function of coverage with CO on protruding iron atoms

Coverage in p(2 × 1)-unit cell	1	2	3	4
Θ_{CO}/nm^{-2}	2.2	4.4	6.5	8.7
E_{ads}*/(eV/CO)	−1.87	−1.72	−1.48	−1.33
E_{ads}†/(eV/Cu)	0.27	0.32	0.40	0.53
$v_{C=O(stretch)}$/cm^{-1}	1876	1854	1862	1851
		1895	1885	1880
			1959	1945
				2010
$d_{C–O}$/Å	1.181	1.182	1.178	1.176
		1.179	1.175	1.176
			1.166	1.163
				1.160
Angle of CO from surface normal/° Å	14.5	9.6	12.0	12.6
		13.9	16.0	4.5
			30.5	17.5
				23.0
$d_{Fe–C\ in\ CO}$/Å	1.774	1.786	1.781	1.841
		1.775	1.805	1.808
			1.808	1.809
				1.824
$\angle_{Fe–C–O}$/°	177.1	175.0	177.6	175.8
		177.6	178.9	171.0
			178.3	178.0
				178.2
$d_{Cu–C\ in\ CO}$/Å	2.771	2.448	2.579	2.189
		2.665	2.506	2.285
			3.297	3.144
				3.929

*Adsorption energy for CO including zero point energy (ZPE) correction with respect to gas phase CO and copper adsorbed on χ-Fe$_5$C$_2$(100)$_{0.25}$

$$E_{ads} = \frac{E_{slab+Cu+nCO} + ZPE_{slab+Cu+nCO} - E_{slab+Cu} - nE_{CO_g} + ZPE_{CO_g}}{n}$$

†Adsorption energy for copper including zero point energy correction with respect to fcc-Cu and CO adsorbed on χ-Fe$_5$C$_2$(100)$_{0.25}$

$$E_{ads} = E_{slab+Cu+nCO} + ZPE_{slab+Cu+nCO} - E_{fcc-Cu} - (E_{slab+nCO} + ZPE_{slab+nCO})$$

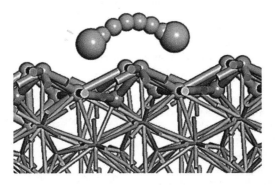

$$\Theta_{CO} = 0$$
$$Ea = 0.62 \text{ eV}$$

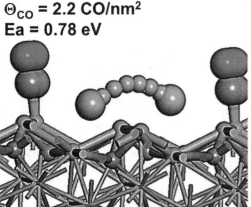

$$\Theta_{CO} = 2.2 \text{ CO/nm}^2$$
$$Ea = 0.78 \text{ eV}$$

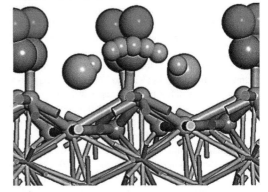

$$\Theta_{CO} = 4.4 \text{ CO/nm}^2$$
$$Ea = 0.76 \text{ eV}$$

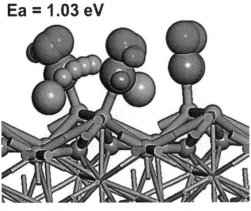

$$\Theta_{CO} = 6.5 \text{ CO/nm}^2$$
$$Ea = 1.03 \text{ eV}$$

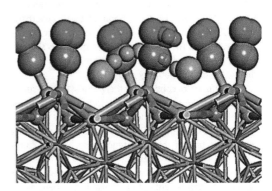

$$\Theta_{CO} = 8.7 \text{ CO/nm}^2$$
$$Ea = 1.08 \text{ eV}$$

Figure 3 Diffusion of atomic copper over ridge of iron atoms on χ-Fe$_5$C$_2$(100)$_{0.25}$ indicating diffusion pathway

associated with a simultaneous movement of the adsorbed CO species on the opposite ridge. In the transition state, the C=O bond is shortened from 1.181 to 1.177 Å, and now making an angle of ~10° with the surface normal. The higher activation energy for diffusion results in slightly lower hopping frequency of 2×10^6 s^{-1} at a coverage of 2.2 CO per nm^2.

Increasing the CO-coverage further to 4.4 molecules CO adsorbed per nm^2 results in similar activation energy for the diffusion of atomic copper over the ridge of the χ-Fe$_5$C$_2$(100)$_{0.25}$-surface as at a coverage of 2.2 molecules of CO per nm^2 (i.e. 0.76 eV). Now the copper atom has to pass over the ridge, on which one CO molecule is adsorbed and the minimum distance between the copper atom and carbon in CO is 2.24 Å. It can be observed that the diffusing copper atom pushes CO out of its way in the transition state. This results in an interesting transition state with a highly activated CO-molecule close to the copper atom with a C=O-bond length of 1.236 Å at an angle of 40° from the surface normal (the Fe–C bond length is 1.776 Å and the angle between Fe–C–O is reduced to 170.5°). The C=O stretching frequency decreased in this transition state to 1536 cm^{-1}. The hopping frequency at 525 K is now drastically reduced to 8×10^3 s^{-1}. The decrease in the hopping frequency can be attributed to a change in the activation energy and in the pre-exponential factor, i.e. both energetic and entropic factors affect the mobility of atomic copper over the ridge on this surface.

At a CO-coverage of 6.5 CO per nm^2 copper may diffuse over the ridge with one CO adsorbed in the top position or with two CO molecules adsorbed in the top position. In the latter case, atomic copper pushes the adsorbed CO-molecules out of its way to minimize its contact. The distance between the copper atom and carbon in the CO-molecules is in the transition state 2.09 ± 0.04 Å. Furthermore, in the transition state the angle of adsorbed CO with the surface normal is increased to 40°. The C=O bond is elongated to 1.190 Å indicating a weakening of the C=O bond (as also shown by a redshift in the C=O stretching frequency by ~100 to 1763 cm^{-1}). The activation barrier for the diffusion of a copper atom over the ridge containing two copper atoms was determined to be 1.03 eV resulting in a hopping frequency for this pathway of the copper atom at 525 K of 3×10^3 s^{-1}.

A similar activation barrier (1.08 eV) was obtained for the diffusion of a copper atom over the ridge on the χ-Fe$_5$C$_2$(100)$_{0.25}$-surface at a coverage of 8.7 CO per nm^2. The CO-molecules adsorbed on the ridge crossed over by the diffusing copper atom are pushed away. In the transition state the carbon atoms in the adsorbed CO molecules are 2.04 ± 0.01 Å from the copper atom. The angle of adsorbed CO with the surface normal is in the transition state increased to 32°. The C=O bond of these adsorbed CO molecules is elongated to 1.189 Å indicating a weakening of the C=O bond. The C=O stretching frequency is shifted to a lower frequency of 1773 cm^{-1}. This results in a hopping frequency for this pathway of 5×10^2 s^{-1}.

The diffusion of copper over the χ-Fe$_5$C$_2$(100)$_{0.25}$-surface is impeded by the presence of co-adsorbed CO. Nevertheless, the high hopping frequency of larger than 10^2 s^{-1} implies that copper on Hägg iron carbide will only be present for a very short time, if this is a static phase. Then it can thus under those circumstances not be considered an electronic promoter for the Fischer–Tropsch synthesis (a similar argument was recently put forward by Tian et al.[46]). However, it was recently argued that iron traverses through a number of different phases during the Fischer–Tropsch synthesis[1,25,26] i.e. iron carbide oxidizes splitting off small magnetite crystallites, which may sinter and subsequently transform back into iron carbide. The effectiveness of copper is then not only a function of its mobility on the catalytically active phase, but also of the average lifetime of the catalytically active phase and its mobility on magnetite. It has recently been shown that the adsorption of atomic

copper on magnetite is rather stable[47] implying that the lifetime of the carbide phase in this cycle may be one of the determining factors in the effectiveness of copper as an electronic promoter in these catalysts. Furthermore, it must be realized that the model studied here presumes unimpeded migration of copper. The presence of carbon in the catalyst may however inhibit the formation of larger copper crystallites.

Conclusions

The stability of atomic copper in or on the iron-rich Fe$_{40}$C$_{16}$(100)$_{0.25}$-surface was investigated using GGA–RPBE. The adsorption of atomic copper on this surface is endergonic with respect to fcc-Cu. The presence of CO stabilizes atomic copper to some extent, but its thermodynamically preferred state is still fcc-Cu. Copper does affect the bonding of CO to the Fe$_{40}$C$_{16}$(100)$_{0.25}$-surface; co-adsorbing CO and atomic copper will result in an elongation of the C=O bond and a redshift in the C=O stretching frequency.

Atomic copper has a high mobility on Fe$_{40}$C$_{16}$(100)$_{0.25}$ and its mobility is only slightly impeded by the presence of co-adsorbed CO. The impediment is only partially energetic in nature and entropic effects also play a role. The latter are mainly caused by the strong redshift in the C=O stretching frequency if atomic copper is near. The high mobility of copper on this surface implies that the average time atomic copper will be on the surface is too short to be an effective, electronic promoter in the Fischer–Tropsch synthesis.

Conflicts of interest

The authors declare no conflicts of interest.

References

1. E. van Steen and M. Claeys: Chem.l Eng. Technol., 2008, **31**, 655.
2. J. W. Niemantsverdriet, A. M. van der Kraan, W. L. van Dijk and H. S. van der Baan: J. Phys. Chem., 1980, **84**, 3363.
3. G. B. Raupp and W. N. Delgass: J. Catal., 1979, **58**, 361.
4. A. K. Datye, Y. Jin, L. Mansker, R. T. Motjope, T. H. Dlamini and N. J. Coville: Stud. Surf. Sci. Catal., 2000, **130**, 1139.
5. E. de Smit, F. Cinquini, A. M. Beale, O. V. Safonova, W. van Beek, P. Sautet and B. M. Weckhuysen: J. Am. Chem. Soc., 2010, **132**, 14928.
6. D.-B. Cao, F.-Q. Zhang, Y.-W. Li and H. Jiao: J. Phys. Chem. B, 2004, **108B**, 9094.
7. D.-B. Cao, F.-Q. Zhang, Y.-W. Li, J. Wang and H. Jiao: J. Phys. Chem. B, 2005, **109B**, 833.
8. P. J. Steynberg, J. A. van den Berge and W. Janse van Rensburg: J. Phys.: Condens. Matters, 2008, **20**, 064238
9. D. C. Sorescu: J. Phys. Chem. C, 2009, **133C**, 9256.
10. M. A. Petersen, J. A. van den Berg and W. Janse van Rensburg: J. Phys. Chem. C, 2010, **114C**, 7863.
11. J. M. Gracia, F. F. Prinsloo and J. W. Niemantsverdriet: Catal. Lett., 2009, **113**, 257.
12. D.-B. Cao, Y.-W. Li, J. Wang and H. Jiao: J. Mol. Catal. A-Chem., 2011, **346A**, 55–69.
13. T. H. Pham, X. Duan. G. Qian, X. Zhou and D. Chen: J. Phys. Chem. C, 2014, **118C**, 10170.
14. C. F. Huo. Y. W. Li, J. Wang and H. Jiao: J. Am. Chem. Soc., 2009, **131**, 14713.
15. M. E. Dry: 'The Fischer-Tropsch synthesis', in 'Catalysis science and technology', (ed. J. R. Anderson and M. Boudart), Vol. 1, 15–255; 1981, Berlin, Springer Verlag.
16. R. P. Mogorosi, N. Fischer, M. Claeys and E. van Steen: J. Catal., 2012, **289**, 327–334.
17. H. Kölbel and M. Ralek: Catal. Rev. – Sci. Eng., 1980, **21**, 225.
18. I. E. Wachs, D. J. Dwyer and E. Iglesia: Appl. Catal., 1984, **12**, 201.

19. D. B. Bukur, D. Mukesh and S. A. Patel: *Ind. Eng. Chem. Res.*, 1990, **29**, 194.
20. S. Li, A. Li, S. Krishnamoorthy and E. Iglesia: *Catal. Lett.*, 2001, **77**, 197.
21. S. Li, S. Krishnamoorthy, A. Li, G. D. Meitzner and E. Iglesia: *J. Catal.*, 2002, **206**, 202–217.
22. Y. N. Wang, W. P. Ma, Y. L. Lu, J. Yang, Y. Y. Xu, H. W. Xiang, Y. W. Li, Y. L. Zhao and B. J. Zhang: *Fuel*, 2003, **82**, 195–213.
23. R. J. O'Brien and B. H. Davis: *Catal. Lett.*, 2004, **94**, 1.
24. E. de Smit, F. M. de Groot, R. Blume, M. Hävecker, A. Knop-Gericke and B. M. Weckhuysen: *Phys. Chem. Chem. Phys.*, 2010, **12**, 667.
25. G. A. Huff Jr and C. N. Satterfield: *Ind. Eng. Chem. Process Des. Dev.*, 1984, **23**, 696.
26. J. Xu and M. Saeys: *J. Catal.*, 2006, **242**, 217.
27. M. Elahifard, E. Fazeli, A. Joshani and M. Gholami: *Surf. Interface Anal.*, 2013, **45**, 1081.
28. J. Cheng, P. Hu, P. Ellis, S. French, G. Kelly and C. M. Lok: *Surf. Sci.*, 2009, **603**, 2752.
29. J. J. Mortensen, B. Hammer and J. K. Nørskov: *Phys. Rev. Lett.*, 1998, **80**, 4333.
30. G. Kresse and J. Haffner: *Phys. Rev. B*, 1993, **47B**, 558.
31. G. Kresse and J. Haffner: *Phys. Rev. B*, 1994, **49B**, 14251.
32. G. Kresse and J. Furthmüller: *Phys. Rev. B*, 1996, **54B**, 11169.
33. G. Kresse and J. Furthmüller: *Comput. Mater. Sci.*, 1996, **6**, 15.
34. B. Hammer, L. B. Hansen and J. K. Nørskov: *Phys. Rev. B*, 1999, **59B**, 7413.
35. P. E. Blöchl: *Phys. Rev. B*, 1994, **50B**, 17953.
36. G. Kresse and D. Joubert: *Phys. Rev. B*, 1999, **59B**, 1758.
37. H. J. Monkhorst and J. D. Pack: *Phys. Rev. B*, 1976, **13B**, 5188.
38. G. Kresse, M. Marsman and J. Furthmüller: VASP the Guide, 2007, available at: http://cms.mpi.univie.ac.at/VASP/
39. H. Jónsson, G. Mills and K. W. Jacobson: in 'Classical and quantum dynamics in condensed phase simulation', (ed. B. Berne, G. Ciocotti and D. F. Coker), 385–404; 1998, Singapore, World Scientific.
40. G. Henkelman, A. Arnaldsson and H. Jónsson: *Comput. Mater. Sci.*, 2006, **36**, 354.
41. D. Henkelman, S. D. Kenny, R. Smith and G. Henkelman: *J. Comput. Chem.*, 2007, **28**, 899.
42. Z. H. Chonco, A. Ferreira, L. Lodya, M. Claeys and E. van Steen: *J. Catal.*, 2013, **307**, 283.
43. Z. H. Chonco, L. Lodya, M. Claeys and E. van Steen: *J. Catal.*, 2013, **308**, 363.
44. D. C. Sorescu: *Surf. Sci.*, 2011, **605**, 401.
45. S. J. Jenkins and D. A. King: *Chem. Phys. Lett.*, 2000, **317**, 372.
46. X. Tian, T. Wang, Y. Yang, Y.-W. Li, J. Wang and H. Jiao: *J. Phys. Chem. C*, 2014, **118C**, 21963.
47. R. M. van Natter, J. S. Coleman and C. R. F. Lund: *J. Mol. Catal. A*, 2009, **311A**, 17.

11

Development and characterization of thermally stable supported V–W–TiO$_2$ catalysts for mobile NH$_3$–SCR applications

Andrew M. Beale*[1,2,3], Ines Lezcano-Gonzalez[1,2], Teuvo Maunula[4] and Robert G. Palgrave[1]

[1]Department of Chemistry, University College London, 21 Gordon Street, WC1H 0AJ, London, UK
[2]UK Catalysis Hub, Rutherford Appleton Laboratory, Research Complex at Harwell, Harwell, Didcot, OX11 0FA, UK
[3]Finden Ltd, Clifton Hampden, Oxfordshire, OX14 3EE, UK
[4]Dinex Ecocat Oy, DET Finland, Catalyst Development, Typpitie 1, FI-90620 Oulu, Finland

Abstract Vanadium based catalysts supported on a mixture of tungsten and titanium oxide (V$_2$O$_5$/WO$_3$–TiO$_2$) are known to be highly active for ammonia selective catalytic reduction (NH$_3$–SCR) of NO$_x$ species for heavy-duty mobile applications. However they are also known to be sensitive to high temperatures which leads to both sintering of the anatase TiO$_2$ support and a first order phase transition to rutile at temperatures >600°C. Here we report our attempts to use SiO$_2$ to stabilize the TiO$_2$ anatase phase and to compare its catalytic activity with that of a

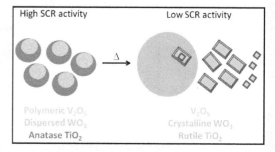

non-stabilized V$_2$O$_5$/WO$_3$–TiO$_2$ catalyst after thermal aging up to 800°C. Detailed characterization using spectroscopic (Raman, UV–vis, X-ray absorption spectroscopy), scattering and techniques providing information on the catalytic surface (Brunauer–Emmet–Teller, NH$_3$ adsorption) have also been performed in order to understand the impact of high temperatures on component speciation and the catalytic interface. Results show that non-stabilized V$_2$O$_5$/WO$_3$–TiO$_2$ catalysts are initially stable after thermal aging at 600°C but on heating above this temperature a marked drop in catalytic activity is observed as a result of sintering and phase transformation of Anatase into Rutile TiO$_2$ and phase segregation of initially highly dispersed WO$_3$ and polymeric V$_2$O$_5$ into monoclinic WO$_3$ and V$_2$O$_3$ crystallites. Similar behavior was observed for the 4–5 wt-% of SiO$_2$-stabilised sample after aging above 700°C, importantly therefore, offset by some ~100°C in comparison to the unstabilised sample.

Keywords V–W–TiO$_2$, NH$_3$–SCR, Thermal deactivation, Stability, Spectroscopy

Introduction

Low-loaded (~2%) V containing catalysts such as V/TiO$_2$ are well known to be highly active catalysts for the selective catalytic reduction (SCR) of NO$_x$ using NH$_3$ as a reductant.[1] Operating temperatures for maximum NO$_x$ conversion range between 300 and 450°C and the technology is typically employed for stationary NO$_x$ abatement but since 2005 in Europe, the technology has been employed commercially in heavy-duty mobile applications. In order to enhance both activity and stability they are normally 'promoted' to a significant extent (10%) with group XVI elements

such as Mo and W rendering the commercially applied catalysts as mixed phase V$_2$O$_5$/(WO$_3$ or MoO$_3$)/TiO$_2$. Much research has been undertaken in order to elucidate both the mechanism and active phase of the catalyst and it appears that there is general consensus on both; from an active phase perspective it has been proposed that both the VO$_x$ and WO$_x$ species are well spread over the TiO$_2$ support with O from the V–O–Ti/V–O–V of polymeric vanadates thought to be critical for activity at low temperatures.[2–7] In addition to enhanced deNO$_x$ activity, the WO$_x$ is proposed to have a number of additional benefits including the prevention of V polymerization and volatilization, the maintenance of the high surface area of the TiO$_2$ anatase support and the offsetting of the first order solid state phase

*Corresponding author, email andrew.beale@ucl.ac.uk

transition to the TiO_2 rutile phase with a much reduced surface area.[2,8]

A major problem in catalytic application is deactivation. V_2O_5/(WO_3 or MoO_3)/TiO_2 catalysts are known to be generally resistant to chemical poisoning by SO_2 since the TiO_2 support is weakly and therefore reversibly sulfated. However thermal deactivation is a major problem for emission control catalysts, particularly when used in mobile applications since temperature spikes caused by running rich in fuel or air can mean that the catalyst experiences temperatures as great as 700–800°C. Such high temperatures are known to cause sintering, phase transformation of TiO_2 anatase to rutile at temperatures >600°C and potentially volatalization of vanadia; the melting point of pure V_2O_5 is 670°C.

In this paper we demonstrate the results of a new, thermally stable V–SCR catalyst based on tailored silica stabilized raw materials and preparation methods when compared against the conventional reference V–SCR catalyst based on pure anatase with the equivalent amount of vanadium and tungsten. We examine then the effects of thermal aging of the two catalysts (a reference and an SiO_2 stabilized thermally durable type) at various temperatures before performing a detailed characterization in order to examine how these temperature extremes lead to deactivation of the catalyst.

Experimental

Catalyst preparation

Two types of V–SCR samples were prepared: a conventional (reference) vanadium SCR catalyst (VSCR1) and a thermally stabilized (VSCR2) catalyst that contains about 4–5 wt-% silica, demonstrated to act as a stabilizer, where the resultant catalyst has been prepared in a way so as to produce a well dispersed, high surface area containing stable catalyst.[9,10] Vanadium SCR catalysts are based on pre-blended amorphous SiO_2 stabilized TiO_2 (anatase) powder material, which is mixed in an aqueous slurry, where tungsten oxide (WO_x) and vanadium oxide (VO_x) are added dissolved in de-ionized water solutions of ammonium metavanadate (NH_4VO_3) and ammonium metatungstate hydrate ($(NH_4)_6H_2W_{12}O_{40}\cdot xH_2O$) precursors and left to homogenize for 1 h. The mixed acidic slurry was coated on a metallic substrate (thickness 50 μm, cell density 600 cells per in²) and the coated catalyst was dried at ~80°C and calcined at 550°C for 4 h. The loadings of vanadium as coating used are 2.2 wt-% (calculated based on V_2O_5) and 10 wt-% tungsten (calculated based on WO_3).

Catalyst aging protocol

Catalysts were aged in situ in static air at 600, 700 and 800°C for 3 h before being removed from the metallic substrate as powder samples for characterization purposes so as to investigate how/whether such temperatures lead to thermally induced phase transformation/sintering and concordant changes in surface area. Therefore, the preparation and aging in this study were based on real structured catalysts, having a direct linkage to real catalysts in production, in contrast to preparing samples as powders only. The fresh honeycomb samples were also hydrothermally aged at 650 and 700°C for 20 h in a flow of 10% water in air and comparative catalytic activity measurements made.

Catalyst characterization

The surface area and pore size distribution were measured using the Brunauer–Emmet–Teller (BET) method with nitrogen adsorption–desorption isotherms (Sorptomatic 1990). NH_3 adsorption–desorption studies were performed by first adsorbing NH_3 first at 200°C via a step exchange experiment (0→500 ppm NH_3 in nitrogen, 42 000 h⁻¹) followed by desorption of NH_3 in the presence of oxygen (Temperature Programmed Oxidation, 20°C min⁻¹, 10% O_2 in N_2). The reactor products were analyzed using a Gasmet Fourier transform infrared spectroscopy (FTIR) analyzer during a thermal ramping up to 600°C. The ammonia adsorption capacity at 200°C was calculated by the integration of NH_3 consumption during the step exchange response.

X-ray absorption spectroscopy data were recorded on the Dutch–Belgian beamline (DUBBLE; BM26A) at the ESRF and B18 @ the Diamond Light Source.[11] Data were collected using a Si(111) double crystal monochromator with harmonic rejection achieved using the appropriate optics. V K-edge (5656 keV) measurements on the samples were performed in fluorescence mode using a nine-element monolithic germanium detector. Transmission measurements were performed on the crystalline oxide reference samples (V_2O_5, VO_2 and V_2O_3) and a 10 μm V foil to calibrate the monochromator position. In a typical experiment, approximately 100 mg sample was pressed to form self-supporting wafers before being mounted in air. Measurements were performed at room temperature in normal step-scanning mode over the range of 5420–5620 eV. A typical X-ray absorption near-edge structure (XANES) spectrum was collected over a period of 30 min and at least three scans were performed for each sample. X-ray absorption near-edge structure data were processed using the using Athena (IFFEFIT software package).[12,13] The XANES spectra were normalized from 30 to 150 eV above the edge energy, using a quadratic polynomial regressed through the data above the edge and extrapolated back to E_0. The extrapolated value of the post edge polynomial at E_0 is used as the normalization constant.

X-ray diffraction data were recorded on a Bruker D8 diffractometer with Bragg–Brentano geometry in flat plate mode using a Co $K_{\alpha+\beta}$ radiation source and rotating sample holder. Data were recorded from 10 to 90° 2θ using a step size of 0.02°/1 s acquisition time. X-ray diffraction profile fitting to extract quantitative phase analysis information of the final phase composition for the various samples was obtained using the Rietveld method. In order to do this the Powdercell programme was used to profile data over a 2θ range of 10–80° with no excluded regions. The background was simulated using a polynomial function whereas the diffraction peaks were profiled using a pseudo-voigt function and the atom positions and B iso values were fixed in order to enable the refinement to converge. Included in the final refinement were scale factors (and therefore the % age of each phase present), profile parameters, cell parameters and the zero-shift. Typical refined R_{wp} values ranged from 5.25 to 7.30%. The obtained compositions were refined with a precision/accuracy of 5–10% (depending on the number of phases present) of their actual weight percent. An example experimental pattern, as well as the fitted profiles and residuals are presented in ESI Fig. 2 (see

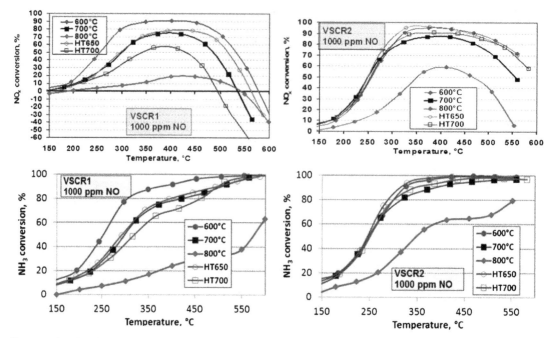

Figure 1 Standard SCR activities and NH_3 conversion data of both thermally and hydrothermally aged VSCR1 and VSCR2 samples, using GHSV of 50 000 h^{-1} and feed composition of 1000 ppm NO, 1000 ppm NH_3, 10%O_2 and 10%H_2O

Supplementary Material www.maneyonline.com/doi/suppl/ 10.1179/2055075814Y.0000000005).[14] For particle size estimations using the Scherrer equation an instrument broadening function was first calculated using crystalline LaB_6 as a reference material.[15]

Raman data were recorded on self-supported wafers using a Kasier Optics instrument with a 5.5 inch objective lense (200 μm spot size) in backscatter mode with a 532 nm laser operating at 15 mW power and a CCD camera to record the data. The acquisition time employed was 10 s/spectra with 10 accumulations performed per sample with a total acquisition time of 5 min (inc. equivalent dark time acquisition). All Raman spectra were offset corrected using Thermo Galactic Grams AI v. 7.0 software.

UV–vis data were recorded in diffuse reflectance mode using a Cary-50 instrument with an acquisition time of 1 min/spectrum. Five spectra were acquired and summed per sample with a dark and MgO reference used for calibration purposes.

X-ray photoelectron spectroscopy (XPS) was carried out with a Thermo K-alpha instrument using monochromated Al K_α (1486.6 eV) radiation. Charge compensation was achieved using a dual mode (electron and Ar^+) flood gun, and spectra were referenced to adventitious Carbon 1s peak set to 285.0 eV. The V 2p region was fitted with Gaussian–Lorentzian line shapes on a Shirly background.

Catalyst activity measurements

The activity of SCR catalysts was investigated by steady state experiments between 150 and 600°C in a quartz reactor tube mounted in an IR furnace.[16] The inlet concentration in the standard mixture was stoichiometric using NH_3/NO_x = 1 : 1000 ppm NO_x (NO only or NO_2/NO=400 : 600 ppm), 10%O_2, 10%H_2O, and balance nitrogen with space velocity

(SV) of 50 000 h^{-1}. The inclusion of NO_2 in the feed simulated the presence of a Pt oxidation catalyst before the SCR unit as is typically encountered in the mobile emissions system.[17,18]

The gas compositions were analyzed with Fourier transform infrared spectroscopy (Gasmet CR 2000) equipped with heated (180°C) sampling lines, which prevents ammonia and water adsorption/condensation on the transfer lines.

Results and discussion

SCR activity

The catalytic performance of both thermally aged VSCR1 and VSCR2 catalysts was examined in the Standard SCR (NO_2/NO_x=0) and Fast SCR (NO_2/NO_x=0.4) reactions, as shown in Figs. 1 and 2, respectively.

After thermal aging at 600°C, VSCR1 still maintained reasonable standard SCR activity (>90%) at 350–450°C, which began to decline however, at higher temperatures, reaching ~38% at 550°C (Fig. 1). In contrast NH_3 conversion increased with increasing reaction temperature (>95% at 450–600°C), resulting in a ratio between converted NO_x and converted NH_3 less than 1. This implies that part of the NH_3 is not consumed during the SCR process and that is converted following a different reaction, probably by direct oxidation with oxygen[19]

$$2NH_3 + 3/2O_2 \rightarrow N_2 + 3H_2O \qquad (1)$$

$$2NH_3 + 2O_2 \rightarrow N_2O + 3H_2O \qquad (2)$$

$$2NH_3 + 5/2O_2 \rightarrow 2NO + 3H_2O \qquad (3)$$

Above 377–427°C, reactions (1)–(3) become competitive with the standard SCR, decreasing the amount of available NH_3 and/or producing undesired NO or N_2O. In line with the

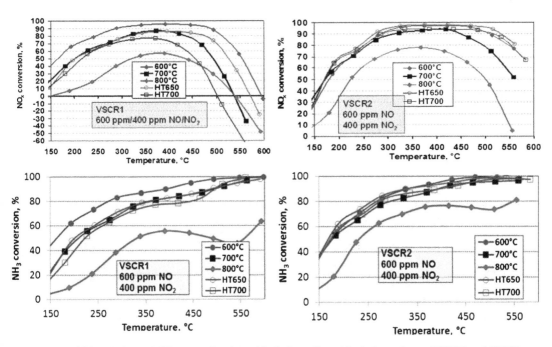

Figure 2   Fast SCR activities and NH$_3$ conversion data of both thermally and hydrothermally aged VSCR1 and VSCR2 samples, using GHSV of 50 000 h^{-1} and feed composition of 600 ppm NO, 400 ppm NO$_2$, 1000 ppm NH$_3$, 10%O$_2$ and 10%H$_2$O

NH$_3$ conversion data, increased amounts of N$_2$O (>6 ppm) were produced over the VSCR1 catalyst at reaction temperatures higher than 450°C, most likely through reaction (2).

Upon more severe aging treatments, VSCR1 catalysts showed a strong decrease in SCR activity, along with a drop in NH$_3$ conversion, especially evident for the sample aged at 800°C. A pronounced imbalanced conversion of NO$_x$ and NH$_3$ was observed at high reaction temperatures (>450°C), more marked with increasing aging temperature, leading to negative NO$_x$ conversions above 500°C. Furthermore, considerable amounts of N$_2$O were formed over the aged VSCR1 catalysts. For the sample treated at 800°C, an important decrease in the ratio between converted NO$_x$ and NH$_3$ was observed (X_{NOx}/X_{NH3}<1 over the whole temperature range) however, only small amounts of N$_2$O (~4% at 600°C) were formed on this catalyst, suggesting that NH$_3$ was mainly oxidized to NO according to equation (3). Hydrothermal treatments had a more significant effect on the catalytic performance of VSCR1. The hydrothermally aged samples exhibited a greater drop in activity, together with a considerable decrease in the ratio between converted NO$_x$ and converted NH$_3$ above 450°C. In accordance, larger amounts of N$_2$O were produced over these catalysts.

VSCR2 samples presented an enhanced stability as compared to VSCR1. As seen in Fig. 1, the VSCR2 catalysts maintained a relatively high activity under standard SCR conditions, together with a higher N$_2$ selectivity than VSCR1 samples. Aging at 800°C also lead to a loss in activity and selectivity, but less pronounced than for VSCR1. Importantly, VSCR2 catalysts exhibited a much better tolerance to hydrothermal treatments, with HT650 and HT700 samples presenting a high level of NO$_x$ conversion (>90%) at 350–450°C. The hydrothermally aged VSCR2 samples showed as

well a reduced NH$_3$ oxidation activity and a less pronounced unbalanced conversion of NO$_x$ and NH$_3$. In contrast to VSCR1 catalysts, VSCR2 were more resistant to hydrothermal than to thermal treatments.

As seen in Fig. 2, both VSCR1 and VSCR2 catalysts showed an enhancement in the activity at low temperatures under fast SCR conditions. This is related with the V redox properties, which lead to a higher deNO$_x$ efficiency.[20] Similarly to the standard SCR, all VSCR2 catalysts presented much better stability than VSCR1, maintaining a high level of NO$_x$ conversion (>90% at 350–450°C) even upon aging at 700°C. Again, VSCR1 samples exhibited a gradual drop in activity with increasing aging temperature, as well as negative NO$_x$ conversions above 500°C (see ESI Fig. 1 in Supplementary Material www.maneyonline.com/doi/suppl/10.1179/2055075814Y.0000000005). Furthermore, as previously observed, VSCR2 catalysts displayed an improved stability towards hydrothermal aging, more evident than under standard conditions.

Surface area and ammonia adsorption capacity

The BET specific surface area studies for the thermally aged samples are given in Fig. 3. The reduction in surface area with increasing temperature indicated that morphological changes, which have an effect on the stability of V and W compounds on the porous TiO$_2$ surface, have taken place. The surface area of a newly prepared sample is typically 80–100 m^2 g^{-1} for both samples but drops then with the aging temperature.[16] Whilst VSCR1 loses a part of the surface area already at 600°C, stabilized VSCR2 retains most of its surface area at this temperature (Fig. 3). A critical surface area is thought to about 50 m^2 g^{-1} which is sufficient to maintain enough surface and to keep V and W compounds stable for SCR reactions.[15] Hydrothermal treatment at 700°C led to

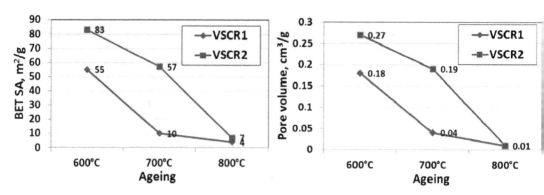

Figure 3 BET surface area and pore volumes of VSCR catalysts by N_2 adsorption–desorption isotherms at −196°C as function of aging temperature

severe loss of surface area (10 m² g⁻¹) for VSCR1, which correlated also to the drop in SCR activity seen in Figs. 1 and 2. The pore volume related to N_2 adsorption–desorption isotherm correlated also to the changes in BET surface.

Ammonia adsorption capacity is a property related to SCR functionality adsorbing on both Brønsted and Lewis acid sites. It has previously been shown that ammonia adsorbed on Brønsted acid sites play a role in the vanadium catalyzed SCR process at low temperatures (<300°C) but less so at temperatures >400°C.[21] Ammonia adsorption measurement at 200°C showed that the total acidity of catalyst surface mimics the drop seen in the surface area shown in Fig. 4 – note that the NH_3 adsorption capacity of both these catalysts are initially >100 µmol g⁻¹.[16] Aging at 800°C destroys the surface (and its acidic properties) on both VSCR catalysts, although a clear effect of stabilization in the VSCR2 catalyst can be seen at the critical temperature at 700°C.

X-ray diffraction data

X-ray diffraction (XRD) data for both catalysts after thermal aging are given in Fig. 5 and ESI Fig. 2 (Supplementary Material www.maneyonline.com/doi/suppl/10.1179/2055075814Y.0000000005) respectively and revealed that all samples are highly crystalline with the main reflections ascribable to the presence of either TiO_2 anatase or rutile (see ESI Table 1 in Supplementary Material www.maneyonline.com/doi/suppl/10.1179/2055075814Y.0000000005). After treatment at 600°C this is the only phase present in both samples. The rather broad but clear reflections corresponding to the (101),

(200) were used for performing a Scherrer analysis and suggested the TiO_2 particles to be ∼16–18 nm – the similarity of the particle size estimation suggesting also that the particles are reasonably isotropic.[22] Heat treatment at 700°C lead to a small degree of sintering (slightly anisotropic) of the TiO_2 anatase phase as evidenced by growth of the (111) reflection (∼30% more intense) than that seen at 600°C. Treatment at 800°C lead to the formation of small amounts (∼20% by weight) of TiO_2 rutile. Sintering of the TiO_2 anatase phase is now severe (particles >100 nm) and in addition peaks due to monoclinic WO_3 are now also present. The results from a multiple phase Lebail extraction performed on the samples treated at 700 and 800°C are shown in Table 1.

For the XRD data of VSCR1 revealed all samples to be highly crystalline. The sample aged at 600°C contained reflections ascribable to TiO_2 anatase possessing anisotropic crystallites greater in size than that seen for VSCR2. The sample aged at 700°C contains also clear evidence of TiO_2 anatase although the intensity of this signal is greater (∼100% more intense) than that seen at 600°C suggesting

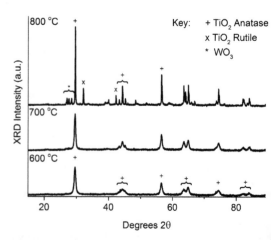

Figure 5 X-ray diffraction data with VSCR2 with increasing temperature of thermal treatment. Important Bragg peaks for respective phases TiO_2 anatase, TiO_2 rutile and monoclinic WO_3 have been identified with a +, x or * accordingly. Table 1 in ESI (Supplementary Material www.maneyonline.com/doi/suppl/10.1179/2055075814Y.0000000005) contains list of the seven most pertinent reflections of each phase present

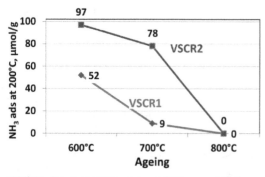

Figure 4 Ammonia adsorption capacity at 200°C on VSCR catalysts as function of aging temperature

further growth/sintering of this phase; the average particle size observed now being $\gg$100 nm in size. Also now present in the pattern are reflections due to TiO_2 rutile as well as the appearance of Bragg peaks due to the formation of monoclinic WO_3 crystallites as a result of phase segregation. VSCR1 aged at 800°C is by contrast completely dominated by the presence of TiO_2 rutile which also undergoes significant sintering between 700 and 800°C. Also present in the pattern are peaks for TiO_2 rutile and again clear evidence for monoclinic WO_3. From ESI Fig. 4 (Supplementary Material www.maneyonline.com/doi/suppl/10.1179/2055075814Y.0000000005), the characteristics of the WO_3 phase do not however change particularly on heating to 800°C. The results from a multiple phase Lebail extraction performed on the samples aged at 800°C are shown in Table 1. The rutile fraction was about 20 wt-% on VSCR1 aged at 700°C and on VSCR2 aged at 800°C, showing an improved thermal stability of about 100°C between these two coatings.

The XRD data and in particular the large TiO_2 crystallites demonstrate that the decrease in the surface area and pore volume seen in Figs. 3 and 4 occur due to severe sintering and phase transformation of the TiO_2 crystallites.

Raman

Raman spectra obtained from the samples correspond well with the XRD data, containing strong well defined bands for TiO_2 anatase phase (see Table 2 and Fig. 6 and ESI Fig. 5 (Supplementary Material www.maneyonline.com/doi/suppl/10.1179/2055075814Y.0000000005) for further details and assignments) or in the VSCR1-800°C sample, rutile.[23] With increased temperature treatment the TiO_2 anatase bands become both more narrow and weaker as the phase sinters and eventually transforms into TiO_2 rutile. The sintering of TiO_2 anatase affects the intensity/peak widths of the A_{1g} and B_{1g} modes more dramatically than the E_g mode and suggests the sintering to be non-uniform. In addition to the TiO_2 phases weak bands between 804 and 807 cm^{-1} can be seen ascribable to the υ (O–W–O) mode of monoclinic WO_3; in instances where the WO_3 phase is particularly crystalline (i.e. after heat treatment at 700–800°C the υ (W–O) band at $\sim$712 cm^{-1} can also be seen).[24] Evidence for the presence of presence of vanadium in the form of polymeric V_2O_5 likely dispersed over the WO_3 surface can also be seen by virtue of the band at $\sim$985 cm^{-1}.[2,25] Weak bands $\sim$930 cm^{-1} also suggest the presence of V_xO_y species.[26] The absence of bands >1000 cm^{-1} suggests that V=O species often associated with isolated V^{5+} species are not present to any great extent; however, it is debateable whether such bands are visible unless measurements are made in the absence of water

(dehydrated).[5] After treatment at higher temperatures the weak bands at $\sim$930 cm^{-1} followed by the 985 cm^{-1} band disappear suggesting disappearance of the V_2O_5 polymeric species. However, no clear evidence for an alternative vanadium containing phase could be seen from either Raman or XRD.

UV–vis data

The presence of TiO_2 (both forms) in large amounts is also evident from the UV–vis spectra (Fig. 7 and ESI Fig. 6 in Supplementary Material www.maneyonline.com/doi/suppl/10.1179/2055075814Y.0000000005). The major features observed comprise the ligand to metal charge transfer (LMCT) bands at 230–350 nm associated with TiO_6 species in both TiO_2 forms.[27] A pale yellow hue to the samples suggests the presence of V^{5+} (possible polymeric V_2O_5 type species as seen by Raman scattering) are also present although the contribution to this component is somewhat difficult to unambiguously discern against the TiO_2 'background'.[28] At higher temperatures the samples become increasingly darker (increased absorption in the visible region) suggestive of both a reduction in the amount of TiO_2 anatase (concurrent increase in the amount of TiO_2

Table 1 Phase composition (mass-%) for crystalline phases present in VSCR1 and VSCR2 determined from full pattern Lebail fitting*

Sample	Anatase	Rutile	WO_3
VSCR1-700	73.2	20.9	5.9
VSCR1-800	6.5	91.5	1.9
VSCR2-800	75.9	19.9	4.2

*The only crystalline phase present in VSCR1-600, VSCR2-600 and VSCR-2-700 is TiO_2 Anatase.

Table 2 Raman bands and fitted position, width and intensities for various samples after thermal treatment

Sample	Band/vibration origin	Position/ cm^{-1}	Band width/ cm^{-1}	Relative band intensity (a.u.)
VSCR2 (600°C)	(TiO_2) Anatase (B_{1g})	396	34	0.84
	(A_{1g})	517	34	0.996
	(E_g)	639	31	1
	υ (O–W–O)	804	46	0.06
	V^{5+}–O–V^{5+}	928	54	0.03
	V_2O_5 polymeric	985	28	0.04
VSCR2 (700°C)	(TiO_2) Anatase (B_{1g})	396	29	0.63
	(A_{1g})	517	28	0.63
	(E_g)	640	29	1
	υ (O–W–O)	804	47	0.07
	V^{5+}–O–V^{5+}	928	55	0.04
	V_2O_5 polymeric	985	103	0.07
VSCR2 (800°C)	(TiO_2) Anatase (B_{1g})	396	27	0.59
	(A_{1g})	517	26	0.44
	(E_g)	640	25	1
	υ (W–O)	712	18	0.09
	υ (O–W–O)	807	30	0.24
VSCR1 (600°C)	(TiO_2) Anatase (B_{1g})	396	26	0.73
	(A_{1g})	517	28	0.76
	(E_g)	639	30	1
	υ (O–W–O)	804	125	0.08
	V^{5+}–O–V^{5+}	930	49	0.04
	V_2O_5 polymeric	985	22	0.05
VSCR1 (700°C)	(TiO_2) Anatase (B_{1g})	396	34	0.55
	(A_{1g}) 516	516	28	0.45
	(E_g) 639	639	28	1
	υ (O–W–O) 804	804	41	0.18
VSCR1 (800°C)	(TiO_2) Rutile (E_g)	449	95	1
	(A_{1g})	610	43	0.69
	υ (W–O)	713	36	0.1
	υ (O–W–O)	807	55	0.25

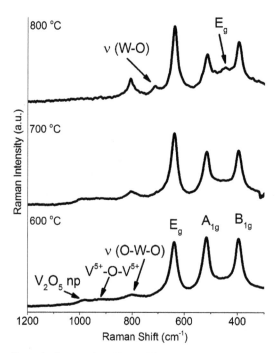

Figure 6 Raman data from VSCR2 with increasing temperature of thermal treatment. Band assignments and results from peak profiling are illustrated and are also listed in Table 2. E_g band in sample aged at 800°C originates from TiO_2 rutile

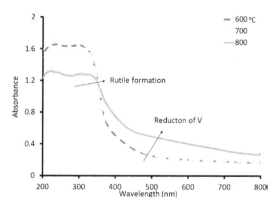

Figure 7 UV–vis spectra for VSCR2 with increasing temperature of thermal aging. The arrows indicate the shifts in either the LMCT band position (due to TiO_2 Rutile formation) or else the increased absorption in the visible region due to reduction of V species. The LMCT bands present in the spectra are listed in Table 3.

rutile as evidenced by the bands ∼350 nm) and reduction of the initial V species. This increased absorption in the visible region between 420 and 700 nm has previously been ascribed to polymerization of V^{5+} containing VO_x species on the catalyst surface. No confirmatory evidence for this could be found in the Raman spectrum for V_2O_5 nanoparticles or bulk species. Most likely then such a change should be interpreted in terms of a reduction of V^{5+} species to either V^{4+} or V^{3+} containing species.

XANES and XPS data

V K-edge XANES data are shown in Fig. 8 for VSCR2 whilst reference spectra for the VSCR1 series and reference spectra for V^{5+} to V^{3+} containing reference oxides are shown in ESI Fig. 7 (Supplementary Material www.maneyonline.com/doi/suppl/10.1179/2055075814Y.0000000005). There appears much debate in the literature as to which features in the XANES spectra can be used to determine the oxidation and coordination state and therefore the nature V species in solid phases.[29–31] Attempts have been made to correlate the position, width and/or intensity of the pre-edge peak much in the same way as has been done successfully for Ti and Fe containing compounds although the richer redox and local structure variation for the V compounds has demonstrated that this is more difficult.[32,33] The key features for the V K-edge XANES spectra have been highlighted with arrows in Fig. 8 and ESI Fig. 7 (Supplementary Material www.maneyonline.com/doi/suppl/10.1179/2055075814Y.0000000005) and include the pre-edge peak due to the 1s–3d dipole transition, the edge position (E_0) and the rising absorption edge dominated by the

dipole allowed 1s–4p transition. It is clear that changes in all three regions occur with increasing aging temperature; broadly these changes involve a reduction in the pre-edge peak intensity, a positive (in energy) shift in the E_0 position and an increase in the rising absorption edge. These key observations and their interpretation from the samples are given in Table 4. A simple comparison of the XANES spectra for the samples with those for the reference compounds (ESI Fig. 7 in Supplementary Material www.maneyonline.com/doi/suppl/10.1179/2055075814Y.0000000005) suggested that the end members of the sample series (i.e. samples treated at 600 and 800°C) to possess V environments similar to V_2O_5 and V_2O_3 respectively. In particular the samples treated at 800°C look almost identical to the V_2O_3 reference spectrum. Furthermore the XANES spectra from the samples treated at intermediate temperatures can be rationalized as comprising a mixture of V_2O_5 and V_2O_3, consistent with the Raman and UV–vis results. We propose to use a straightforward approach of comparison of edge position (E_0 typically found around ∼5480 eV and in this case we determined the edge position from 50% of the normalized intensity) as a method to determine average oxidation state in the samples since the likely V species present will probably resemble the oxide reference compounds V_2O_5, VO_2 and V_2O_3. These reference compounds have also been measured and used to create a calibration curve (ESI Fig. 8 in Supplementary Material www.maneyonline.com/doi/suppl/10.1179/2055075814Y.0000000005) from which the average

Table 3 Centroid band position of LMCT bands from UV–vis spectra shown in Fig. 3 and ESI Fig. 5 (Supplementary Material www.maneyonline.com/doi/suppl/10.1179/2055075814Y.0000000005)

Sample	LMCT band positions/nm
VSCR1 600°C	237 and 296 nm
VSCR1 700°C	234 and 304 nm
VSCR1 800°C	244 and 348 nm
VSCR2 600 °C	241 and 296 nm
VSCR2 700°C	238 and 298 nm
VSCR2 800°C	224 and 308 nm

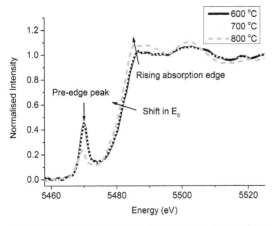

Figure 8 Spectra (XANES) for VSCR2 after aging at three different temperatures. Key features including the pre-edge peak, E_0 position and rising absorption edge and their tendency with increasing treatment temperature have been highlighted using arrows

oxidation state of V can be estimated. Whilst this approach works well for oxidation state determination, it does not tell us much about the coordination state of the species present. Here we need to consider the full XANES spectrum to determine whether multiple V^{5+} species species/environments are present in the sample. Using a combination of the XANES data recorded for the oxide references (and observations made with Raman and UV–vis) and the previous work of Boyesen et al. we conclude that the initial VSCR1/VSCR2 catalysts contain predominantly polymeric V_2O_5 species but that a reduced pre-edge peak intensity (0.41/0.45) in comparison to the V_2O_5 (0.46) suggests that there are either a small amount of octahedrally coordinated monomeric V^{5+} species or else that these monomeric V^{5+} species undergo a 'structural distortion' as a result of a coordinating water ligand (see ESI Fig. 8 in Supplementary Material www.maneyonline.com/doi/suppl/10.1179/2055075814Y.0000000005 for a comparative plot).[28,31] The lower pre-edge peak intensity for the VSCR1 sample suggests that there are slightly more of these octahedral species present in comparison to VSCR2. Both the pre-edge peaks in VSCR1 and VSCR2 are slightly wider than that of the V_2O_5 crystalline reference

Table 4 Pre-edge peak position, intensity, edge position (50% of normalized intensity) and oxidation state determined for various V-containing oxide references and for series of hydrothermally aged samples

Sample	Pre-edge peak position/eV	Pre-edge peak intensity	E_0/eV	Oxidation state (average)
V_2O_5	5470.0	0.46	5481.1	5+
VO_2	5470.3	0.61	5480.6	4+
V_2O_3	5469.0	0.23	5479.0	3+
VSCR1 600°C	5470.0	0.41	5481.0	5+
VSCR1 700°C	5469.6	0.29	5480.6	4+
VSCR1 800°C	5469.0	0.17	5479.3	3.2+
VSCR2 600°C	5470.0	0.45	5481.0	5+
VSCR2 700°C	5470.0	0.45	5480.8	4.5+
VSCR2 800°C	5469.5	0.24	5480.4	3.9+

sample suggesting a wider V–O bond distance variation, consistent with the presence of non-crystalline V_2O_5 species being present.[34] Although there appears little evidence for VO_2 in the samples from the bulk measurements it appears from the XPS data shown in Table ESI 3 (Supplementary Material www.maneyonline.com/doi/suppl/10.1179/2055075814Y.0000000005) that V^{4+} species may be present in the sample although concentrated at the surface. In fact all samples contained surfaces rich in reasonably equivalent amounts of $V^{4+/5+}$ species even though the bulk of the samples tended to be dominated by either V^{5+} or V^{3+} species.

Summary

Both V_2O_5/WO_3–$TiO_2(SiO_2)$ catalysts after thermal aging at 600°C still possessed the essential characteristics that render them active for NH_3–SCR; Raman spectroscopy and V K-edge XANES measurements (and to some extent the absence of Bragg diffraction) point towards the presence of polymeric V_2O_5 being the predominant V^{5+} containing species present. These results are consistent with previous observations that these polymeric V species are more active for NO_x conversion than isolated V sites, although it has been reported that are somewhat less selective towards N_2, especially at high temperatures.[35,36] Critically it appears that SCR activity increases with increasing V content up to monolayer surface coverage, decreasing after that owing to the formation of microcrystalline V_2O_5 particles.[35–37] Selectivity to N_2 is also affected by the formation of bulk V_2O_5.[37–39] Raman spectroscopy also reveals that well dispersed WO_3 species are present in these samples most likely supporting/stabilizing the well dispersed V_2O_5 species according to recent research. Finally the V_2O_5 polymeric species/WO_3 are well dispersed over the Anatase TiO_2 and SiO_2 stabilized TiO_2 crystallites <20 nm in size.[25,40]

The major impact on the system when heating to temperatures >700°C, particularly for the VSCR1 catalyst is first sintering of the TiO_2 (as determined by a drop in surface area and observation of larger crystallites (>100 nm) by XRD) followed by the anatase to rutile phase transformation. This sintering, severe already after aging at 700°C, leads to phase segregation of the catalyst and in particular WO_3 which subsequently crystallizes in the monoclinic form. Vanadium species are partially reduced (average valence from V K-edge XANES of 4.0) most likely as a result of partial phase segregation of the polymeric V_2O_5 species from the sintered catalyst and subsequent autothermal reduction. Heating to 800°C sees further sintering of the TiO_2 anatase and rutile phases whilst further autothermal reduction takes place leading to the eventual formation of V_2O_3 species (according to UV–vis and V K-edge XANES data). The presence of bands due to V_2O_3 in the Raman spectrum will be obscured by those due to TiO_2 anatase and rutile species, whilst the V loading is too low to detect crystallites. Although not characterized here we propose that it is likely that the hydrothermally aged samples undergo a similar mechanism of deactivation, i.e. phase transformation/sintering process as previous results suggest.[16]

Whilst the SiO_2 stabilized VSCR2 catalyst exhibits similar sintering and phase segregation behavior to VSCR1 crucially

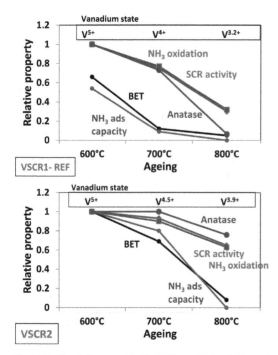

VSCR1- REF

VSCR2

Figure 9 Correlation graph illustrating effect of thermal aging treatment against relative change in measured variable (surface area, NH₃ adsorption capacity, crystalline phase, SCR activity and V oxidation state) of VSCR1 and VSCR2 catalysts. Note that values for relative property are given with respect to optimal value for VSCR2 catalyst

this behavior is offset. From the results of the characterization performed here there are very few obvious differences in the nature of the V, WO₃ and TiO₂ species present in either VSCR1 or VSCR2 although, consistent with previous studies, this small amount of SiO₂ offsets sintering and the anatase to rutile phase transformation by ~100°C.9 In this previous work by Hirano *et al.* studying a range of SiO₂:TiO₂ composites, evidence was found particularly in the highest loaded (48.2 mol.-%) SiO₂ composites, of an amorphous SiO₂ shell surrounding the outside of the anatase particles. Considering that the loadings used here are similar to the lower loadings used in the study by Hirano *et al.* it is therefore likely the same amorphpous SiO₂ shell is present in our samples.

Specifically VSCR2 is more resistant to treatment at 700°C (even in the presence of water) retaining much of the surface area present at 600°C whilst also maintaining a significant proportion of Brønsted acidity and active polymeric V₂O₅ species (although an average oxidation state of 4.5 suggests some autothermal reduction takes place). Treatment at 800°C does however lead to notable sintering, phase transformation, segregation and autothermal V reduction as seen in VSCR1 all leading to a reduced overall catalytic performance.

Figure 9 contains an overall summary of the impact of thermal treatment on the physical properties of the catalysts and their overall activity. With regards to the SCR reaction mechanism, we note that thermal aging at 700°C of the VSCR1 catalyst had a dramatic effect on both the ammonia adsorption capacity and the surface area of the TiO₂ support

although NO$_x$ conversion remained at ~80% of the conversion seen for VSCR1 treated at 600°C. This suggested that the primary effect of thermal aging on NO conversion concerns the vanadium species and in particular that the loss of polymeric V₂O₅ species and formation of inactive bulk V₂O₃ probably leads to deactivation via a reduction in the number of available active sites and a hiatus in the solid state redox cycle.

Conclusions

The addition of small (~%) amounts of stabilizing agent together with the developed preparation methods has a significant effect on the thermal stability of an active V-SCR catalyst enabling it to withstand high operation temperatures and retain much of its catalytic activity. This study also demonstrates the importance of using multiple characterization techniques in conjunction with detailed catalytic testing to fully appreciate and understand the evolution of the structure of the supported catalyst with its catalytic function and thermal durability.

Conflicts of interest

The authors declare no conflict of interest.

Acknowledgements

The authors are grateful to TEKES Finland in the FCEP research project for funding. The authors acknowledge the UK Catalysis BAG access to B18 of Diamond Light Source, EPSRC for funding support to Andrew M. Beale and I. Lezcano-Gonzalez and ESRF for access to BM26A.

References

1. F. Liu, W. Shan, X. Shi and H. He: *Prog. Chem.*, 2012, **24**, 445–455.
2. P. G. W. A. Kompio, A. Brueckner, F. Hipler, G. Auer, E. Loeffler and W. Gruenert: *J. Catal.*, 2012, **286**, 237–247.
3. L. J. Alemany, L. Lietti, N. Ferlazzo, P. Forzatti, G. Busca, E. Giamello and F. Bregani: *J. Catal.*, 1995, **155**, 117–130.
4. N. Y. Topsoe, J. A. Dumesic and H. Topsoe: *J. Catal.*, 1995, **151**, 241–252.
5. I. E. Wachs: *Dalton Trans.*, 2013, **42**, 11762–11769.
6. B. M. Weckhuysen and D. E. Keller: *Catal. Today*, 2003, **78**, 25–46.
7. I. Nova, L. D. Acqua, L. Lietti, E. Giamello and P. Forzatti: *Appl. Catal. B: Environ.*, 2001, **35B**, 31–42.
8. A. Kubacka, A. Iglesias-Juez, M. di Michiel, M. Isabel Becerro and M. Fernandez-Garcia: *Phys. Chem. Chem. Phys.*, 2014, **16**, 19540–19549.
9. M. Hirano, K. Ota and H. Iwata: *Chem. Mater.*, 2004, **16**, 3725–3732.
10. A. Li, Y. Jin, D. Muggli, D. T. Pierce, H. Aranwela, G. K. Marasinghe, T. Knutson, G. Brockman and J. X. Zhao: *Nanoscale*, 2013, **5**, 5854–5862.
11. S. Nikitenko, A. M. Beale, A. M. J. van der Eerden, S. D. M. Jacques, O. Leynaud, M. G. O'Brien, D. Detollenaere, R. Kaptein, B. M. Weckhuysen and W. Bras: *J. Synchrot. Radiat.*, 2008, **15**, 632–640.
12. M. Newville: *J. Synchrot. Radiat.*, 2001, **8**, 322–324.
13. B. Ravel and M. Newville: *J. Synchrot. Radiat.*, 2005, **12**, 537–541.
14. P. Ngene, P. Adelhelm, A. M. Beale, K. P. de Jong and P. E. de Jongh: *J. Phys. Chem. C*, 2010, **114C**, 6163–6168.
15. U. Holzwarth and N. Gibson: *Nat. Nanotechnol.*, 2011, **6**, 534.
16. T. Maunula, T. Kinnunen, K. Kanniainen, A. Viitanen and A. Savimaki: 'Thermally durable vanadium-SCR catalysts for diesel applications', SAE Technical Paper 2013-01-1063, SAE International, 2013.
17. G. Madia, M. Koebel, M. Elsener and A. Wokaun: *Ind. Eng. Chem. Res.*, 2002, **41**, 3512–3517.
18. T. Maunula, R. Lylykangas, A. Lievonen and M. Härkönen: 'NOx reduction by urea in the presence of NO2 on metal substrated SCR catalysts for heavy-duty vehicles', SAE Technical Paper 2003-01-91, SAE International, 2003.
19. G. Busca, L. Lietti, G. Ramis and F. Berti: *Appl. Catal. B: Environ.*, 1998, **18**, 1–36.

20. E. Tronconi, I. Nova, C. Ciardelli, D. Chatterjee and M. Weibel: *J. Catal.*, 2007, **245**, 1–10.

21. N. Y. Topsoe, H. Topsoe and J. A. Dumesic: *J. Catal.*, 1995, **151**, 226–240.

22. E. K. Gibson, M. W. Zandbergen, S. D. M. Jacques, C. Biao, R. J. Cernik, M. G. O'Brien, M. Di Michiel, B. M. Weckhuysen and A. M. Beale: *ACS Catal.*, 2013, **3**, 339–347.

23. H. C. Choi, Y. M. Jung and S. B. Kim: *Vibrat. Spectrosc.*, 2005, **37**, 33–38.

24. R. F. Garcia-Sanchez, T. Ahmido, D. Casimir, S. Baliga and P. Misra: *J. Phys. Chem. A*, 2013, **117A**, 13825–13831.

25. C. A. Carrero, C. J. Keturakis, A. Orrego, R. Schomaecker and I. E. Wachs: *Dalton Trans.*, 2013, **42**, 12644–12653.

26. A. Bruckner and E. Kondratenko: *Catal. Today*, 2006, **113**, 16–24.

27. J. G. Li, T. Ishigaki and X. D. Sun: *J. Phys. Chem. C*, 2007, **111C**, 4969–4976.

28. X. T. Gao, S. R. Bare, J. L. G. Fierro and I. E. Wachs: *J. Phys. Chem. B*, 1999, **103B**, 618–629.

29. J. Wong, F. W. Lytle, R. P. Messmer and D. H. Maylotte: *Phys. Rev. B*, 1984, **30B**, 5596–5610.

30. P. Chaurand, J. Rose, V. Briois, M. Salome, O. Proux, V. Nassif, L. Olivi, J. Susini, J.-L. Hazemann and J.-Y. Bottero: *J. Phys. Chem. B*, 2007, **111B**, 5101–5110.

31. K. L. Boyesen and K. Mathisen: *Catal. Today*, 2014, **229**, 14–22.

32. M. Wilke, F. Farges, P. E. Petit, G. E. Brown and F. Martin: *Am. Mineral.*, 2001, **86**, 714–730.

33. F. Farges, G. E. Brown and J. J. Rehr: *Phys. Rev. B*, 1997, **56B**, 1809–1819.

34. K. L. Boyesen, F. Meneau and K. Mathisen: *Phase Trans.*, 2011, **84**, 675–686.

35. G. T. Went, L. J. Leu and A. T. Bell: *J. Catal.*, 1992, **134**, 479–491.

36. I. E. Wachs, G. Deo, B. M. Weckhuysen, A. Andreini, M. A. Vuurman, M. deBoer and M. D. Amiridis: *J. Catal.*, 1996, **161**, 211–221.

37. S. Djerad, L. Tifouti, M. Crocoll and W. Weisweiler: *J. Mol. Catal.-Chem.*, 2004, **208**, 257–265.

38. G. Madia, M. Elsener, M. Koebel, F. Raimondi and A. Wokaun: *Appl. Catal. B-Environ.*, 2002, **39B**, 181–190.

39. O. Kroecher and M. Elsener: *Appl. Catal. B-Environ.*, 2008, **77**, 215–227.

40. A. Burkardt, W. Weisweiler, J. A. A. van den Tillaart, A. Schafer-Sindlinger and E. S. Lox: *Top. Catal.*, 2001, **16**, 369–375.

12

Support effects in the gas phase hydrogenation of butyronitrile over palladium

Yufen Hao, Xiaodong Wang, Noémie Perret, Fernando Cárdenas-Lizana and Mark A. Keane*

Chemical Engineering, School of Engineering and Physical Sciences, Heriot-Watt University, Edinburgh EH14 4AS, Scotland, UK

Abstract The role of the support in the gas phase hydrogenation of butyronitrile over Pd/Al$_2$O$_3$ and Pd/C (2.5–3.0 nm mean Pd size) has been studied, taking bulk Pd as benchmark. Catalyst activation by temperature programmed reduction was monitored and the metal and acid functions characterized by H$_2$

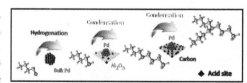

and NH$_3$ chemisorption/temperature programmed desorption and electron microscopy (STEM/TEM). Butyronitrile hydrogenation was stable with time on-stream to deliver butylamine where consecutive condensation with the intermediate butylidenimine generated dibutylamine and tributylamine. Condensation can occur on bulk Pd but selectivity is influenced by the support and reaction over Pd/Al$_2$O$_3$ generated dibutylamine as principal product. Preferential tertiary amine formation was observed over Pd/C and attributed to greater surface acidity that favors the condensation step. Increased hydrogen spillover and acidity (associated with Pd/C) elevated butyronitrile consumption rate.

Keywords Pd/Al$_2$O$_3$, Pd/C, Bulk Pd, Butyronitrile hydrogenation, Surface acidity, Spillover hydrogen

List of symbols

d_{H2} Pd particle size diameter from H$_2$ chemisorption measurements, nm
d_{TEM} mean Pd particle size diameter from TEM analysis, nm
F reactant inlet flow rate, mmol h^{-1}
$GHSV$ gas hourly space velocity, h^{-1}
i.d.   internal diameter, mm
n moles of palladium, mol
N_i stoichiometric coefficient for product 'i'
S_i selectivity to product 'i', %
SSA   specific surface area, m^2 g^{-1}
T_{max} temperature maximum for H$_2$ (or NH$_3$) released during H$_2$ (or NH$_3$) TPD, K
x_i molar fraction of reactant/product 'i'
X_{BT} fractional butyronitrile conversion
σ mean deviation

Introduction

The catalytic hydrogenation of nitriles is an established route to amines, widely used as intermediates in the production of agrochemicals and pharmaceuticals.[1] This reaction is typically conducted in batch liquid phase at elevated H$_2$

pressure (20–45 atm)[2–6] with alkane solvents (e.g. hexane,[5,7] heptane[3,4,6] and octane[7,8]). A move from batch to continuous processes has, however, been highlighted by the fine chemical/pharmaceutical sector as a priority to achieve higher throughput and sustainable production.[9] Nitrile hydrogenation has been conducted over supported metal (Ni,[2,8,10] Co,[5,8,10] Ru,[3,4,6,10] Cu,[4,10] Rh,[3,4] Pt[3,4,10] and Pd[3,4,7,10,11]) catalysts where primary amines are preferentially produced over Ru, Ni and Co, whereas Cu and Rh promote the formation of secondary amines, and Pd and Pt exhibit higher selectivity to tertiary amines. It is striking that Pd, although extensively used in hydrogenation applications, has not been employed to any significant extent in nitrile reduction and the work to date has primarily considered the performance of bimetallic (Pd–Ni,[3,4] Pd–Ag,[4] Pd–Cu,[4] Pd–Pt[12]) or (PdZn, PdGa$_5$, Pd$_5$Ga$_2$ and Pd$_{0.48}$In$_{0.52}$) alloy[11] catalyst formulations.

Metal oxides have been used as support in the hydrogenation of butyronitrile (BT),[3–6,10,13] benzylcyanide,[3] acetonitrile[4–6,12] and lauronitrile.[2] Use of carbon as metal carrier has focused on reactions promoted by Ni[14–16] with limited work on supported Pd.[13,17] Catalytic hydrogenation has been shown to be influenced by support acid-base character[18] with conflicting results for nitrile reduction. In the hydrogenation of acetonitrile[19–22] over oxide (MgO,[19,20] Al$_2$O$_3$,[19–21] Cr$_2$O$_3$,[19] SiO$_2$,[19,20,22] TiO$_2$,[19] ZrO$_2$,[19] ThO$_2$[19] and

UO$_2$[19]) supported Ni, surface acidity was proposed to contribute to condensation step(s). In contrast, no apparent selectivity dependence on support acidity was observed for reaction over oxide[23] (Al$_2$O$_3$, TiO$_2$, SiO$_2$–Al$_2$O$_3$ and SiO$_2$) and zeolite (NaY[6]) supported Ru,[6] Ni,[6,23] Rh[6] and Pt.[6] Given the available literature, it is difficult to establish any explicit link between catalyst performance and surface acid properties. In this report, we set out to decouple the effect of metal and support in determining catalyst performance and evaluate the role of surface acidity in the gas phase continuous hydrogenation of BT, as a model aliphatic nitrile reactant, over Pd/C and Pd/Al$_2$O$_3$, taking bulk Pd as benchmark.

Experimental methods

The alumina support (Puralox, Condea Vista Co.) was used as received, (1 wt-%) Pd/C, (1.2 wt-%) Pd/Al$_2$O$_3$ and PdO were obtained from Sigma-Aldrich. The samples were sieved into a batch of 75 μm average diameter and activated in 60 cm^3 min^{-1} H$_2$ (BOC, >99.99%) at 10 K min^{-1} to 573 K, which was maintained for 1 h. Reduction conditions to convert PdO to zero valent Pd have been established elsewhere.[24] Samples for off-line analysis were passivated in 1% v/v O$_2$/He at ambient temperature.

Catalyst characterization

Palladium content was measured by inductively coupled plasma optical emission spectrometry (Vista-PRO, Varian Inc.) from the diluted extract in HF. Temperature programmed reduction (TPR), H$_2$ and NH$_3$ chemisorption/temperature programmed desorption (TPD) and specific surface area (SSA) measurements were conducted using the commercial CHEM-BET 3000 (Quantachrome) unit. The samples were loaded into a U-shaped Quartz cell (3.76 mm i.d.) and heated in 17 cm^3 min^{-1} (Brooks mass flow controlled) 5% v/v H$_2$/N$_2$ at 10 K min^{-1} to 573 ± 1 K. The effluent gas passed through a liquid nitrogen trap and changes in H$_2$ consumption monitored by a thermal conductivity detector with data acquisition/manipulation using the TPR Win software. The reduced samples were maintained at the final temperature in H$_2$/N$_2$ until the signal returned to baseline, swept with 65 cm^3 min^{-1} N$_2$ for 1.5 h, cooled to ambient temperature and subjected to H$_2$ (or NH$_3$) chemisorption using a pulse (50–1000 μL) titrations. Samples were thoroughly flushed in N$_2$ with TPD at 10–50 K min^{-1} to 923–1173 K. The resultant profile was corrected using the TPD recorded in parallel directly following TPR to explicitly determine H$_2$ (or NH$_3$) release. Specific surface area was determined in a 30% v/v N$_2$/He flow using undiluted N$_2$ as internal standard. At least two cycles of N$_2$ adsorption–desorption were employed using the standard single point BET method. Specific surface area and H$_2$/NH$_3$ uptake/desorption were reproducible to ±5% and the values quoted represent the mean. Supported Pd particle morphology (size and shape) was determined by transmission (TEM, JEOL JEM 2011 unit) and scanning transmission (STEM, JEOL 2200FS field emission gun equipped unit) electron microscopy, employing Gatan DigitalMicrograph 1.82 for data acquisition/manipulation. The samples were crushed and deposited (dry) on a holey carbon/Cu grid (300 Mesh). Up to 800 individual Pd particles

were counted for each catalyst to determine the surface area weighted Pd diameter as described elsewhere.[25]

Catalysis procedure

Reactions were conducted (1 atm, 473 K) in situ, following catalyst activation, in a fixed bed vertical glass reactor (i.d.=15 mm) under conditions that ensured minimal mass or heat transfer limitations. The BT reactant was delivered at a fixed calibrated flow rate via a glass/teflon air tight syringe and teflon line using a microprocessor controlled infusion pump (Model 100 kd Scientific). A layer of borosilicate glass beads served as preheating zone where the reactant was vaporized and reached reaction temperature before contacting the catalyst. Isothermal conditions (±1 K) were maintained by diluting the catalyst bed with ground glass (75 μm) and the temperature was continuously monitored by a thermocouple inserted in a thermowell within the catalyst bed. A co-current flow of butyronitrile (<1% v/v) and H$_2$ was maintained at GHSV=1.0 × 10^4 h^{-1} with an inlet flow rate (F) of 6.9 mmol h^{-1} where H$_2$ was in excess (by a factor of 24) of the stoichiometric requirement for the formation of the butylidenimine intermediate. The molar palladium (n) to F ratio spanned the range 0.3 × 10^{-4}–1.3 × 10^{-3} h. The reactor effluent was frozen in a liquid nitrogen trap for subsequent analysis, which was made using a Perkin–Elmer Auto System XL chromatograph equipped with a flame ionization detector, employing a DB-1 capillary column (i.d.=0.33 mm, length=30 m, film thickness=0.20 μm). Data acquisition/manipulation was performed using the TotalChrom Workstation Version 6.1.2 (for Windows) chromatography data system and reactant/product molar fractions (x$_i$) were obtained using detailed calibration plots (not shown). Fractional BT hydrogenation (X$_{BT}$) was obtained from

$$X_{BT} = \frac{[BT]_{in} - [BT]_{out}}{[BT]_{in}} \qquad (1)$$

where selectivity to product i (S$_i$, %) is given

$$S_i = \frac{N_i x_i}{[BT]_{in} - [BT]_{out}} \times 100 \qquad (2)$$

where, [BT]$_{in}$ and [BT]$_{out}$ represent inlet and outlet BT concentration, respectively, and N$_i$ is the stoichiometric coefficient for product 'i'. Repeated reactions with different samples from the same batch of catalyst delivered raw data reproducibility better than ±6%.

Results and discussion

Catalyst characterization

The critical physicochemical properties of the catalytic systems considered in this study are recorded in Table 1.

Pd/Al$_2$O$_3$

The TPR profile generated for Pd/Al$_2$O$_3$ is presented in Fig. 1(A) where the occurrence of a negative peak (H$_2$ release) at 355 K can be attributed to Pd hydride decomposition.[26] The hydride is generated by H$_2$ absorption, which is known to proceed at H$_2$ partial pressures >0.02 atm;[27] a pressure of 0.05 atm during TPR was used in this work. The absence of any H$_2$ consumption (positive signal) before

Table 1 Specific surface area (SSA), Pd mean particle size (from TEM/STEM (d_{TEM}) and H_2 chemisorption (d_{H2} measurements), H_2 uptake and release during TPD (with associated temperature of maximum release (T_{max})) and NH_3 uptake and release (with T_{max})

| Catalyst | SSA/ $m^2\ g^{-1}$ | Pd size/nm | | H_2 uptake/ mmol g_{Pd}^{-1} | H_2 TPD | | NH_3 uptake/mmol g^{-1} | NH_3 TPD | |
		$d_{TEM}\pm\sigma$	d_{H2}		H_2 desorbed/ mmol g_{Pd}^{-1}	T_{max}/K		NH_3 desorbed/ mmol g^{-1}	T_{max}/K
Pd/Al_2O_3	145	3.0±1.3	2.4	2.1	9.6	767	0.52	0.51	464
Al_2O_3	160	...	...	...	...	...	0.71	0.69	450
Pd	3		131	0.04	0.05	782			
$Pd+Al_2O_3$	133			0.04	0.09	782			
$Pd/Al_2O_3+Al_2O_3$	152			2.3	46.6	767			
Pd/C	870	2.5±1.1	2.2	2.5	79.8	785, 1173	0.94	0.92	480

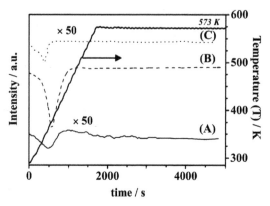

Figure 1 Temperature programmed reduction (TPR) profiles for (A) Pd/Al_2O_3 (solid line), (B) PdO (dashed line) and (C) Pd/C (dotted line)

hydride decomposition suggests the presence of metallic Pd in advance of the temperature ramp. Palladium particle size was determined by electron microscopy and verified by H_2 chemisorption on the basis of dissociative adsorption (Pd/H stoichiometry=1:1).[28] Representative TEM images presented in Fig. 2(A) and (B) reveal pseudo-spherical Pd particles at the nanoscale. The associated Pd size distribution histogram (Fig. 2(C)) delivered a mean diameter (d_{TEM}=3.0±1.3 nm) in good agreement with the value obtained from H_2 chemisorption (d_{H2}=2.4 nm). We should note that chemisorption titration measurements were conducted at H_2 partial pressure=0.001 atm, circumventing any contribution due to hydride formation. Subsequent TPD generated the profile presented in Fig. 3(Aa) with H_2 release over 665–890 K. Hydrogen desorption far exceeded the amount chemisorbed (Table 1). This suggests the involvement of spillover hydrogen, i.e. migration of atomic

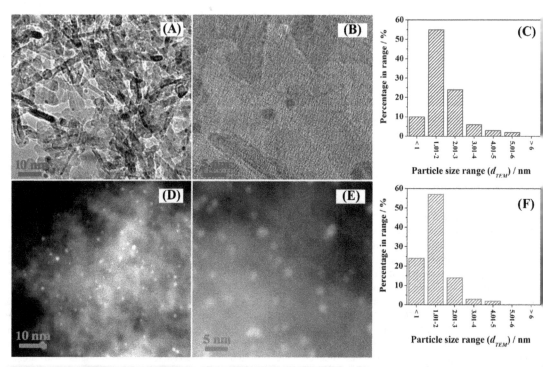

Figure 2 Representative (A,D) medium and (B,E) higher magnification TEM and STEM images with associated (C,F) Pd particle size distributions for (A–C) Pd/Al_2O_3 and (D–F) Pd/C

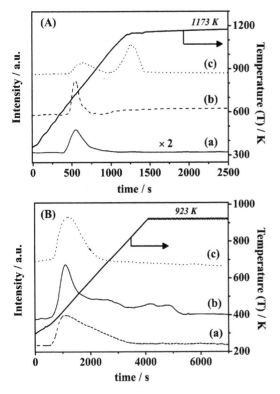

Figure 3 (A) Hydrogen temperature programmed desorption (TPD) profiles for (a) Pd/Al$_2$O$_3$ (solid line), (b) Pd/Al$_2$O$_3$+Al$_2$O$_3$ (dashed line) and (c) Pd/C (dotted line); (B) Ammonia TPD profiles for (a) Al$_2$O$_3$ (dashed line), (b) Pd/Al$_2$O$_3$ (solid line) and (c) Pd/C (dotted line)

hydrogen to the support following dissociation on Pd sites during TPR.[29] Alumina has been demonstrated to accommodate spillover through the action of surface hydroxyl groups.[30] The temperature related maximum H$_2$ release (T_{max}=767 K) is consistent with publications demonstrating spillover desorption from alumina at $T \geqslant 503$ K.[24,31]

There is some controversy regarding the predominant role of metal or support in nitrile hydrogenation, particularly regarding condensation steps.[4,21,22,32] In order to decouple these effects, we examined bulk Pd (generated by reducing PdO) and a Pd+Al$_2$O$_3$ physical mixture, where the Pd content was equivalent to that in Pd/Al$_2$O$_3$. We should note that there was no detectable H$_2$ uptake or release on or from Al$_2$O$_3$, a response that is expected and in agreement with the literature.[26] The SSA recorded for Pd+Al$_2$O$_3$ (Table 1) is a composite with additive contributions from both components. Temperature programmed reduction of PdO also generated a negative peak (Fig. 1(B)) with a T_{max} (=373 K) and associated hydride composition (H/Pd=0.67) which differed from that measured for Pd/Al$_2$O$_3$ (T_{max}=355 K; H/Pd=0.05). This agrees with the reported shift in hydride decomposition to higher temperatures and increased H/Pd for larger Pd particles.[26] The appreciably lower (by a factor greater than 50) H$_2$ uptake on bulk Pd relative to Pd/Al$_2$O$_3$ can be related to the lower specific Pd surface area. While H$_2$ chemisorption was unchanged with Al$_2$O$_3$ addition, TPD

from the physical mixture was measurably higher (twofold) than that from Pd alone, suggesting the occurrence of spillover, as noted elsewhere.[29] Hydrogen spillover in catalyst+support physical combinations where the two components are well mixed[33] or present as discrete layers[34] has been demonstrated with a reported[35] spillover transport across non-contiguous surfaces. This effect was also observed for the Pd/Al$_2$O$_3$+Al$_2$O$_3$ mixture (see TPD profile in Fig. 3(Ab)) where H$_2$ released was (5 times) greater than that recorded for Pd/Al$_2$O$_3$ (Table 1).

Surface acidity was probed by NH$_3$ adsorption coupled with TPD; the TPD profiles for Al$_2$O$_3$ (a) and Pd/Al$_2$O$_3$ (b) are given in Fig. 3(B). The conventional approach (as documented in the literature) to quantifying acidity has been based solely on a measurement of NH$_3$ desorption.[22,36,37] This can, however, be subject to inaccuracy, notably as a result of contributions due to thermal degradation of surface functionalities, particularly dehydroxylation.[38] In this study, total acidity obtained from integration of the TPD signal matched NH$_3$ uptake measured in pulse chemisorption (Table 1). The reported acid site data for alumina show some disparity depending on sample pre-treatment and experimental desorption conditions, e.g. gas flow and heating rates.[39] The NH$_3$ TPD profile for Al$_2$O$_3$ (Fig. 3(Ba)) is characterized by T_{max}=450 K where the profile shape is similar to that recorded by Skotak et al.[40] Surface acidity associated with Al$_2$O$_3$ is predominantly attributable to Lewis sites (Al^{3+})[41,42] with a secondary -OH group contribution.[42] The TPD profile generated for Pd/Al$_2$O$_3$ (Fig. 3(Bb)) exhibited a similar maximum (at 464 K) but the NH$_3$ adsorbed (and released) was measurably lower than Al$_2$O$_3$ (see Table 1). A similar effect has been reported previously[40] and explained on the basis of Pd interaction with surface acid sites, following Pd incorporation on Al$_2$O$_3$, that results in a decrease in total acidity. The NH$_3$ measurements recorded in this study for Pd/Al$_2$O$_3$ (0.51 ± 0.01 mmol g^{-1}, see Table 1) are very close (0.54 mmol g^{-1}) to that reported by Nam et al.[37]

Pd/C

In order to assess further the possible role of the metal carrier, Pd/C was examined where the SSA (870 m^2 g^{-1}, Table 1) far exceeded that of Pd/Al$_2$O$_3$ and is typical of activated carbon supported metal catalysts.[25] The TPR profile of Pd/C (Fig. 1(C)) exhibits a Pd hydride decomposition peak at 353 K with associated H/Pd=0.04, close to that obtained for Pd/Al$_2$O$_3$ and suggesting an equivalent Pd size.[26] The STEM images (Fig. 2(D) and 2(E)) reveal near spherical Pd particles with a mean diameter (d_{TEM}=2.5 ± 1.1 nm) from the size distribution (Fig. 2(F)) that agrees with the H$_2$ chemisorption (d_{H2}=2.2 nm, Table 1) measurement. We must stress the convergence of Pd loading (1.1 ± 0.1 wt-%), size distribution and mean (d_{TEM}=2.9 ± 0.1 nm) for Pd/Al$_2$O$_3$ and Pd/C, which facilitates an explicit analysis of support effects. Hydrogen desorption from Pd/C was significantly greater (by a factor of over 30) than that chemisorbed (Table 1), again suggesting spillover contributions. The H$_2$ TPD profile for Pd/C Fig. 3(Ac) shows two stages of H$_2$ desorption with associated T_{max} at 785 and 1173 K and a greater total H$_2$ release (8-fold) relative to Pd/Al$_2$O$_3$. Differences in the amount of spillover hydrogen associated with the same

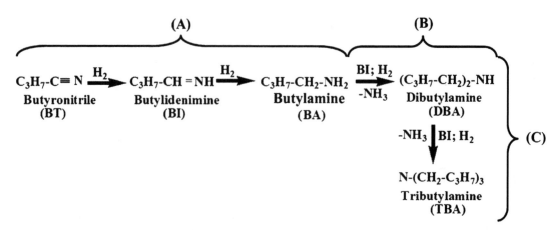

Figure 4 Reaction pathways for hydrogenation of butyronitrile (BT) to (A) primary (butylamine, BA), (B) secondary (dibutylamine, DBA) and (C) tertiary (tributylamine, TBA) amine products

metal (and size) on different carriers have been noted in the literature.[43] Spillover can be influenced by the concentration of initiating and acceptor sites and degree of contact between the participating phases.[44] We can link increased H_2 desorption from Pd/C to the greater SSA that can accommodate more spillover. Surface carboxylic and phenolic groups associated with carbonaceous materials can act as spillover acceptor sites.[34] A wide variation in acidity of carbon supported transition metal catalysts has been observed and is sensitive to carbon source and pretreatment conditions.[45,46] The total acidity of Pd/C from NH_3 chemisorption matched the desorption measurement (0.93 ± 0.1 mmol g^{-1}) and exceeded that of Pd/Al_2O_3 (Table 1). The NH_3 TPD profile for Pd/C (Fig. 3(Bc)) is characterized by a $T_{max}=480$ K, equivalent to that (473–483 K) reported elsewhere for carbon supported systems[45] and ascribed to the presence of weak acid sites.[47,48]

Gas phase hydrogenation of butyronitrile

Reaction pathways in BT hydrogenation that have been reported in the literature[3,49] are presented in Fig. 4. Nitrile

reduction generates a reactive aldimine (butylidenimine, BI) intermediate that is hydrogenated to the primary amine (butylamine, BA, step (A)). Butylamine can undergo condensation with the imine in the presence of hydrogen, releasing ammonia to generate a secondary amine (dibutylamine, DBA, step (B)) which, in turn, can be transformed into a tertiary amine (tributylamine, TBA, step (C)). The reactivity of BI, in terms of hydrogenation or condensation, governs product selectivity. Fractional BT conversion (X_{BT}) over Pd/Al_2O_3 was time invariant and representative time on-stream plots are shown in Fig. 5(A). This response is significant given that (liquid and gas phase) hydrogenation of nitriles over supported Pt,[6,50] Ru[6] and Ni[14,16,22] was accompanied by catalyst deactivation and a temporal decline in activity. This was attributed to:

(i) agglomeration of metal particles during reaction[6,15]

(ii) occlusion of active sites by the amine product(s)[6,12,15]

(iii) catalyst coking due to the formation of dehydrogenated surface species and carbides.[14,22]

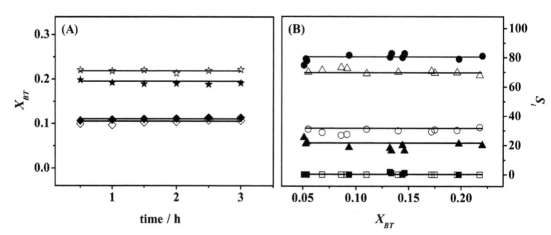

Figure 5 (A) Temporal dependence of butyronitrile fractional conversion (X_{BT}) at varying n/F (0.8×10^{-3} h, ◆; 1.3×10^{-3} h, ★; 0.4×10^{-3} h, ◇; 0.8×10^{-3} h, ☆) and (B) Selectivity (S_i, %) to butylamine (BA, ■, □), dibutylamine (DBA, ●, ○) and tributylamine (TBA, ▲, △ as a function of X_{BT} for reaction over Pd/Al_2O_3 (solid symbols) and Pd/C (open symbols). Reaction conditions: $T=473$ K, $P=1$ atm

These deleterious effects do not appear to apply in our system. Selectivity as a function of conversion over Pd/Al$_2$O$_3$ is shown in Fig. 5(B). The secondary amine was the principal product with TBA as byproduct and negligible BA formation (S_{BA}<2%); selectivity was independent of conversion. The results suggest that once formed, BA participates in a condensation step with the aldimine to generate DBA as the main product. The secondary amine that is produced undergoes further reaction with BI to give the tertiary amine but this step was not promoted to the same extent. Our findings are in line with the work of Huang and Sachtler[4] who reported the formation of secondary amine as the main reaction product in the conversion of BT over Pd/NaY in both liquid and gas phase operation. We should, however, note that Iwasa and co-workers[11] have reported preferential tertiary amine formation in gas phase acetonitrile hydrogenation over Pd/Al$_2$O$_3$.

An unambiguous correlation of metal and support in governing nitrile hydrogenation selectivity has yet to be established. It has been proposed that support acidity can affect product distribution[16,19,22] and Lewis and Brønsted acid sites on alumina are known to contribute to condensation reactions.[19,49] Weak adsorption of BT on Al$_2$O$_3$ with surface coordination through the nitrogen (of the -C≡N group) has been demonstrated by infrared spectroscopy.[51] Hegedűs and Máthé[52] proposed that the Pd phase determines selectivity while the alumina carrier only influences activity. In order to decouple these contributions, we conducted the reaction over Al$_2$O$_3$ and bulk Pd. There was no detectable conversion using the alumina support alone, which can be ascribed to the inability of Al$_2$O$_3$ to dissociate H$_2$[26] and generate the reactive intermediate. The nitrile consumption rate over bulk Pd was appreciably lower (by a factor of 65) than Pd/Al$_2$O$_3$ (Table 2), which can be attributed to lower H$_2$ uptake/release (Table 1). Comparison of selectivity is only meaningful at a common conversion and the selectivity data presented in Table 2 were obtained at X_{BT}=0.15. Bulk Pd generated near equivalent butylamine and DBA with no detectable tertiary amine formation. This suggests that the condensation reaction to generate secondary (but not tertiary) amine can proceed over Pd sites without the involvement of the support. Drawing on the pathway in Fig. 4, the nitrile can be activated on bulk Pd, react with the available surface hydrogen to form imine with further hydrogenation to BA and condensation to DBA. In

contrast to Pd/Al$_2$O$_3$, the secondary amine formed on bulk Pd must desorb without further reaction. Enhanced selectivity to DBA (and formation of TBA) exhibited by Pd/Al$_2$O$_3$ suggests a contribution due to the support and/or metal/support interface to promote condensation. Gluhoi et al.[19] have proposed that the metal-support interface is critical in nitrile hydrogenation. To develop this further, the reaction was conducted over a Pd + Al$_2$O$_3$ physical mixture, which delivered a measurably higher reaction rate than Pd alone (Table 2). The H$_2$ TPD results presented in Table 1 establish the occurrence of spillover hydrogen on Al$_2$O$_3$ in the mixture which can account for the increased rate. Moreover, selectivity to DBA was enhanced with the isolation of TBA in the product stream. This suggests a direct contribution of Al$_2$O$_3$ to primary (and secondary) amine condensation, where activated nitrile reacts with spillover hydrogen generated by Pd. Any contribution due to hydrogen spillover in Pd/Al$_2$O$_3$ + Al$_2$O$_3$ was negligible and reaction rate/product distribution was essentially the same as that obtained with Pd/Al$_2$O$_3$ (Table 2). The intrinsic activity of Pd/Al$_2$O$_3$ is such that spillover due to addition of Al$_2$O$_3$ (Table 1) did not influence performance.

Pd/C delivered a significantly higher nitrile consumption rate than Pd/Al$_2$O$_3$ (Table 2), which can be linked to greater available surface reactive hydrogen (Table 1), given the equivalency of Pd particle size in both catalysts. Structure sensitivity has been proposed for gas phase nitrile hydrogenation over supported Ni[14,15] and Pt[50] with higher specific activity for larger metal particles in the 10–18 nm range. The basic character of BT (electron lone pair on nitrogen in -C≡N) must be considered in possible nitrile interaction(s) with the support.[49,51] The greater surface acidity exhibited by Pd/C favors nitrile/amine activation and can contribute to higher reaction rate. This is consistent with literature that has linked activity to support acidity.[16,52] Catalyst stability with time on-stream also applies to Pd/C (Fig. 5(A)) where selectivity was independent of conversion (Fig. 5(B)) with TBA as predominant product, DBA as byproduct and no detectable BA. This deviates from the selectivity response for Pd/Al$_2$O$_3$ and enhanced tertiary amine selectivity over Pd/C can be ascribed to greater surface acidity that promotes condensation.[19,49] Chen et al.[2] reported increased conversion of lauronitrile over more acidic catalysts (Ni/SiO$_2$, Ni/Al$_2$O$_3$ and Ni/SiAlO) with low selectivity to the primary amine. Moreover, given the decreasing amine basicity, in the order TBA>DBA>BA,[53] increased surface acidity enhances amine interaction, facilitating sequential reaction (BA→DBA→TBA, see Fig. 4) leading to predominant tertiary amine formation.

Conclusions

This work has set out to decouple the role of metal (Pd) and support in nitrile hydrogenation. The results presented support the following:

1. In solvent-free continuous gas phase BT hydrogenation over Pd/C and Pd/Al$_2$O$_3$ with equivalent mean Pd size, Pd/C exhibited appreciably higher activity that can be attributed to increased total surface hydrogen (chemisorbed on Pd and spillover, from H$_2$ TPD) coupled with

Table 2 Butyronitrile consumption rate and product (butylamine (BA), dibutylamine (DBA) and tributylamine (TBA)) selectivity (S_i) for reaction over bulk and supported Pd at equal fractional conversion (X_{BT}=0.15); reaction conditions: T=473 K, P=1 atm

Catalyst	Rate/mol h^{-1} mol$_{Pd}^{-1}$	S_i/% BA	DBA	TBA
Pd/Al$_2$O$_3$	110	0	84	16
Al$_2$O$_3$	…	…	…	…
Pd	1.7	55	45	0
Pd + Al$_2$O$_3$	2.5	43	52	5
Pd/Al$_2$O$_3$ + Al$_2$O$_3$	112	0	82	18
Pd/C	183	0	30	70

greater surface acidity (from NH_3 chemisorption/TPD). Nitrile conversion over both catalysts was constant with time on-stream.

2. Reaction over bulk Pd generated primary and secondary amines, indicating that condensation can proceed over Pd without support. Alumina alone was inactive but in combination with Pd (physical mixture) an increase in activity (relative to Pd) was observed and attributed to the involvement of spillover hydrogen. Addition of Al_2O_3 to bulk Pd shifted reaction to higher amines, demonstrating direct contribution of Al_2O_3 to condensation.

3. Pd/Al_2O_3 and Pd/C generated distinct product distributions with secondary amine as principal product over Pd/Al_2O_3 and preferential tertiary amine formation over Pd/C; product selectivity was independent of conversion. The greater surface acidity of Pd/C facilitates surface interaction(s) with BA and DBA, promoting sequential reaction with the BI intermediate, favoring tertiary amine formation.

Conflicts of interest

There are no conflicts of interest associated with this work.

Acknowledgements

The authors acknowledge EPSRC support for free access to the TEM/SEM facility at the University of St Andrews and financial support to Y. Hao and X. Wang through the Overseas Research Students Award Scheme (ORSAS).

References

1. S. Gomez, J. A. Peters and T. Maschmeyer: Adv. Synth. Catal., 2002, 344, (10), 1037–1057.
2. H. Chen, M. Xue, S. Hu and J. Shen: Chem. Eng. J., 2012, 181–182, 677–684.
3. Y. Huang and W. M. H. Sachtler: Appl. Catal. A: Gen., 1999, 182, (2), 365–378.
4. Y. Huang and W. M. H. Sachtler: J. Catal., 1999, 188, (1), 215–225.
5. P. Schärringer, T. E. Müller and J. A. Lercher: J. Catal., 2008, 253, (1), 167–179.
6. Y. Huang, V. Adeeva and W. M. H. Sachtler: Appl. Catal. A: Gen., 2000, 196, (1), 73–85.
7. L. Hegedűs, T. Máthé and T. Kárpáti: Appl. Catal. A: Gen., 2008, 349, (1–2), 40–45.
8. A. Chojecki, M. Veprek-Heijman, T. E. Müller, P. Schärringer, S. Veprek and J. A. Lercher: J. Catal., 2007, 245, (1), 237–248.
9. C. Jiménez-González, P. Poechlauer, Q. B. Broxterman, B.-S. Yang, D. am Ende, J. Baird, C. Bertsch, R. E. Hannah, P. Dell'Orco, H. Noorrnan, S. Yee, R. Reintjens, A. Wells, V. Massonneau and J. Manley: Org. Proc. Res. Dev., 2011, 15, (4), 900–911.
10. D. J. Segobia, A. F. Trasarti and C. R. Apesteguía: Appl. Catal. A: Gen., 2012, 445–446, 69–75.
11. N. Iwasa, M. Yoshikawa and M. Arai: Phys. Chem. Chem. Phys., 2002, 4, (21), 5414–5420.
12. M. C. Carrión, B. R. Manzano, F. A. Jalón, I. Fuentes-Perujo, P. Maireles-Torres, E. Rodríguez-Castellón and A. Jiménez-López: Appl. Catal. A: Gen., 2005, 288, (1–2), 34–42.
13. J. Krupka, J. Drahonsky and A. Hlavackova: React. Kinet. Mech. Cat., 2013, 108, (1), 91–105.
14. A. Nieto-Márquez, D. Toledano, J. C. Lazo, A. Romero and J. L. Valverde: Appl. Catal. A Gen., 2010, 373, (1–2), 192–200.
15. A. Nieto-Márquez, D. Toledano, P. Sánchez, A. Romero and J. L. Valverde: J. Catal., 2010, 269, (1), 242–251.
16. A. Nieto-Márquez, V. Jiménez, A. M. Raboso, S. Gil, A. Romero and J. L. Valverde: Appl. Catal. A Gen., 2011, 393, (1–2), 78–87.
17. X.-F. Guo, Y.-S. Kim and G.-J. Kim: Catal. Today, 2010, 150, (1–2), 22–27.
18. S. Santiago-Pedro, V. Tamayo-Galván and T. Viveros-García: Catal. Today, 2013, 213, 101–108.
19. A. C. Gluhoi, P. Mărginean and U. Stănescu: Appl. Catal. A: Gen., 2005, 294, 208–214.
20. M. J. F. M. Verhaak, A. J. van Dillen and J. W. Geus: Catal. Lett., 1994, 26, (1–2), 37–53.
21. P. Braos-García, P. Maireles-Torres, E. Rodríguez-Castellón and A. Jiménez-López: J. Mol. Catal. A: Chem., 2001, 168, (1–2), 279–287.
22. P. Braos-García, P. Maireles-Torres, E. Rodríguez-Castellón and A. Jiménez-López: J. Mol. Catal. A: Chem., 2003, 193, (1–2), 185–196.
23. C. V. Rode, M. Arai, M. Shirai and Y. Nishiyama: Appl. Catal. A Gen., 1997, 148, (2), 405–413.
24. C. Amorim, X. Wang and M. A. Keane: Chin. J. Catal., 2011, 32, (5), 746–755.
25. C. Amorim and M. A. Keane: J. Colloid Interf. Sci., 2008, 322, 196–208.
26. S. Gómez-Quero, F. Cárdenas-Lizana and M. A. Keane: Ind. Eng. Chem. Res., 2008, 47, (18), 6841–6853.
27. C. Shi and B. W.-L. Jang: Ind. Eng. Chem. Res., 2006, 45, (17), 5879–5884.
28. G. Prelazzi, M. Cerboni and G. Leofanti: J. Catal., 1999, 181, (1), 73–79.
29. C. Amorim and M. A. Keane: J. Hazard. Mater., 2012, 211–212, 208–217.
30. J. T. Miller, B. L. Meyers, F. S. Modica, G. S. Lane, M. Vaarkamp and D. C. Koningsberger: J. Catal., 1993, 143, (2), 395–408.
31. B. Lin, R. Wang, X. Yu, J. Lin, F. Xie and K. Wei: Catal. Lett., 2008, 124, (3–4), 178–184.
32. A. Infantes-Molina, J. Mérida-Robles, P. Braos-García, E. Rodríguez-Castellón, E. Finocchio, G. Busca, P. Maireles-Torres and A. Jiménez-López: J. Catal., 2004, 225, (2), 479–488.
33. A. D. Lueking and R. T. Yang: Appl. Catal. A: Gen., 2004, 265, (2), 259–268.
34. W. C. Conner and J. L. Falconer: Chem. Rev., 1995, 95, (3), 759–788.
35. P. Baeza, M. S. Ureta-Zañartu, N. Escalona, J. Ojeda, F. J. Gil-Llambías and B. Delmon: Appl. Catal. A: Gen., 2004, 274, (1–2), 303–309.
36. M. Turco, G. Bagnasco, C. Cammarano, P. Senese, U. Costantino and M. Sisani: Appl. Catal. B: Environ., 2007, 77, (1–2), 46–57.
37. I. Nam, K. M. Cho, J. G. Seo, S. Hwang, K.-W. Jun and I. K. Song: Catal. Lett., 2009, 130, (1–2), 192–197.
38. L. Rodríguez-González, F. Hermes, M. Bertmer, E. Rodríguez-Castellón, A. Jiménez-López and U. Simon: Appl. Catal. A: Gen., 2007, 328, (2), 174–182.
39. M. Moreno-Bravo, M. Hernández-Luna, J. Alcaraz-Cienfuegos and A. Rosas-Aburto: Appl. Catal. A: Gen., 2003, 249, (1), 35–52.
40. M. Skotak, D. Łomot and Z. Karpiński: Appl. Catal. A: Gen., 2002, 229, (1–2), 103–115.
41. P. J. Chupas and C. P. Grey: J. Catal., 2004, 224, (1), 69–79.
42. C. Morterra and G. Magnacca: Catal. Today, 1996, 27, (3–4), 497–532.
43. S. T. Srinivas and P. K. Rao: J. Catal., 1994, 148, (2), 470–477.
44. G. L. Xu, K. Y. Shi, Y. Gao, H. Y. Xu and Y. D. Wei: J. Mol. Catal. A: Chem., 1999, 147, 47.
45. X. Tan, W. Deng, M. Liu, Q. Zhang and Y. Wang: Chem. Commun., 2009, (46), 7179–7181.
46. W. Deng, X. Tan, W. Fang, Q. Zhang and Y. Wang: Catal. Lett., 2009, 133, (1–2), 167–174.
47. C. Fang, D. Zhang, L. Shi, R. Gao, H. Li, L. Ye and J. Zhang: Catal. Sci. Technol., 2013, 3, (3), 803–811.
48. D. Zhang, L. Zhang, L. Shi, C. Fang, H. Li, R. Gao, L. Huang and J. Zhang: Nanoscale, 2013, 5, (3), 1127–1136.
49. I. Ortiz-Hernandez and C. T. Williams: Langmuir, 2007, 23, (6), 3172–3178.
50. M. Arai, Y. Takada and Y. Nishiyama: J. Phys. Chem. B, 1998, 102, (11), 1968–1973.
51. M. R. Strunk and C. T. Williams: Langmuir, 2003, 19, (22), 9210–9215.
52. L. Hegedűs and T. Máthé: Appl. Catal. A: Gen., 2005, 296, (2), 209–215.
53. N. L. Allinger, M. P. Cava, D. C. de Jongh, C. R. Johnson, N. A. Lebel and C. L. Stevens: Organic chemistry, 572–573; 1974, New York, Worth Publishers.

13

Stabilization of iron by manganese promoters in uniform bimetallic FeMn Fischer–Tropsch model catalysts prepared from colloidal nanoparticles

M. Dad*[1], H. O. A. Fredriksson[1], J. Van de Loosdrecht[2], P. C. Thuene[1] and J. W. Niemantsverdriet[1]

[1]Laboratory for Physical Chemistry of Surfaces, Eindhoven University of Technology, P.O. Box 513, 5600 MB, Eindhoven, The Netherlands
[2]Sasol, Group Technology, P.O. Box 1, Sasolburg 1947, South Africa

Abstract A systematic study was carried out to investigate the response of monodisperse supported Fe and FeMn nanoparticles to treatments in O_2, H_2 and H_2/CO at temperatures between 270 and 400°C. Uniform size (7–14 nm), Fe and mixed FeMn nanoparticles were synthesised by applying thermal decomposition of Fe- and Mn-oleate complexes in a high boiling point solvent. By combining X-ray photoelectron spectroscopy (XPS), transmission electron microscopy (TEM) and energy-dispersive X-ray (EDX) analysis, the phase composition and morphology of the model catalysts were studied. Energy-dispersive X-ray analysis shows that the catalyst particles have the expected composition of Fe and Mn. Well-defined crystallite phases [maghemite (γ-Fe_2O_3) and mixed FeMn-spinel] were observed after calcination at 350°C in Ar/O_2 using XPS analysis. Upon subsequent treatments in H_2 and H_2/CO the crystal phases changed from maghemite (γ-Fe_2O_3) to metallic Fe, Fe carbide and graphitic C. Using Mn as a promoter influences the nanoparticle size achieved during the fabrication of Fe nanoparticles and improves their stability against morphological change and agglomeration during reduction and Fischer–Tropsch synthesis conditions.

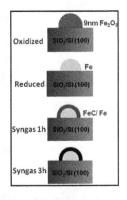

Keywords Fischer–Tropsch, Model catalyst, Iron-manganese oxide, Nanoparticles, X-ray photoelectron spectroscopy, Transmission electron microscopy

Introduction

Fischer–Tropsch synthesis (FTS) is of profound interest because it provides an alternative route for production of a great variety of products, such as short-chain alkenes, diesel fuel, short-chain alkenes and oxygenates from syngas ($CO + H_2$). When the syngas is produced from coal, iron-based materials are used industrially to catalyse this process,[1–6] where structural and chemical promoters, such as silica, potassium, sodium, copper, calcium, zinc, zirconium and chromium,[5,7–9] are added to improve the catalyst selectivity, activity and stability toward sintering. Mn is widely used as the promoter of choice for iron FTS catalysts, particularly when used for producing lower olefins.[10–12] It has also been reported that manganese acts not only as a chemical promoter to alter the chemisorption of the reactants on the catalyst but also as a structural promoter to enhance dispersion of active iron and to stabilise the catalyst during the FTS process.[13–15] Stabilising the catalysts in their active state for a longer time is often an important challenge and understanding how structural promoters enhance catalyst stability is of great importance.

Detailed characterisation of conventional porous heterogeneous catalysts is usually difficult, because the relevant structural, morphological and chemical changes take place on the active metal nanoparticles, situated inside the pores of the support. Therefore, they cannot easily be studied with electron microscopy or surface sensitive characterisation techniques, such as X-ray photoelectron spectroscopy (XPS). The surface science approach, to prepare well-defined flat models of these

*Corresponding author, email emad.dodd@gmail.com

complex, porous, supported catalysts is a good way to overcome these limitations. Although measuring catalytic activity of such systems is not easy, a planar system has the great advantage that it can be characterised by a host of surface sensitive techniques and that the active catalytic material is efficiently exposed to the ambient gas atmosphere. A typical model catalyst approach, using well-defined nanoparticles deposited on the flat silicon oxide membranes,[16] can be used to study the details of Fe catalysts used for various reactions like FTS, CO and methane oxidation reactions.[17–19]

The aim of our work was to synthesise uniform iron manganese oxide model catalysts for clarifying the possible roles of Mn in Fe Fischer–Tropsch catalysts. Colloidal synthesis methods enable one to obtain particles of uniform size and composition.[20] A combination of advanced surface characterisation techniques, such as XPS, transmission electron microscopy (TEM) and energy-dispersive X-ray (EDX) analysis, were applied. We are particularly focus on changes in morphology and surface composition of silica-supported Fe and mixed FeMn nanoparticles that occur during calcination in air, reduction in H_2 and subsequent syngas CO/H_2 treatment. The results show that Mn promotion helps to preserve the structural integrity of the catalysts, which show enhanced stability against agglomeration during hydrogen reduction and against disintegration in synthesis gas.

Experimental

Synthesis of the FeO_x and mixed $FeMnO_y$ model catalysts

The Fe oxides and mixed FeMn oxide nanoparticles investigated in this study were prepared by thermal decomposition of iron oleate and manganese oleate complexes using 1-octadecene as a high boiling point solvent, as described by Park et al.[21]

Synthesis of the Fe-Oleate Precursor

The Fe-oleate complex was prepared by dissolving 10.8 g of iron chloride ($FeCl_3·6H_2O$, 40 mmol, Sigma-Aldrich Chemie B.V., The Netherlands, 98%) and 36.5 g of sodium oleate (120 mmol, TCI EUROPE N.V., Belgium, 95%) in a solvent mixture consisting of 80 mL ethanol, 60 mL distilled water and 140 mL heptane. The solution was heated to 70°C and maintained at this temperature for 4 h. When the reaction was completed, the upper organic layer containing the Fe-oleate complex was washed three times with 30 mL of distilled water in a separatory funnel. After washing, the solvents were evaporated off using a rotary evaporator (1×10^{-2} mbar) at 100–150°C for 2 h resulting in Fe-oleate complex in a waxy solid form.

Synthesis of the Mn-oleate precursor

The Mn-oleate was prepared according to the procedure reported in Ref. 22 with some modifications: 7.94 g of manganese chloride tetrahydrate ($MnCl_2·4H_2O$, 40 mmol, Aldrich, 98%) and 22.60 g of oleic acid ($C_{18}H_{34}O_2$, 80 mmol, Aldrich, 90%) were dissolved in 200 mL of methanol. A solution of 3.2 g (80 mmol) of sodium hydroxide in 200 mL of methanol was added dropwise to the stirred $MnCl_2$/oleic acid solution over a period of 1 h. The initially clear colorless mixture turned pink and a deep red oily substance precipitated. After being stirred for another hour, the solvent was removed and the product was washed with acetone, ethanol and water. After evaporating the solvent, using the rotary evaporator (1×10^{-2} mbar) at 100–150°C for 1 h, a deep red waxy solid Mn-oleate was formed.

Synthesis of Fe oxide and FeMn nanoparticles

To synthesis of Fe oxide nanoparticles, 1.8 g (2 mmol) of the Fe-oleate complex and 0.28 g of oleic acid (1 mmol) were dissolved in 16 g of 1-octadecene (Aldrich, 90%) at room temperature. The reaction mixture was heated to 320°C with a constant heating rate of 3.3°C min^{-1} and then kept at that temperature for 30 min. When the reaction temperature reached 320°C, a severe reaction occurred and the initial transparent solution became turbid and brownish black. The resulting solution, containing the nanocrystals, was then cooled to room temperature and 28 mL of ethanol was added to the solution to precipitate the nanocrystals. The nanoparticles were separated by centrifugation at 8000 rev min^{-1} for 12 min and re-dispersed in toluene (1:10 dilution of nanoparticles in toluene). In order to synthesise mixed FeMn oxide nanoparticles, iron oleate and manganese oleate were mixed in the desired ratios following the same procedure as for making iron oxide nanoparticles.

Uniform colloidal particles on planar support

Solutions of the Fe and mixed FeMn nanoparticles in toluene were sonicated using a horn sonicator (SonicVibracell VC750, Newtown, CT, USA) with a cylindrical tip (6 mm end cap diameter) delivering 3000 J min^{-1}. The solutions were then deposited onto 10×10 mm^2 (for the XPS studies) and 20×20 mm^2 (for the TEM studies) flat silica supports, by applying a technique called 'interfacial self-assembly' as described by Brinker and co-workers.[23]

As schematically shown in Figure 1a, a mixture of 40 μL of toluene with 3 mmol nanoparticles and 15 mg polystyrene mL^{-1} was added to a distilled water bath. As soon as the mixture hit the water, an oily thin film was formed on the water surface (Fig. 1b). After 1 or 2 min, the toluene evaporated and a stable polymer/nanoparticle film was left floating on the water (Fig. 1c). After draining the water via a valve at the bottom of the water bath (Fig. 1d), the film was deposited onto the surface of the sample, located at the bottom of the water bath.

SiO_2/SiN_x TEM window

For TEM studies, the nanoparticles were deposited onto custom-made TEM substrates.[24] As shown in Figure 2a, these substrates consist of a silicon wafer with a thin silicon nitride (SiN_x) layer on top. Part of the silicon wafer underneath the SiN_x layer is etched away to create a 100×100 μm^2 wide, 15–20 nm thick 'membrane window' through which an electron beam can pass (Fig. 2b). A 3-nm thick surface layer of silicon oxide was formed on top of the SiN_x after calcination at 750°C for 24 h in dry air.[24] These TEM membranes were used as a support for the synthesised Fe and mixed FeMn particles because they can withstand both high reaction temperatures and gas flows and since the terminating, inert SiO_2-layer mimics a frequently used catalysts support.

Characterization

X-ray photoelectron spectroscopy was measured with a Kratos AXIS Ultra spectrometer, equipped with a

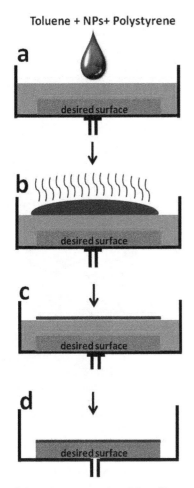

Toluene + NPs+ Polystyrene

a

desired surface

b

desired surface

c

desired surface

d

desired surface

Figure 1 Schematic representation of deposition method *a* Adding the nanoparticle/polymer solution onto the distilled water solution; *b* toluene evaporation; *c* formation of the stable nanoparticles containing polymer film and *d* deposition of the polymer film on the flat support. The procedure represents a modification of a method introduced by Brinker and co-workers[23]

monochromated Al Kα source (1486.6eV) and a delay-line detector (DLD). The spectra were obtained with the X-ray source operating at 10mA, 15kV and 40eV pass energy, 0.1eV step size was used. The background pressure in the analysis chamber during measurements was typically 1×10^{-8} mbar. Casa XPS software, Version 2.3.16 Pre-rel 1.4 was used. Binding energies were calibrated with the standard Si 2s=154.6eV peak in SiO_2. For quantitative analysis, the Si 2s, C 1s, Fe 2p, Mn 2p, and O 1s and the Wagner relative sensitivity factors peaks were used. The TEM studies were carried out using a FEI Tecnai G^2 Sphera microscope operating with a 200kV LaB_6 filament and a bottom mounted 1024×1024 Gatan CCD camera. All images were measured using the TEM in bright field mode. Electron diffraction patterns of species were obtained to determine the related crystalline phases present in the samples. The S/TEM characterisation was carried out using a FEI Talos F200X equipped with a high brightness Schottky FEG and the SuperX EDS system which includes four SDD detectors EDS symmetrically placed around the sample and a 16MP CMOS camera, the FEI Ceta 16M. The mapping experiments were performed with different beam currents, optimised to minimise beam damage and with a dwell time of 10ms to further protect the specimen.

Sample treatments

X-ray photoelectron spectroscopy

Sample treatments were performed in a high-temperature gas reaction cell (Kratos Analytical Ltd, Manchester, U.K.), consisting of a dome shaped quartz chamber, allowing for *in vacuo* transfer into the measurements chamber. The samples were supported on a fused silica stub and the chamber pressure was kept at 1.5bar during all treatments. For oxidation treatments, an Ar/O_2 mixture (10% O_2 in Ar, ~50mL$_n$min^{-1}) was used. The samples were heated up to 350°C at a rate of 5°Cmin^{-1} and held at this temperature for 30min. This treatment was performed before each reduction treatment, to ensure the catalyst was always in a reproducible, fully oxidized state. After the oxidation treatment, the reaction cell was cooled down to below 70°C. For the

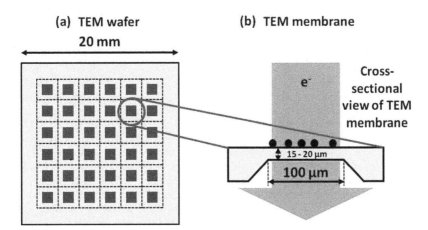

(a) TEM wafer

20 mm

(b) TEM membrane

e⁻

Cross-sectional view of TEM membrane

15 - 20 μm

100 μm

Figure 2 *a* Schematic representation of custom-made TEM wafer, consisting of 36 TEM membranes with windows; *b* crosssectional view of the individual TEM membrane with the electron transparent membrane suspended on the silicon frame. Figure adapted from Thune *et al.*[24]

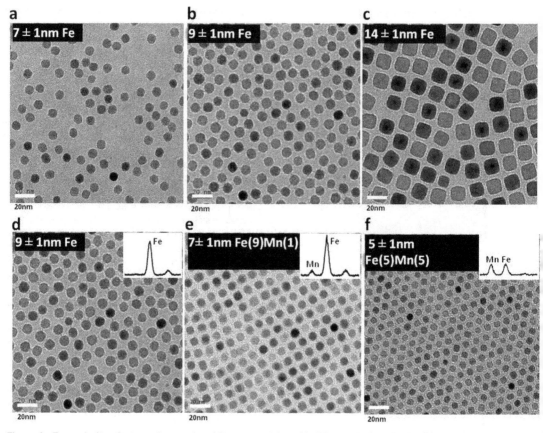

Figure 3 Transmission electron micrographs of Fe nanoparticles with different diameters *a* 7 nm Fe; *b* 9 nm Fe and *c* 14 nm Fe; *d–f* show Fe nanoparticles with various amount of Mn incorporated; *d* pure Fe; *e* Fe$_9$Mn$_1$; and *f* Fe$_5$Mn$_5$ nanoparticles dispersed on a SiO$_2$/SiN$_x$ membrane after calcination. Energy-dispersive X-ray spectra are added as insets to show the relative Fe/Mn contents

subsequent reduction treatments, pure H$_2$, 6.0 was used, without further filtration or purification. Before these treatments, the reaction cell and the gas lines were evacuated to 5×10^{-6} mbar, then back-filled with H$_2$ and once again evacuated to 5×10^{-6} mbar to get rid of contaminants (in particular oxygen). The samples were then heated under H$_2$-flow (80 mL min^{-1}) to 400°C at a rate of 5°C min^{-1}. The samples were held at the targeted temperature for 30 min, 1 and 1.5 h and then the gas flow was stopped, the reaction cell immediately evacuated and the samples rapidly cooled to room temperature. After treatment in reactive gases the samples were transferred, *in vacuo*, to the measurement chamber of the XPS. The pure H$_2$, 6.0 as well as CO, 4.7 were used for the subsequent syngas treatments. After reduction at 400°C, the reaction cell cooled down in H$_2$-flow, to 200°C to avoid the catalyst re-oxidisation. The gas flow was then switched to syngas (CO/H$_2$, 10/10 mL min^{-1}) and the fully reduced sample was heated to 270°C at a rate of 5°C min^{-1} and kept there for 1 and 3 h. Then the samples were transferred to the measurement chamber under vacuum conditions.

Transmission electron microscopy

For the TEM studies, similar treatments were applied in a quartz tube flow-reactor, i.e. oxidation at 350°C for 30 min, followed by reduction in H$_2$ at 400°C for 1 h and subsequent

syngas treatment at 270°C for 1 h. The quartz tube was then cooled down to 90°C and transferred to glove box and opened there in order to gently re-oxidise the catalyst.

Results and discussion

Catalyst preparation and properties

Figure 3 shows a set of supported Fe and FeMn-oxide nanoparticles prepared by thermal decomposition of metal oleate. The TEM images confirm the stability of these nanoparticles against sintering upon calcination at 350°C. Figure 3a–c shows pure Fe particles with diameters between 7 and 14 nm.

The various diameters were achieved by varying the molar ratios of oleic acid to Fe-oleate precursor (0.1, 0.5 and 1.5). After depositing the particles onto SiO$_2$/Si (100) or SiO$_2$ terminated Si$_3$N$_4$ TEM-window supports, a calcination in dry Ar/O$_2$ at 350°C for 1 h was applied, in order to remove residues of the polymer used for deposition as well as the oleic acid, serving to stabilize the particles while in solution. Figure 3a shows that the 7-nm Fe nanoparticles are sparsely spread over the support, whereas the 9- and 14-nm particles (Fig. 3b and c) are well distributed over the whole support. Figure 3d–f shows monodispersed Fe and mixed FeMn-oxide nanoparticles. For all of them, uniform size and well-defined crystallite phases were observed, as well as particle distribution over the surface.

Moreover, precise control over Fe and Mn ratios was obtained after using different relative amounts of Fe- and Mn-oleate, as indicated by the inset EDX spectra. Interestingly, we observed that the addition of Mn also influences the size of particles, i.e. mixed FeMn particles have smaller diameters than Fe particles grown under identical conditions. Figure 3d shows well-distributed Fe nanoparticles with average diameter of 9 nm, whereas Figure 3e and f presents the Fe nanoparticles promoted with 10 and 50 mol.-% Mn, respectively. The addition of Mn modifies the dispersion of nanoparticles and decreases its crystallite size.[25] The diameter of 9-nm Fe particles decreases to 7 nm in Figure 3e for Fe_9Mn_1 and 5 nm in Figure 3f for the Fe_5Mn_5 sample. As can be seen in the insets of Figure 3d–f, the EDX spectroscopy graphs demonstrate the expected composition of Fe and Mn (i.e. 89:11 for the Fe_9Mn_1 and 49.5:50.5 for the Fe_5Mn_5 sample).

X-ray photoelectron spectroscopy measurements were performed in order to specify the oxidation state of the outer few nanometres of the particles after calcination. For all six samples, the Fe $2p_{3/2}$ peak is located at 710.7 eV[26] and exhibits a satellite peak at 719 eV, characteristic for Fe^{3+} species. The Mn 2p spectra of the Mn-promoted samples exhibit two peaks at 641.6 and 653.4 eV, which can be attributed to the Mn $2p_{3/2}$ and Mn $2p_{1/2}$, respectively, with binding energies corresponding to MnO_2.[27]

Catalyst response to gas treatment

Promotion of Fe FTS-catalysts with an appropriate amount of Mn (typically 8–12 mol.%) is known to increase the catalyst performance in terms of activity, selectivity to lower olefins (C_2–C_4) and stability. Therefore, we selected the Fe_9Mn_1 in comparison with the Fe sample for further investigations. Figure 4 shows TEM images of Fe- and mixed FeMn-oxide nanoparticles after calcination in Ar/O_2 at 350°C (Fig. 4a and d), after reduction in dry hydrogen at 400°C (Fig. 4b and e) and after syngas treatment at 270°C (Fig. 4c and f). The insets show the corresponding electron diffractograms.

As pointed out in the previous section, calcined Fe and Fe_9Mn_1 nanoparticles (Fig. 4a and d) are homogeneously distributed over the silica support with no sign of particle agglomeration. Furthermore, comparison of the electron diffraction pattern of the calcined Fe particles (Fig. 4a) with the reference patterns[26,28] indicates that Fe particles resemble maghemite (γ-Fe_2O_3) more closely than haematite (α-Fe_2O_3), whereas the spinel phases $MnFe_2O_4$ and γ-Fe_2O_3 were observed for the Fe_9Mn_1 particles (Fig. 4d). For the Fe particles reduced in dry H_2 at 400°C for 1 h, we found a significant increase in the average particle diameter as well as considerable morphology changes. Moreover, no obvious diffraction pattern rings were observed, indicating an amorphous species. This is compatible with spontaneous re-oxidation at low temperature as described in the literature, where reduced Fe-nanoparticles that have been exposed to air at room temperature are described as core–shell structures[29] with a metallic core encapsulated in an oxide shell as a result of surface oxidation. Wang et al. have also reported that the oxide layer is a mixture of magnetite and maghemite.[30] The corresponding change for mixed FeMn particles after

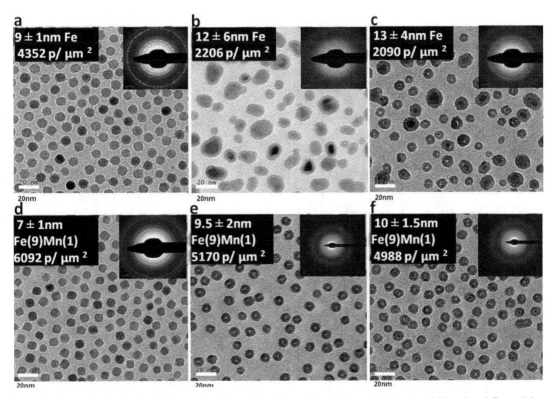

Figure 4 Transmission electron micrographs of *a* calcined 9-nm Fe particles; *b* calcined and H_2 reduced Fe particles; *c* calcined, H_2 reduced and syngas treated Fe particles; *d* calcined 7-nm Fe_9Mn_1 particles; *e* calcined and H_2 reduced Fe_9Mn_1 particles and *f* calcined, H_2 reduced and syngas treated Fe particles dispersed on a SiO_2/SiN_x membrane

H_2 reduction (Fig. 4e to d) is that the average particle size increases only slightly. However, strong diffraction ring is seen after reduction (Fig. 4e) with a d spacing of 2.0 Å, characteristic for the high intensity 110 plane of metallic Fe. Comparing the reduced Fe and FeMn nanoparticles (Fig. 4b and e), it appears that adding 10 mol.-% Mn both helps preserving nanoparticles size dispersion by preventing sintering after reduction and helps retarding re-oxidation of the Fe upon contact with air during transfer to the microscope. Figure 4c shows the reduced Fe sample after a subsequent 1 h syngas treatment at 270°C. It seems that particles experienced a slight further increase in diameter, probably because of the formation of FeC, which can also be confirmed by the single distinct ring in the TEM electron diffraction pattern. This ring corresponds to the d spacing of 2.0 Å, which can be attributed to the high intensity 510 plane of Hägg carbide (χ-Fe$_5$C$_2$),[31] the high intensity 031 plane of cohenite (Fe$_3$C) or the high intensity 110 plane of metallic Fe. The core–shell structure observed for both the Fe and the Fe$_9$Mn$_1$ particles in Figure 4c and f might consist of either a metallic Fe or Fe carbide core surrounded by FeMn oxides. In contrast with syngas treated Fe particles in Figure 4c, Fe particles promoted with Mn shows better size dispersion. In conclusion, Mn as a promoter exhibits a strong interaction with Fe, which modifies the size of nanoparticles during fabrication[32] and retards the agglomeration of Fe oxide during reduction and syngas treatment.

Figure 5a and b shows the XPS in the Fe 2p and C 1s regions for the Fe catalyst after different gas treatments. As pointed out above, the calcined sample exhibits a Fe 2p spectrum, which is consistent with the Fe^{3+} oxidation state. Combined with the electron diffraction pattern shown in Figure 4a, we can thus conclude that (γ-Fe$_2$O$_3$) is the dominant phase for Fe after calcination. The corresponding carbon spectra (Fig. 5b) show a small peak at 284.6 eV because of a small amount of remaining carbon from the sample preparation process. However, applying 1 h reduction in H_2 removes almost all impurities from the surface. After 30 min reduction in H_2, the first sign of a peak at 706.5 eV, corresponding to metallic Fe was observed indicating partial reduction of the nanoparticles. Treatment for 1 h in H_2 results in the disappearance of the Fe^{3+} related peak, validating the complete reduction of Fe$_2$O$_3$ to metallic Fe. The peak at 706.4 eV remains after introducing syngas (H$_2$/CO = 1) for 1 and 3 h. However, the appearance of a broad C-peak at 282–283 eV indicates the formation of iron carbide.[33,34]

Figure 6a–c shows Fe 2p, Mn 2p and C 1s XPS of the mixed FeMn (9:1) sample after similar treatments. As can be seen in Figure 6a, the Fe 2p peaks of the mixed FeMn particles present largely the same trend as was observed for the Fe sample in Figure 5a. From the narrow peak at 706.6 eV, appearing after reduction in H_2, it can be concluded that the particles first transform from maghemite (Fe^{3+}) to a mixed state of metallic

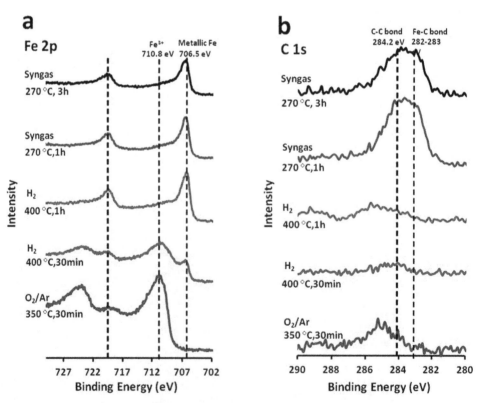

Figure 5 *a* Fe 2p spectra and *b* C 1s spectra of 9-nm Fe/SiO$_2$ model nanoparticles at different stages of catalyst treatment. Ar/O$_2$ denotes catalyst after calcination at 350°C for 30 min; H_2 denotes subsequent reduction at 400°C for 30 min and 1 h; and syngas denotes spectra taken after a final treatment in (H$_2$/CO = 1) at 350°C for 1 and 3 h. The dashed vertical lines indicate peak positions reported in the literature for metallic Fe ($\sim$707 eV), Fe^{3+} in γ-Fe$_2$O$_3$ (710.8 eV), the Fe^{3+}-satellite (719 eV), Graphitic C (284 eV) and carbidic C (282.8 eV)

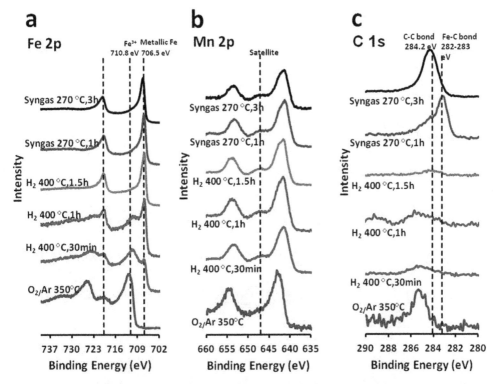

Figure 6 *a* Fe 2p spectra; *b* Mn 2p spectra; *c* C 1s spectra of 7-nm Fe$_9$Mn$_1$/SiO$_2$ model nanoparticles after different stages of catalyst treatments. O$_2$/Ar denotes catalyst after calcination at 350°C for 30 min; H$_2$ reduction for 30 min, 1 and 1.5 h; subsequent syngas treatment (H$_2$/CO = 1) at 350°C for 1 and 3 h. The dash vertical lines show metallic Fe ($\sim$707 eV), Fe^{3+} in γ-Fe$_2$O$_3$ (710.8 eV), the Fe^{3+}-satellite (719 eV), carbidic C (282–283 eV) and graphitic C ($\sim$284.6 eV)

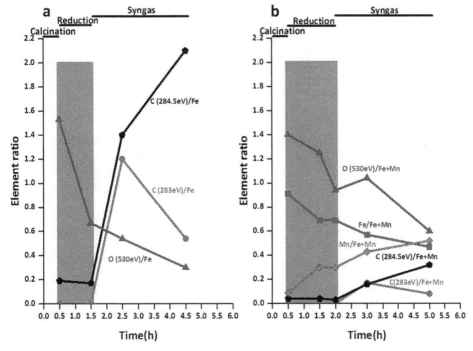

Figure 7 Effect of different treatments on element ratio at the surface of the sample with *a* 9-nm Fe particles and *b* 7-nm Fe$_9$Mn$_1$ particle

Fe, magnetite and FeO and then to metallic Fe. However, in order to fully reduce Fe in the mixed FeMn particles, exposure to H_2 gas flow for 1.5 h was required. This suggests that the presence of Mn as a promoter suppresses the reduction of Fe oxide to metallic Fe. This result can be explained by the fact that formed MnO (during reduction) segregates on the surface of FeO. As a result, FeO_x is surrounded by MnO,[35] thereby slowing down the reduction from (Fe,Mn)O to Fe and MnO.[36] Based on the absence of a satellite peak in between the two main Mn 2p peaks after calcination (Fig. 6b), the dominant phase for the calcined FeMn particles is MnO_2. After treatments in H_2 and syngas, a satellite peak appears at 647.5 eV and the Mn $2p_{3/2}$ shifted to 641.5 eV, which is significant for MnO. No further changes in the Mn oxidation state are observed after the initial reduction treatment. Figure 6c shows only trace amounts of surface C after calcination and reduction in H_2. The subsequent 1 h syngas treatment at 270°C results in

the formation of Fe carbide, as indicated by the peak at 283 eV. Prolonged treatment in syngas for 3 h results in a shift towards higher binding energies (284.5 eV). This can be interpreted as formation of graphite-like surface species.

In addition to the qualitative information provided by the peak shapes and positions, quantification of the elements associated with the nanoparticles gives important information about changes in their surface constitution. Figure 7 shows the quantification of these elements, including the metals (Fe and Mn), oxygen and carbon, after the various gas treatments. The oxygen peak is split into two peaks, one at 533 eV corresponding to the SiO_2 substrate and the other at 530 eV associated with the metal oxide in the nanoparticles (only the latter is considered here). Similarly, the C 1s peak is deconvoluted into two separate peaks, one corresponding to graphitic carbon (284.5 eV) and the other corresponding to carbide (283 eV). The quantities are expressed as the ratio

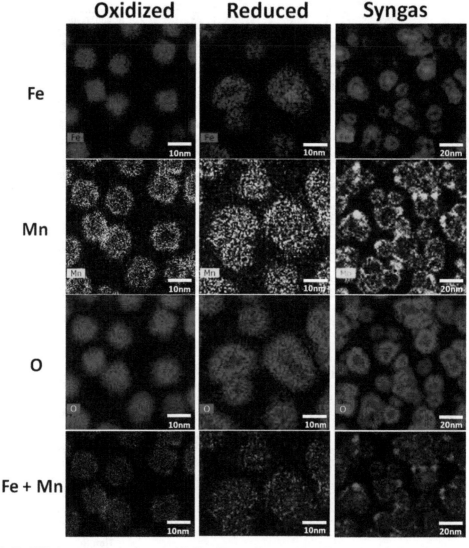

Figure 8 The TEM elemental mapping images of Fe_9Mn_1: first row Fe; second row Mn; third row O; Fe + Mn forth row, after calcination at 350°C for 30 min (first column); H_2 reduction for 1 h (second column) and subsequent Syngas treatment ($H_2/CO = 1$) at 350°C for 1 h (third column)

between each element and the total amount of metal. As can be seen in both graphs (Fig. 7a and b), the amount of oxygen shows a decreasing trend during conversion of Fe oxide to metallic Fe as well as MnO_2 to MnO (after H_2-treatment), and after H_2/CO mixture-treatment. In both cases, a dramatic increase in the C (283 eV) signal intensity is the clear evidence of the formation of Fe carbide after treating the reduced Fe particles by H_2/CO gas mixture for 1 h, as well as graphitic C after 3 h exposure. This indicates that the particles are covered by a graphitic surface layer. In the case of Fe_9Mn_1 sample (Fig. 7b), the increase of Mn/Fe+Mn ratio during reduction and syngas treatment indicates that Mn migrates to the catalyst surface, while a slight decrease in Fe/Fe + Mn ratio during conversion of Fe oxide to metallic Fe (after H_2-treatment) and during formation of FeC and C-deposition (after H_2/CO mixture-treatment) was observed. Figure 8 shows TEM elemental mapping images for the Fe_9Mn_1 catalyst surface after three different gas treatments, intending to demonstrate the spatial distribution of the elements in the individual particles (note that the images are not taken from the same area and in the same magnification). The figures in the left column show the distribution of Fe, Mn and O after calcination at 350°C for 1 h. For clarity, the Fe and Mn distribution is also shown as superimposed figures in the bottom row. Fe and O are homogeneously mixed in the particles, whereas Mn is predominantly situated at the catalyst's surface. Subsequent reduction in H_2 at 400°C for 1 h (Fig. 8, middle column) results in a homogeneous mixture of Fe and Mn, whereas the particle size appeared a bit bigger. However, the superimposed image (lowest row) reveals that Mn is at the surface of the particle in this case too. In addition, an absence of O at the center of the particles can be seen validating the creation of non-oxidic particle cores. After syngas treatment for 1 h (Fig. 8, right column), Mn is still predominantly present at the surface of the particles, whereas no O is observed in the interior. Furthermore, Mn appears to have segregated out of the particles, forming domains of pure MnO on the outside of the iron-rich particles, probably at the interface with the support. We tentatively propose that such an arrangement would help to stabilise the iron particles against agglomeration or disintegration during carburisation.

Although the TEM elemental mapping is largely in agreement with XPS results, it should be kept in mind that the former was acquired after prolonged exposure to atmosphere, whereas in the latter case, the samples were transferred in vacuo and the measurements done immediately after the treatments. The XPS results therefore give a more accurate picture of the state of the catalysts at the different gas temperature conditions.

Conclusion

The effect of gas treatment on the supported Fe and mixed FeMn nanoparticles on flat silica surfaces was studied to investigate the effect of manganese on morphology and composition of iron under conditions relevant for FTS. To reduce the complexity associated with commercial catalysts and to facilitate the use of advanced surface characterisation techniques, such as XPS, TEM and EDX analysis, flat model catalysts consisting of colloidal particles deposited on planar SiO_2/Si(001) or SiO_2/Si_4N_3 supports were used. The main conclusion of the

work is that Mn, present in the form of MnO, serves as a structural promoter, which by exhibiting a strong interaction with iron stabilises the composite particles against agglomeration during reduction and exposure to synthesis gas.

Acknowledgements

The authors gratefully acknowledge Dr Mauro Porcu from FEI Company in Eindhoven for providing TEM elemental mapping images.

References

1. E. de Smit and B. M. Weckhuysen: Chem. Soc. Rev., 2008, **37**, 2758–2781.
2. A. Demirbas: Prog. Energy Combust. Sci., 2007, **33**, 1–18.
3. H. Schulz, E. van Steen and M. Claeys: Stud. Surf. Sci. Catal., 1994, **81**, 455–460.
4. Y. Yang, H. Xiang, R. Zhang, B. Zhong and Y. Li: Catal. Today, 2005, **106**, 170–175.
5. T. Herranz, S. Rojas, F. J. P. érez-Alonso, M. Ojeda, P. Terreros and J. L. G. Fierro: Appl. Catal. A Gen., 2006, **311**, 66–75.
6. H. Hayakawa, H. Tanaka and K. Fujimoto: Appl. Catal. A Gen., 2006, **310**, 24–30.
7. S. Li, S. Krishnamoorthy, A. Li, G. D. Meitzner and E. Iglesia: J. Catal., 2002, **206**, 202–217.
8. N. Lohitharn, J. G. Goodwin Jr. and E. Lotero: J. Catal., 2008, **255**, 104–113.
9. Y. Jin and A. K. Datye: J. Catal., 2000, **196**, 8–17.
10. G. C. Maiti, R. Malessa and M. Baerns: Appl. Catal., 1983, **5**, 151–170.
11. C. K. Dasa, N. S. Dasa, D. P. Choudhury, G. Ravichandranb and D. K. Chakrabartyb: Appl. Catal. A Gen., 1994, **111**, 119–132.
12. C. Wang, Q. Wang, X. Sun and L. Xu: Catal. Lett., 2005, **105**, 93–101.
13. K. B. Jensen and F. E. Massoth: J. Catal., 1985, **92**, 98–108.
14. J. Barrault and C. Renard: Appl. Catal., 1985, **14**, 133–143.
15. J. J. Venter, A. Chen and M. A. Vannice: J. Catal., 1989, **117**, 170–187.
16. P. Moodley, F. J. E. Scheijen, J. W. Niemantsverdriet and P. C. Thüne: Catal. Today, 2010, **154**, 142–148.
17. G. Hutchings, M. Hall, A. Carley, P. Landon, B. Solsona, C. Kiely, A. Herzing, M. Makkee, J. A. Moulijn, A. Overweg, J. C. Fierro-Gonzalez, J. Guzman and B. C. Gates: J. Catal., 2006, **242**, 71–81.
18. S. T. Daniells, A. R. Overweg, M. Makkee and J. A. Moulijn: J. Catal., 2005, **230**, 52–65.
19. R. J. H. Grisel, C. J. Weststrate, A. Goossens, M. W. J. Crajé, A. M. van der Kraan and B. E. Nieuwenhuys: Catal. Today, 2002, **72**, 123–132.
20. J. Park, J. Joo, S. G. Kwon, Y. Jang and T. Hyeon: Angew. Chem. Int. Edit., 2007, **46**, 4630–4660.
21. J. Park, K. An, Y. Hwang, J. G. Park, H. J. Noh, J. Y. Kim, J. H. Park, H. M. Hwan and T. Hyeon: Nat. Mater., 2004, **3**, 891–894.
22. T. D. Schladt, T. Graf and W. Tremel: Chem. Mater., 2009, **21**, 3183–3190.
23. S. Xiong, X. Miao, J. Spencer, C. Khripin, T. S. Luk and C. J. Brinker: Small, 2010, **6**, 2126–2129.
24. P. C. Thüne, C. J. Weststrate, P. Moodley, A. M. Saib, J. van de Loosdrecht, J. T. Millerc and J. W. Niemantsverdriet: Catal. Sci. Technol., 2011, **1**, 689–697.
25. D. Das, G. Ravichandran and D. K. Chakarbarty: Catal. Today, 1997, **36**, 285–293.
26. T. Yamashita and P. Hayes: Appl. Surf. Sci., 2008, **254**, 2441–2449.
27. M. C. Biesinger, B. P. Payne, A. P. Grosvenor, L. W. M. Lau, A. R. Gerson, R. St and C. Smart: Appl. Surf. Sci., 2011, **257**, 2717–2730.
28. X. Gou, G. Wang, J. Park, H. Liu and J. Yang: Nanotechnology, 2008, **19**, 125606.
29. T. Hyeon, S. S. Lee, J. Park, Y. Chung and H. B. Na: J. Am. Chem. Soc., 2001, **123**, 12798–12801.
30. C. M. Wang, D. R. Baer, J. E. Amonette, M. H. EngelhardL and Y. Qiang: J. Am. Chem. Soc., 2009, **131**, 8824–8832.
31. V. D. Blank, B. A. Kulnitskiy, I. A. Perezhogin, Y. L. Alshevskiy and N. V. Kazennov: Sci. Technol. Adv. Mater., 2009, **10**, 1–5.
32. Z. Tao, Y. Yang, C. Zhang, T. Li, M. Ding, H. Xiang and Y. Li: J. Nat. Gas Chem., 2007, **16**, 278–285.
33. M. Ding, Y. Yang, B. Wu, T. Wang, H. Xiang and Y. Li: Fuel Process. Technol., 2011, **92**, 2353–2359.
34. S. Airaksinen, M. Banares and A. O. I. Krause: J. Catal., 2005, **230**, 507–513.
35. T. Grzybek, H. Papp and M. Baerns: Appl. Catal., 1987, **29**, 330–335.
36. T. Li, Y. Yang, C. Zhang, X. An, H. Wan, Z. Tao, H. Xiang, Y. Li, F. Yi and B. Xu: Fuel, 2007, **86**, 921–928.

14

Single step combustion synthesized Cu/Ce$_{0.8}$Zr$_{0.2}$O$_2$ for methanol steam reforming: structural insights from *in situ* XPS and HRTEM studies

Dipak Das[1], Jordi Llorca[2], Montserrat Dominguez[2], and Arup Gayen*[1]

[1]Department of Chemistry, Jadavpur University, Kolkata 700032, India
[2]Institut de Tècniques Energètiques and Centre for Research in Nanoengineering, Universitat Politècnica de Catalunya, 08028 Barcelona, Spain

Abstract Single step combustion synthesized Cu (5–15 at.-%)/Ce$_{0.8}$Zr$_{0.2}$O$_2$ materials containing highly dispersed copper have been assessed for methanol steam reforming (MSR). The activity patterns suggest Cu (10 at.-%)/Ce$_{0.80}$Zr$_{0.20}$O$_2$ as the most active formulation, converting ~51% methanol at 300 °C at a gas hourly space velocity of 40,000 h^{-1} (W/F = 0.09 s). The *in situ* XPS experiments carried over the most active sample show a sharp falloff of Cu-surface concentration from a considerably high value of 26% before to 7.4% after the *in situ* MSR tests and it is associated with the complete reduction of oxidized Cu-species (Cu^{2+}) to metallic copper (Cu^0). These findings point to the sintering of copper during MSR which is attributed to be responsible for the deactivation observed with time on stream. Interestingly, the MSR activity is shown to be regenerated nearly completely through an intermediate *in situ* oxidation step in the consecutive cycles of methanol reforming.

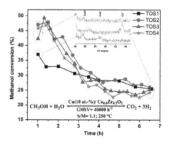

Keywords Copper, Ceria–zirconia, Solution combustion, Methanol steam reforming, Cu-sintering, Regeneration

Introduction

Hydrogen energy is proven to be a promising alternative to fossil fuels. The most potential environment-friendly technology for the production of clean electrical power can be developed by the help of hydrogen-powered fuel cells.[1] The on-board reforming of hydrocarbons has gained renewed interest of researchers as it allows hydrogen production *in situ*, which can solve the problems of storage.[2,3] The production of fuel cell grade hydrogen remains a challenging issue. The low temperature methanol steam reforming (MSR) is a simple and convenient route to produce nearly CO-free hydrogen.[4]

The literature reports several Cu-based[5–9] and Pd-based[4,10] catalysts for MSR. The Cu-based catalysts are comparatively more active and selective at lower temperature. The commercially used MSR catalyst is Cu/ZnO/Al$_2$O$_3$ which contains a high loading of copper and ZnO and a relatively low amount of Al$_2$O$_3$.[11] The usefulness of ZnO and Al$_2$O$_3$ is due to several factors that are widely discussed in the literature.[12,13] But finding a suitable alternative to Al$_2$O$_3$ to increase the reducibility of copper is highly needed. Addition of ZrO$_2$ to Cu-based

Al$_2$O$_3$-supported catalysts is shown to enhance the MSR activity.[14,15] It also increases the reducibility of CuO. On the other hand, CeO$_2$ improves the thermal stability of Al$_2$O$_3$-supported Cu-catalysts and it also reduces CO selectivity through the water gas shift (WGS) reaction.[16,17] The formation of CO from CO$_2$ occurs mainly via the reverse WGS reaction.[5,14,18,19] Whereas the Cu(111) model system is more selective toward CO$_2$ and H$_2$, the Pd(111) model system is more selective toward CO and H$_2$.[20] The nature of active site, particularly the oxidation state of copper is as debated as the mechanistic aspects. Nevertheless, there is a general consensus that the MSR activity depends very much on Cu$^+$ to Cu^0 molar ratio.[7,21,22]

There are limited literature reports on Cu-loaded ceria–zirconia for MSR activity. Mastalir *et al.* have shown long-term stability in MSR over a 15 mol % Cu-loaded CeO$_2$–ZrO$_2$ (1:1 mol ratio).[6] Oguchi *et al.* have reported a CuO/ZrO$_2$/CeO$_2$ catalyst (weight ratio of 8/1/1) which exhibits enhanced catalytic performance in MSR through stabilization of Cu$_2$O on the catalyst surface.[7] Pojanavaraphan *et al.* have established the beneficial effect of Au addition to Cu/Ce$_{0.75}$Zr$_{0.25}$O$_2$ for this reaction.[23,24] In a very recent report, we have studied the MSR behavior of copper impregnated over various Ce–Zr oxides.[25]

*Corresponding author, email: agayen@chemistry.jdvu.ac.in

In this work, we report the synthesis of Cu (5–15 at.-%)/ $Ce_{1-x}Zr_xO_2$ with varied Ce to Zr molar ratio as catalyst materials for MSR via a single step solution combustion method. Two formulations, namely Cu (10 at.-%)/$Ce_{0.8}Zr_{0.2}O_2$ and Cu (15 at.-%)/ $Ce_{0.8}Zr_{0.2}O_2$, have been thoroughly investigated. The MSR studies have also been carried out in situ and monitored by XPS.

Experimental

Preparation of materials

We have synthesized Cu-loaded $Ce_xZr_{1-x}O_2$ ($x = 0.6, 0.7$, and 0.8) solid solutions by a single step solution combustion method. In this method, the metal nitrate salts and an organic fuel taken in stoichiometric amounts are first dissolved in the minimum volume of double distilled water in a borosilicate dish through warming. The resulting solution is then transferred to a preheated muffle furnace maintained at the ignition temperature for combustion. Initially, the solution boils with frothing and foaming followed by complete dehydration when the surface gets ignited and burns to yield a solid product within few minutes.

Typically, the preparation of Cu (10 at.%)/$Ce_{0.8}Zr_{0.2}O_2$ involves combustion of 0.3060 g of $Cu(NO_3)_2 \cdot 3H_2O$ (Merck, GR, 99%), 5.0 g of $(NH_4)_2Ce(NO_3)_6$ (Merck, GR, 99%), and 0.5272 g of $ZrO(NO_3)_2 \cdot H_2O$ (Loba Chemie, ZrO_2 assay >44.5%) with 3.007 g of oxalyl dihydrazide ($C_2H_6N_4O_2$, ODH) corresponding to the molar ratio of 0.10:0.72:0.18:2.01 dissolved in ~30 mL of double distilled water at ~350 °C. The ceria–zirconia oxides (with or without copper loading) exhibit flaming-type combustion and the combustion was completed within a minute. The color of the as-synthesized sample is gray.

Characterization of materials

The powder X-ray diffraction (XRD) data were collected on a Bruker D8 Advance X-ray diffractometer using Cu Kα radiation ($\lambda = 1.5418$ Å) generated at 40 kV and 40 mA. Data were collected in the 2θ range of 10–100° using a Lynxeye detector with 0.02° step size and scan time of 0.4 s per step and analyzed by ICDD (International Centre for Diffraction Data) database for phase identification. Average particle sizes were calculated from the line-broadening of the XRD peaks using the Scherrer equation.

The Brunauer-Emmett-Teller (BET) specific surface areas (SAs) were determined using nitrogen in an Autosorb 1 C SA analyzer (Quantachrome, USA). Prior to analysis, the samples were degassed at 250 °C in vacuum for 2 h.

High-resolution transmission electron microscopy (HRTEM) was performed at an accelerating voltage of 200 kV in a JEOL 2010F instrument equipped with a field emission source. The point-to-point resolution was 0.19 nm, and the resolution between lines was 0.14 nm. The magnification was calibrated against an Si standard. No induced damage of the samples was observed under prolonged electron beam exposure. Samples were dispersed in alcohol in an ultrasonic bath, and a drop of supernatant suspension was poured onto a holey carbon-coated grid. Images were not filtered or treated by means of digital processing, and they correspond to raw data.

X-ray photoelectron spectroscopy (XPS) surface characterization was done on a SPECS system equipped with an Al anode XR50 source operating at 150 mW and a Phoibos 150

MCD-9 detector. The analysis chamber was maintained at a pressure always below 10^{-7} Pa. The area analyzed was about 2 mm × 2 mm setting the pass energy of the hemispherical analyzer at 25 eV and the energy step at 0.1 eV. Charge stabilization was achieved by using a SPECS Flood Gun FG 15/40. The sample powders were pressed to self-consistent disks. The spectra were recorded in the following sequence: survey spectrum, C 1s, Cu 2p, Ce 3d, Zr 3d, Cu LMM Auger, and C 1s again to check for charge stability as a function of time and the absence of degradation of the sample during the analyses. The data processing was performed with the CasaXPS program (Casa Software Ltd., UK). The binding energy values have been centered using the u''' peak of Ce 3d at 916.9 eV (because C 1s signal is partly masked by Ce 4s signal). The atomic fractions (%) were calculated using peak areas normalized on the basis of acquisition parameters after background subtraction, experimental sensitivity factors, and transmission factors provided by the manufacturer.

The in situ MSR experiments were performed in a reaction chamber connected to the XPS analysis chamber that allowed treatments up to 600 °C at atmospheric pressure and sample transfer without exposure to air. The temperature of the sample was measured with a thermocouple in contact with the sample holder, which was heated with an IR lamp. The evolution of products during the in situ experiments (0–100 amu) was followed by a mass spectrometer. Gases were introduced by means of mass flow controllers and liquids were introduced through bubbling the appropriate amount of carrier gas (Ar) to reach the required steam/methanol (S/M) ratio of 1.1 used in our experiments.

Test of reforming activity

The MSR was performed in a continuous-flow fixed-bed downflow quartz microreactor (ID = 6 mm) in the temperature range from 200 to 300 °C under atmospheric pressure. In each experiment, 0.1 g of sample (85–100 mesh) was placed on a bed of quartz wool in a vertical tube furnace. The temperature of the furnace was maintained by thyristor-powered Eurotherm PID controller (model 2416) and a K-type thermocouple (Omega) was inserted in the reactor in close contact with the catalyst bed to measure the actual reaction temperature. A premixed water (Millipore) and methanol (Spectrochem, 99.9%) solution (water/methanol = 1.1) was introduced by a KD100 syringe pump (Cole Parmer) at a rate of 0.4 mL h⁻¹. The liquid was passed through stainless steel tube and traveled a very short distance during which it was evaporated by heating tapes maintained at sufficiently high temperature (>150 °C) to ensure a single phase flow. Flow of nitrogen (23.5 mL min⁻¹) through the system was controlled precisely by Bronkhorst High-Tech B V thermal mass flow controller (model F-201CB). The samples were tested at a gas hourly space velocity (GHSV) of 40,000 h⁻¹. The product gases (e.g. CO, CO_2 and H_2) were analyzed by an online gas chromatograph, Agilent 7890A equipped with a polar Porapak Q and a molecular sieve 5 Å columns, and a thermal conductivity detector and a flame ionization detector (FID). To increase the sensitivity toward CO and CO_2, the FID is associated with a methanizer which converts these gases into CH_4 (though CH_4 was not detected in our experiments).

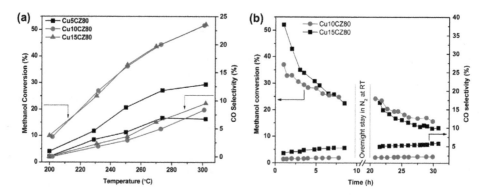

Figure 1 MSR behavior of combustion synthesized catalysts at GHSV of 40,000 h⁻¹: (a) effects of copper loading on CZ80 and (b) time-on-stream behavior of Cu10CZ80 and Cu15CZ80 catalysts at 250 °C (S/M = 1.1; balance is N_2 in each case)

Results and Discussion

Selection of materials

The preliminary MSR experiments (data not included here) carried out over 10 at.-% copper over the $Ce_{1-x}Zr_xO_2$ (Ce–Zr) supports with varied Ce/Zr molar ratio show the 10 at.-% Cu-loaded $Ce_{0.8}Zr_{0.2}O_2$ (named as CZ80), defined as Cu10CZ80, to be the most active formulation exhibiting ~51% methanol conversion at 300 °C at the GHSV of 40,000 h⁻¹ (W/F = 0.09 s). It is to be noted that none of the pure Ce–Zr oxides show any MSR activity over the whole range of temperatures investigated. This indicates that the MSR activity reported here is due to the loading of copper on the Ce–Zr oxides.

Fig. 1a represents the effect of copper loading (5, 10 and 15 at.-%) on the MSR behavior of CZ80. As expected, the conversion of methanol is the lowest in the case of the lowest Cu-loaded sample, 5 at.-% Cu on CZ80 (designated as Cu5CZ80), which shows a conversion of 29% at 300 °C with the highest CO selectivity up to 275 °C. The methanol conversions recorded over Cu10CZ80 and the highest copper-containing sample, 15 at.-% Cu on CZ80 (represented as Cu15CZ80), are found to be similar. But the CO selectivity exhibited by the Cu15CZ80 catalyst is higher than recorded for the Cu10CZ80 catalyst in the complete range of temperatures. The activity behaviors of different Cu-loaded samples led us to choose Cu10CZ80 and Cu15CZ80 as the important formulations for further studies.

The BET-specific SA of the CZ80 oxide is 22 m² g⁻¹. For the lower Cu-loaded sample, Cu10CZ80, the SA increases marginally to 25 m² g⁻¹ and it remains similar to the support for the higher Cu-loaded sample Cu15CZ80. So, the presence of copper appears to have an insignificant role on the SA of the single step combustion synthesized catalysts.

Durability test

The time-on-stream (TOS) activity patterns of Cu10CZ80 and Cu15CZ80 samples at 250 °C (other reaction conditions remaining the same) were studied for a total of 20 h in two consecutive days leaving behind the catalyst overnight at room temperature (RT) in N_2. The conversion behaviors are shown in Fig. 1b. Both the samples show a continuous decrease in conversion with TOS. Though the Cu15CZ80 sample shows

a higher conversion during the initial hours than exhibited by the Cu10CZ80 sample, it decreases rapidly to 22% within first 8 h and beyond that the TOS conversion profile of the Cu15CZ80 sample actually lies below that of the Cu10CZ80 sample exhibiting a methanol conversion of just 13% at the end of the test. On the contrary, the Cu10CZ80 catalyst exhibits an initial methanol conversion of ~36% at 250 °C that decreases to ~25% after 8 h and it is ~16% after 20 h (see Fig. 1b). Thus, the Cu10CZ80 is the best formulation made via the single step combustion route in this study.

A detailed surface analysis of the as-prepared Cu10CZ80 sample and the same obtained after *in situ* MSR reaction (at 300 °C, the temperature at which the sample showed the highest reforming activity) in the XPS reaction chamber was subsequently carried out in order to gain insight(s) about the change(s) in surface composition that prevails during MSR in the microreactor.

In situ XPS studies

Analysis of the XPS data revealed that the as-prepared sample has a surface atomic Ce/Zr ratio of 4, which is exactly the stoichiometric (theoretical) value corresponding to $Ce_{0.8}Zr_{0.2}O_2$. Approximately, the same ratio is recorded after MSR (Ce/Zr = 4.5). This indicates that the ceria–zirconia surface composition remains essentially unaffected apart from marginal cerium enrichment on the surface during the MSR tests.

Fig. 2a and b corresponds to the Cu 2p core level regions of Cu10CZ80 for the as-prepared sample and the sample after *in situ* MSR, respectively. These spectral features together with the respective Auger lines (see Fig. 2c and d) reveal that the as-prepared sample consists exclusively of oxidized copper (the $2p_{3/2}$ peaks at binding energies higher than 934 eV with characteristic satellite features correspond to Cu^{2+}). After MSR, the oxidized Cu-species transforms completely into metallic copper (the $2p_{3/2}$ peak at 932.4 eV is due to Cu^0, which is in accordance with the absence of satellite lines). The dispersion of Cu is impressively high in the as-prepared sample giving an atomic Cu/(Ce+Zr) ratio of about 26%, which is a value much higher than that expected for a 10 at.-% Cu loading. After MSR, the Cu/(Ce+Zr) ratio is calculated to be 7.4%, which means that the dispersion of Cu drops, probably by sintering.

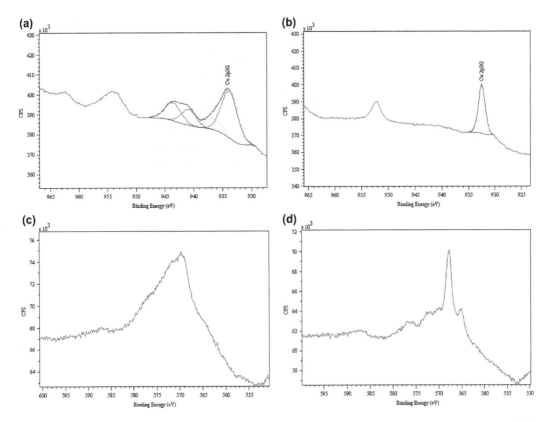

Figure 2   XPS of Cu 2p regions of (a) as prepared and (b) after MSR samples of Cu10CZ80; Cu LMM Auger spectra of (c) as prepared and (d) after MSR samples of Cu10CZ80.

This would explain the decay in catalytic activity during TOS reported above.

XRD studies

The powder XRD patterns of pure CZ80 and the three Cu-loaded CZ80 materials are shown in Fig. 3. All the peaks correspond to the cubic fluorite structure, as expected. The average size of the crystallites calculated using the Scherrer equation is ~22 nm. The peaks observed at 35.5° and 38.7° correspond to CuO(002) and CuO(111) planes in the as-synthesized materials. The intensity of these CuO-related peaks increases with the increase of Cu loading (see the inset of Fig. 3 for a better view). The CuO(111)/CeO$_2$(111) peak area ratio increases from 0.006 in Cu5CZ80 to 0.014 in Cu10CZ80 and 0.023 in Cu15CZ80. The least-square-refined lattice parameters of the pure CZ80 is 5.3687 (5) Å and those for the Cu-loaded oxides are 5.3719 (6) Å for Cu5CZ80, 5.378 (1) Å for Cu10CZ80, and 5.371 (1) for Cu15CZ80. After subtracting the Kα_2 contribution, the least-square-refined lattice parameters are calculated to be 5.372 (1) Å, 5.374 (1) Å, 5.378 (2) Å, and 5.3734 (3) Å, respectively, for the CZ80, Cu5CZ80, Cu10CZ80, and Cu15CZ80 samples. Thus, the lattice parameters of the Cu-loaded samples are marginally higher than that of the pure oxide and, taking into account that a decrease in lattice parameter is expected upon copper substitution [Ce^{4+} (0.097 nm), Ce^{3+} (0.1143 nm), Zr^{4+} (0.084 nm), and Cu^{2+} (0.073 nm)], it is difficult

to comment conclusively about the presence or absence of substitutional Cu in the Ce–Zr solid solution phase. It appears that a large fraction of the copper is present as finely dispersed CuO nanoparticles and the rest minor fraction of copper may have the possibility to remain as interstitial punctual defects in the Ce–Zr lattice owing to the small size of the Cu^{2+} ion.[26]

HRTEM studies

Under low magnification (image not included), the CZ80 support is constituted by crystallites ranging from 10 to 50 nm. All the support particles are well crystallized, as shown in the lattice fringe image depicted in Fig. 4a. Considering the poly-crystalline nature of the crystallites, the size range is consistent with the Scherrer size of 22 nm. Unlike the TEM with highly localized imaging, the information obtained from Scherrer's formula from the XRD data is the information from the bulk material. Thus, the particles size obtained by local imaging (HRTEM) will be averaged out using Scherrer's formula.[27]

Similar characteristics are encountered in the samples loaded with copper in their as-prepared forms, Cu10CZ80 and Cu15CZ80. A lattice fringe image of Cu10CZ80 is depicted in Fig. 4b. It should be stressed out that no copper-containing particles are recognized in the HRTEM images of the Cu10CZ80 catalyst, which suggest that copper may be incorporated into the support nanoparticles or it is so highly dispersed that it escapes TEM detection. Although the lattice fringes do not

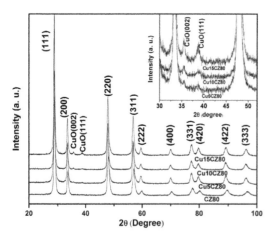

Figure 3   Powder XRD patterns of pure CZ80 and the different Cu-loaded CZ80 oxides. Inset shows the expanded view of the Cu-region ($2\theta = 30°–55°$)

show any variation to the nominal Ce–Zr support, the absence of segregated Cu-containing particles at the surface of the Ce–Zr support points to the incorporation of a part of copper into the Ce–Zr lattice. This result is in agreement with the XPS data, which showed an impressive high copper dispersion. It is also to be noted that the XRD analyses (see inset of Fig. 2) reveal formation of highly dispersed CuO on CZ80 support in all the Cu-loaded CZ80 samples.

In general view (image not included), the structure of the sample with the highest Cu loading, Cu15CZ80, is essentially

identical to the lower Cu-loaded sample. However, in this case, several isolated CuO particles can be recognized, as it is shown in its high-resolution image in Fig. 4c in accordance with XRD data. These CuO particles are about 4–5 nm in size.

The HRTEM images of the aged (after MSR in the microreactor) samples show the presence of metallic copper (see Fig. 4d and e). This is in accordance with the *in situ* XPS experiment (see Fig. 2) and also the *ex situ* characterization by XRD (see Fig. 4f that shows the XRD of the aged samples spanning the Cu-region; the peak at 43.5° corresponds to Cu⁰ (111)). HRTEM image of the aged catalyst of Cu10CZ80 (Cu10CZ80aged) shows the presence of particles with high electron contrast (see Fig. 4d). These particles show the characteristic lattice fringes of metallic Cu. The Cu nanoparticles measure about 2 nm in diameter.

In the lattice fringe image of the aged sample of Cu15CZ80 (Cu15CZ80aged; see Fig. 4e), a well-developed metallic Cu nanoparticle is shown. The size of the metallic Cu nanoparticles in this sample is larger (about 3–4 nm) with respect to the aged Cu10CZ80 catalyst, in accordance with its higher copper loading (see also Fig. 4f). For the aged samples, the CuO(111)/CeO$_2$(111) peak area ratios are similar, 0.025 for Cu10CZ80, and 0.026 for Cu15CZ80. The least-square-refined lattice parameters of Cu10CZ80aged and Cu15CZ80aged samples are found to be 5.376 (1) Å and 5.375 (1) Å, respectively, which are again similar to the respective as-prepared samples. It is thus not possible to infer on the existence of ionically substituted copper species in the Ce–Zr support in the aged samples, so the existence of Cu²⁺ in the Ce–Zr lattice cannot be completely ruled out.

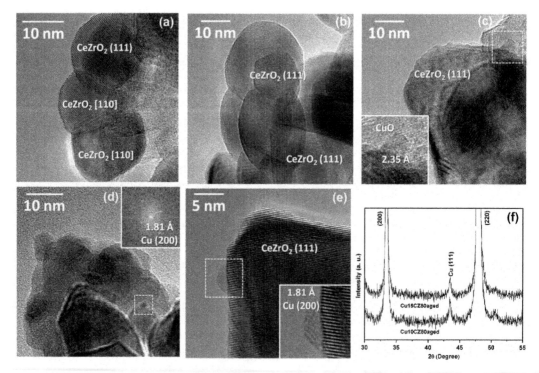

Figure 4   HRTEM images of (a) pure CZ80, (b) Cu10CZ80, (c) Cu15CZ80, (d) Cu10CZ80aged, (e) Cu15CZ80aged, and (f) slow scan XRD data of Cu10CZ80 and Cu15CZ80 catalysts in their aged forms in the 2θ range 30°–55°

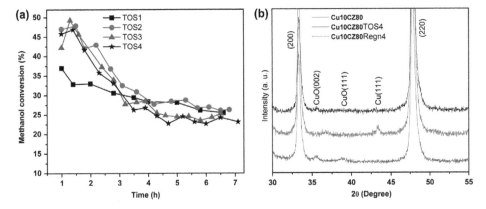

Figure 5 (a) Time-on-stream MSR behavior of Cu10CZ80 catalyst at 250 °C after *in situ* oxidative regeneration treatments for three consecutive cycles (TOS2 to TOS4) along with that in as-prepared form (TOS1) (S/M = 1.1; balance is N$_2$ in each case to reach a GHSV of 40,000 h^{-1}) and (b) XRD patterns of the samples after the fourth TOS cycle (Cu10CZ80TOS4) and after its *in situ* regeneration (Cu10CZ80Regn4 sample) along with that of the as-prepared sample in the 2θ range 30°–55°.

As a complementary chemical characterization tool, we have also conducted electron energy loss spectroscopy (EELS) measurements over the Ce–Zr crystallites to check for the presence of substitutional Cu in the Cu10CZ80 catalyst in its as-prepared form. In all cases, the signals of Cu were absent, suggesting that the Ce–Zr crystallites do not contain Cu at all in their structure. It should be stressed out, however, that the sensitivity of Cu to EELS is not very high and that Cu signals are very close to Ce signals, so the fact that we do not see Cu does not necessarily exclude that there is a small amount of Cu in Ce–Zr oxide of the present combustion synthesized catalyst Cu10CZ80.

Now, if we combine the outcomes of XRD, TEM, and EELS studies with the XPS studies of as-prepared and *in situ* MSR-treated samples, it can be proposed that in both the as-prepared samples of Cu10CZ80 and Cu15CZ80, a part of copper is present as finely dispersed CuO (major Cu-phase) on the Ce–Zr support and the rest portion of copper remains as substitutional ions (minor Cu-phase) in the Ce–Zr lattice essentially on the surface layers that is credited to the high Cu-surface concentration obtained for all the as-prepared and regenerated Cu10CZ80 samples. This is a reasonable proposition since combustion synthesis of base and noble-metal-loaded oxide samples is generally composed of a surface solid solution phase containing substitutional metal ion (here copper) sites and the rest of the metal component remains as finely dispersed metal/metal oxide crystallites.[28,29] The reductive MSR atmosphere transforms the oxidized copper species to metallic copper. This is why both the aged catalysts have been characterized to contain copper metal nanocrystallites.

The nature and ratio of the various Cu-related phases and its surface composition decides the on-stream MSR activity pattern. The initial methanol conversion can be correlated to the presence of oxidized copper species—lattice-incorporated Cu^{2+} as well as dispersed Cu^{2+} in the form of CuO on the Ce–Zr support. As the time progresses, both these two types of oxidized copper species are reduced to metallic copper which subsequently undergoes sintering to form larger Cu-metal nanocrystallites. The proportion

of the oxidized copper species thus decreases with TOS leaving behind a catalyst surface which is less active for the MSR. Here, we can speculate that the finely dispersed CuO on the surface are reduced at a faster pace than the lattice-incorporated Cu^{2+} surface species. A rapid initial fall in methanol conversion followed by a steady decrease in conversion is thus observed in the on-stream MSR pattern of the Cu10CZ80 and Cu15CZ80 catalysts.

From the TOS tests, it is evident that the initial activity of the Cu15CZ80 catalyst is more than that of the Cu10CZ80 catalyst that can be attributed to the larger loading of copper in the former (6.2 wt-% vs. 4.1 wt-%). But with the progress of reaction, the methanol conversion exhibited by the former catalyst decreases below that of the latter and is associated with a larger amount of CO formation. The deactivation of catalyst is mainly associated with the thermal sintering of the small copper nanoparticles leading to a decrease of the active SA[30-32] and or the reduction of oxidized copper to its metallic form.[3] The phase analyses as well as structural studies have already proved that metallic copper is present in both the aged samples, the size of Cu particles being larger in the higher Cu-loaded sample Cu15CZ80. Owing to the larger loading of copper in it, the extent of sintering is more in Cu15CZ80 than in Cu10CZ80. So, the TOS behavior of the Cu10CZ80 and Cu15CZ80 can be accounted for by noting the reduction of oxidized copper to metallic copper followed by its sintering. The sintering becomes more evident in the second day of test, bringing the TOS conversion values of Cu15CZ80 below that of Cu10CZ80.

A rough estimate of the turnover frequencies (TOFs) of the two finalized catalysts have been calculated by considering the mole of hydrogen produced per mole of total copper (nominal amount taken during synthesis) per second at 250 °C after 1 h of reaction. The TOFs are calculated to be 0.034 and 0.025 s^{-1}, respectively for the Cu10CZ80 and Cu15CZ80 catalysts. We have also calculated the TOF values of the as-prepared and aged samples of Cu10CZ80 by considering the surface concentration of copper from XPS and Cu particle size from HRTEM as reported recently.[25] The TOF values of the

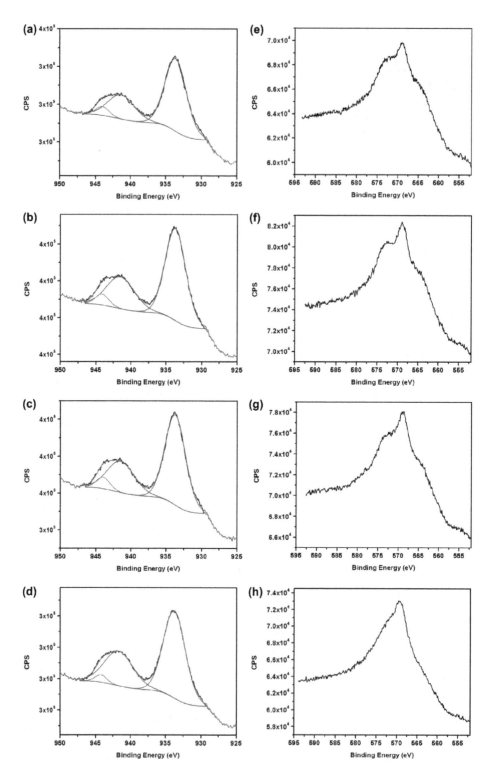

Figure 6 XPS of Cu 2p$_{3/2}$ regions of (a) Regn1, (b) Regn2, (c) Regn3, and (d) Regn4 samples of Cu10CZ80; Cu LMM Auger spectra of (e) Regn1, (f) Regn2, (g) Regn3, and (h) Regn4 samples of Cu10CZ80

as-prepared and aged samples calculated thus are 0.15 and 0.16 s^{-1}, respectively.

Finally, we looked into the aspect of regeneration characteristics of Cu10CZ80. After the first cycle of TOS test (at 250 °C for 6 h; TOS1), the catalyst was purged with nitrogen to RT and left in this flow for overnight and the next day it was regenerated *in situ* in 20% O_2 in N_2 (flowing at 30 mL min^{-1} from RT to 350 °C, at the heating rate of 10 °C min^{-1} and dwell there for 1 h). This was followed by cooling in N_2 to 250 °C and the TOS behavior was assessed again for further 6 h (TOS2). The catalyst was subsequently cooled in nitrogen and left overnight at RT and the cycle of TOS tests was continued in the following days (TOS3 and TOS4). Fig. 5a shows the conversion values recorded for the regenerated samples along with the TOS data of the as-prepared sample (reproduced from Fig. 1b). The regeneration behavior is quite impressive up to ~4 h showing a higher conversion than recorded over the as-prepared form of the catalyst, beyond which the conversions observed are similar to each other. Thus, the combustion synthesized catalyst of this study can be regenerated easily with an intermediate oxidation treatment. We have already explained the loss in activity during TOS to the formation of Cu^0 followed by sintering based on XRD, HRTEM, and *in situ* XPS findings. In the following texts, we put forward possible explanations for the TOS behavior of the regenerated samples based on XRD and XPS analysis.

We first attempt to explain the TOS behavior recorded for the Cu10CZ80 catalyst on regeneration by analyzing the powder XRD patterns of the sample obtained at the end of these TOS tests (Cu10CZ80TOS4 sample) and the one after its *in situ* regeneration (Cu10CZ80Regn4 sample). Fig. 5b shows the XRD patterns (slow scan data) of the Cu region (2θ from 25° to 55°) of these samples along with the as-prepared catalyst. The sample after the fourth cycle of TOS is characterized by Cu^0 presence, as expected. The CuO(111)/CeO$_2$(111) peak area ratio of this sample (after four consecutive cycles) is calculated to be 0.013 which is half of the aged catalyst after 20 h of TOS. This can be attributed to the presence of finer crystallites of copper in the cycled sample (Cu10CZ80TOS4 sample). Interestingly, for the regenerated sample, no CuO-related diffraction peaks could be identified. Thus, the oxidative treatment generates too small copper species (below the detection limit of XRD) and redistributes them in an efficient manner on the catalyst surface. The least-square-refined lattice parameters of the TOS4 and Regn4 samples remain similar (5.3780(7) Å and 5.3740(1) Å, respectively) to the as-prepared sample that again point to no major structural change due to these treatments.

In order to gain further insights about the surface properties of the Cu10CZ80 catalyst subsequent to regeneration in the microreactor, we have carried out XPS analysis (*ex situ* samples) of all the regenerated samples (Regn1 to Regn4 followed by the catalyst name). Fig. 6a–d shows the Cu 2p$_{3/2}$ region of all these samples. It is clearly evident that all the regenerated samples are very similar. They exhibit rather constant Cu/ (Ce+Zr) atomic ratios (0.19–0.21), except sample Regn4 that is enriched in Cu (exhibiting an atomic ratio of 0.32). On the other hand, it is clear from both the photoemitted electrons and the Auger lines (see Fig. 6e–h) that in all the regenerated samples, copper exists only as Cu(II), which nicely fits with the reaction scheme that we have proposed based on the *in situ* MSR studies in the XPS chamber. It should be kept in mind that the samples have been analyzed *ex situ*, so may be the sample contains reduced Cu under reaction conditions (as we saw in the *in situ* MSR experiment) and the surface oxidizes upon exposure to air. We did actually note a visible change in color

of the samples (from black to gray, the color of the as-prepared sample) collected after the regeneration tests when exposed to air in the subsequent days which we could not avoid before performing the XPS measurements. This natural tendency of the combustion-made sample to get spontaneously oxidized in air can possibly explain the differences in the Cu atomic percentages obtained from the *in situ* and *ex situ* XPS data of the various Cu10CZ80 samples. The surface Ce/Zr atomic ratio of the regenerated samples maintains the same value of about 4.5 to that of the sample after *in situ* MSR. The surface enriched highly with nanosized copper species is proposed to be responsible for the overall gain in the MSR activity during initial stages (first 4 h of MSR). Afterward, the catalyst surface of the regenerated sample resembles that of the as-prepared sample and thus shows similar reforming activity.

Conclusions

In the present study, we have shown that the solution combustion synthesis is a simple one-step route to prepare Cu-based ceria–zirconia oxides that are active for MSR. The Cu/Ce$_{0.8}$Zr$_{0.2}$O$_2$ catalyst (Cu10CZ80) possessing a low SA of 25 m^2 g^{-1} and containing 4.1 wt-% Cu (nominal value) shows the best MSR activity exhibiting a methanol conversion of ~51% at 300 °C when the GHSV is maintained at 40,000 h^{-1}. The copper is present both as substitutional ion (minor Cu-phase) and as dispersed CuO crystallites (major Cu-phase) over ceria–zirconia in the as-prepared catalysts. The loss of activity during on-stream MSR tests (aging treatment) results from the complete reduction of copper followed by its sintering. Introduction of an intermediate *in situ* oxidation step leads to nearly complete regeneration of the MSR behavior in the consecutive TOS cycles of methanol reforming.

Acknowledgments

DD thanks JNMF for a research fellowship. Financial support from the Department of Science and Technology, Government of India, by a grant (SR/S1/PC-28/2010) to AG and DST Special Grant to the Department of Chemistry of Jadavpur University in the International Year of Chemistry 2011 is gratefully acknowledged. AG is grateful to Dr. Parthasarathi Bera of National Aerospace Laboratories, India, for useful discussion. JL is Serra Húnter Fellow and is grateful to ICREA Academia program (Generalitat de Catalunya) and MINECO grant ENE2012-36368.

References

1. D. Ramirez, L. F. Beites, F. Blazquez and J. C. Ballesteros: *Int. J. Hydrogen Energy*, 2008, **33**, (16), 4433–4443.
2. D. R. Palo, R. A. Dagle and J. D. Holladay: *Chem. Rev.*, 2007, **107**, (10), 3992–4021.
3. Y. Choi and H. G. Stenger: *Appl. Catal. B Environ.*, 2002, **38**, (4), 259–269.
4. S. Sá, H. Silva, L. Brandão, J. M. Sousa and A. Mendes: *Appl. Catal. B Environ.*, 2010, **99**, (1–2), 43–57.
5. S. D. Jones, L. M. Neal and H. E. Hagelin-Weaver: *Appl. Catal. B Environ.*, 2008, **84**, (3–4), 631–642.
6. A. Mastalir, B. Frank, A. Szizybalski, H. Soerijanto, A. Deshpande, M. Niederberger, R. Schomäcker, R. Schlögl and T. Ressler: *J. Catal.*, 2005, **230**, (2), 464–475.
7. H. Oguchi, T. Nishiguchi, T. Matsumoto, H. Kanai, K. Utani, Y. Matsumura and S. Imamura: *Appl. Catal. A Gen.*, 2005, **281**, (1–2), 69–73.

8. L. Zhang, L. Pan, C. Ni, T. Sun, S. Zhao, S. Wang, A. Wang and Y. Hu: *Int. J. Hydrogen Energy*, 2013, **38**, (11), 4397–4406.

9. Z. Lei, P. Li-wei, N. Chang-jun, S. Tian-jun, W. Shu-dong, H. Yong-kang, W. An-jie and Z. Sheng-sheng: *J. Fuel Chem. Technol.*, 2013, **41**, (7), 883–888.

10. A. Haghofer, K. Föttinger, F. Girgsdies, D. Teschner, A. K. Knop-Gericke, R. Schlögl and G. Rupprechter: *J. Catal.*, 2012, **286**, 13–21.

11. M. Behrens, S. Zander, P. Kurr, N. Jacobsen, J. Senker, G. Koch, T. Ressler, R. W. Fischer and R. Schlögl: *J. Am. Chem. Soc.*, 2013, **135**, (16), 6061–6068.

12. R. Figueiredo, A. Martínez-Arias, M. Granados and J. L. Fierro: *J. Catal.*, 1998, **178**, (1), 146–152.

13. L. C. Wang, Y. M. Liu, M. Chen, Y. Cao, H. Y. He, G. S. Wu, W. L. Dai and K. N. Fan: *J. Catal.*, 2007, **246**, (1), 193–204.

14. J. Agrell, H. Birgersson, M. Boutonnet, I. Melián-Cabrera, R. M. Navarro and J. L. G. Fierro: *J. Catal.*, 2003, **219**, (2), 389–403.

15. H. Jeong, K. Kim, T. Kim, C. Ko, H. Park and I. Song: *J. Power Sources*, 2006, **159**, (2), 1296–1299.

16. X. Zhang and P. Shi: *J. Mol. Catal. A Chem.*, 2003, **194**, (1–2), 99–105.

17. R. Maache, R. Brahmi, L. Pirault-Roy, S. Ojala and M. Bensitel: *Top. Catal.*, 2013, **56**, (9–10), 658–661.

18. B. A. Peppley, J. C. Amphlett, L. M. Kearns and R. F. Mann: *Appl. Catal. A Gen.*, 1999, **179**, (1–2), 31–49.

19. H. Purnama, T. Ressler, R. E. Jentoft, H. Soerijanto, R. Schlögl and R. Schomäcker: *Appl. Catal. A Gen.*, 2004, **259**, (1), 83–94.

20. X. K. Gu and W. X. Li: *J. Phys. Chem. C*, 2010, **114**, (49), 21539–21547.

21. A. Szizybalski, F. Girgsdies, A. Rabis, Y. Wang, M. Niederberger and T. Ressler: *J. Catal.*, 2005, **233**, (2), 297–307.

22. H. Oguchi, H. Kanai, K. Utani, Y. Matsumura and S. Imamura: *Appl. Catal. A Gen.*, 2005, **293**, 64–70.

23. C. Pojanavaraphan, A. Luengnaruemitchai and E. Gulari: *Appl. Catal. A Gen.*, 2013, **456**, 135–143.

24. C. Pojanavaraphan, A. Luengnaruemitchai and E. Gulari: *Int. J. Hydrogen Energy*, 2013, **38**, (6), 1384–1362.

25. D. Das, J. Llorca, M. Dominguez, S. Colussi, A. Trovarelli and A. Gayen: *Int. J. Hydrogen Energy*, 2015, **40**, (33), 10463–10479.

26. E. Moretti, L. Storaro, A. Talon, M. Lenarda, P. Riello and R. Frattini: *Appl. Catal. B Environ.*, 2011, **102**, (3–4), 627–637.

27. M. Abdullah and K. Khairurrijal: *J. Nano Saintek.*, 2008, **1**, (1), 28–32.

28. M. S. Hegde, G. Madras and K. C. Patil: *Acc. Chem. Res.*, 2009, **42**, (6), 704–712.

29. P. Bera, K. R. Priolkar, P. R. Sarode, M. S. Hegde, S. Emura, R. Kumashiro and N. P. Lalla: *Chem. Mater.*, 2002, **14**, (8), 3591–3601.

30. B. Frank, F. C. Jentoft, H. Soerijanto, J. Kröhnert, R. Schlögl and R. Schomäcker: *J. Catal.*, 2007, **246**, (1), 177–192.

31. M. Kurtz, H. Wilmer, T. Genger, O. Hinrichsen and M. Muhler: *Catal. Lett.*, 2003, **86**, (1/3), 77–80.

32. M. V. Twigg and M. S. Spencer: *Top. Catal.*, 2003, **22**, (3/4), 191–203.

15

Study of ethanol reactions on H₂ reduced Au/TiO₂ anatase and rutile: effect of metal loading on reaction selectivity

Muhammad A. Nadeem[1], Imran Majeed[2], Geoffrey I. N. Waterhouse[3] and Hicham Idriss*[1]

[1]SABIC- Corporate Research and Innovation (CRI) at KAUST, Thuwal 23955, Saudi Arabia
[2]Department of Chemistry, Quid-i-Azam University, Islamabad 4200, Pakistan
[3]School of Chemical Sciences, University of Auckland, Private Bag 92019, Auckland, New Zealand

Abstract The effect of Au particle size and loading (over TiO₂ anatase and rutile) on the reaction selectivity and conversion of ethanol has been studied using temperature programmed desorption. The addition of Au onto TiO₂ had three main effects on the reaction. First, a gradual decrease is observed in the reaction selectivity of the dehydration (to ethylene) in favor of dehydrogenation (to acetaldehyde) with increasing Au loading on both polymorphs of TiO₂. Second, a gradual decrease is seen in the desorption temperature of the main reaction products also with increasing Au loading. Third, secondary reaction products [mainly C4 (crotonaldehyde, butene, furan) and C6 (benzene) hydrocarbons] increased considerably with increasing

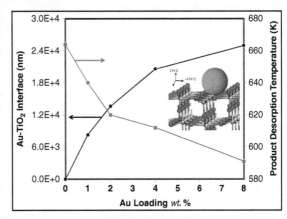

Au loading reaching about 60% for benzene for the 8 wt-%Au/TiO₂ anatase. An inverse relationship between the interface lengths of Au particles on TiO₂ and desorpion temperatures of reaction products is found.

Keywords Gold loading, TiO₂ anatase, TiO₂ rutile, Particle size effect, Ethanol-TPD, Benzene production

Introduction

Since the discovery of supported Au as an active metal for low temperature CO oxidation,[1,2] a considerable amount of research has been directed towards understanding of reaction mechanisms and active sites required for oxidation reactions.[3,4] After the work of Bond et al. who reported the hydrogenation of olefins over supported Au catalysts,[5] Au catalysts for other applications have been continuously explored[6,7] including oxidative decomposition, selective oxidation and selective hydrogenation.[8] Since then the volume of work on this catalyst is steadily increasing with a view that gold catalytic systems have the potential to serve as one of the most energy efficient catalysts.[9] Gold catalytically very active when supported on reducible oxides, i.e. TiO₂, CeO₂, and Fe₂O₃.[10,11] There are many properties explaining the catalytic activity of gold systems and these

include, quantum size effect,[12,13] Au oxidation states,[14] oxygen spill over to and from the support,[15,16] charge transfer to and from the support[17,18] and the presence of very low coordinated gold atoms.[19] Although there is reasonable progress in understanding the CO oxidation reaction,[19] more work is still needed to understand the relationship between particle size and particle density (or loading) for Au supported catalysts on reaction products.

Ethanol because of its renewable origin has been studied for the last decade either as a source for hydrogen production or as reactant for other chemicals. We have studied the application of Au supported TiO₂ catalyst for the photocatalytic activity towards hydrogen production from ethanol.[20–26] These catalysts have shown promising results and have led to some progress in understanding the role of Au during photo-catalytic reactions. Since then there is a growing interest in the photocatalytic activity of Au supported TiO₂ catalysts.[27–32] Apart from photocatalytic hydrogen production, ethanol can also be used to form other valuable chemicals. Earlier, we have reported results indicating the possibility to form other organic chemicals including benzene and furan. However, our study was limited to reduced TiO_2 and 8 wt-%Au supported on TiO_2 anatase.[33] As such, this has been seen for other noble metals on other reducible oxides.[34,35] It was found that reactions of ethanol on bare TiO_2 were dominated by ethanol dehydration to ethylene while Au deposition resulted in shifting the reaction selectivity to acetaldehyde, which further reacts to give other products including to benzene via a complex set of intermediate steps.

The reactions of ethanol have been investigated previously over a wide range of model and real catalytic surfaces.[36–45] These studies indicated that ethanol undergoes dehydrogenation and dehydration to give acetaldehyde and ethylene. Many reasons are invoked and were attempted to give further explanations and these include acid/basic site density,[36,46] bond energy,[36] electronegativity difference,[36] Madelung potential of oxygen anions and metal cations,[47] and the oxygen electronic polarisability of the oxides.[39] Other products mostly resulting from acetaldehyde have also been observed for the reaction of ethanol over metal oxides. These include: crotonaldehyde (formed by β-idolization of acetaldehyde), ethyl acetate (formed by dimerization of two acetaldehyde molecules), acetates (formed by direct oxidation of acetaldehyde),[37] benzene (formed via a series of condensation steps (β-idolization) to give a C_6 unsaturated compounds that undergo dehydration followed by cyclization steps),[35] furan[48] and butenes (formed by reductive coupling of acetaldehyde).[36] The reaction of ethanol was compared over TiO_2 and Pd/TiO_2 using temperature programmed desorption (TPD) by others.[49] It was found that the presence of Pd suppressed the dehydration to ethylene and resulted in carbon–carbon bond dissociation. This observation is actually in line with similar results for Pd, Rh and Pt over CeO_2 where a shift in the reaction selectivity during ethanol TPD towards carbon–carbon bond dissociation was noticed[35,39] when compared to CeO_2 alone. In this study, we further investigate the changes in dehydration and dehydrogenation capability of TiO_2 surface as a function of Au loading, polymorphs in an effort to probe possible contribution of both on the reaction selectivity for higher hydrocarbons.

Experimental

Catalyst preparation

TiO_2 anatase nanoparticles were prepared by the sol–gel hydrolysis of Ti(IV) isopropoxide. Briefly, TiO_2-sol was prepared by dissolving Ti(IV) isopropoxide (284.4 g) in iso-propanol (1 L) at 293 K. Under vigorous stirring, milli Q water (1 L) was added slowly drop wise to Ti(IV) isopropoxide solution resulting in the hydrolysis of the alkoxide and precipitation of hydrous titanium oxides. The final molar ratio of water: Ti(IV) isopropoxide in the reaction mixture was 55.5⊗1. The suspension was then left stirring for 24 h. The particles were subsequently collected by vacuum filtration, washed repeatedly with isopropanol and then air dried for two days at 293 K. Anatase nanoparticles were obtained upon calcination of the dried powders at 673 K for two

hours. Rutile nanofibers were obtained from Sigma-Aldrich (99.5%; $W \times L = 10 \times 40$ nm) and used as such.

TiO_2 supported gold nanoparticle catalysts (Au loading = 1, 2, 4 and 8 wt-%) were prepared by the deposition–precipitation method with urea. A detailed account of synthesis procedures for these materials can be found in our previous publication.[33] Briefly, under vigorous stirring, titania (2.5 g) was added to an appropriate amount of aqueous solution (1.1 mM) containing $HAuCl_4.3H_2O$ (250 mL for Au loadings of 2%) and urea (0.42M). The suspensions of TiO_2 particles were then heated to 360 K and kept at this temperature under continuous stirring for eight hours. The initial pH was 2 which increased with increasing temperature up to 9. The Au (III) impregnated titania were collected by vacuum filtration, washed repeatedly with milli Q water, dried for two days at 393 K in a desiccator over silica gel and then calcined at 573 K for one hour to thermally reduce surface Au (III) cations to Au metal.

Catalyst characterization

The Au/TiO_2 catalysts were characterized by N_2 physisorption, X-ray diffraction (XRD), X-ray photoelectron spectroscopy (XPS) and TEM measurements. The Brunauer–Emmett–Teller (BET) surface areas for all catalysts after adding Au did not deviate from that of TiO_2 anatase nanoparticles (105 m^2 g_{Catal}^{-1}) and rutile nanofibers (170 m^2 g_{Catal}^{-1}). The amount of Au loaded was measured independently by X-ray fluorescence analysis and was almost identical with the nominal loading. X-ray photoelectron spectroscopy analyses were performed on a Kratos Axis Ultra spectrometer. TEM data for sample characterization was collected at ANSTO (Sydney, Australia) using a JEOL 2010F TEM. Powder XRD patterns were taken on a Siemens D5000 Diffractometer equipped with a Cu anode X-ray tube and a curved graphite filter monochromator. X-ray diffraction data was collected from $2\theta = 10$–$90°$ (step 0.02°, scan rate 2° min^{-1}) using Cu K_α X-rays ($\lambda = 1.5418$ Å, 40 mA, 40 kV). X-ray photoelectron spectroscopy and TEM parameters during data collection, analyses procedure and results of 2–8 wt-%Au/TiO_2 (anatase and rutile) catalysts have been discussed in detail elsewhere.[20,33] In general Au particle sizes increase with increasing Au loading from 2 to 8 wt-% while no change in the XPS Au4f binding energies are seen when compared to that of bulk Au metal in agreement with a recent work using environmental XPS.[50]

Temperature programmed desorption

A detailed account of experimental procedures and the equipment used in the TPD studies can be found elsewhere.[21] Relative yields of all desorption products were determined by adopting the method described previously[37] while mass spectrometer sensitivity factor was calculated using the method described by Ko et al.[51] The relative yields were calculated for individual desorption products by quantitatively analyzing the desorption spectra for the respective mass fragments. Due to the possibility of more than one products desorbing at the same temperature and same mass fragment signal, extra care is needed when analyzing the result of the desorption profile. Every peak area from the profile was cautiously subtracted according to the contribution of possible species present from the fragmentation pattern. Two factors were employed when determining the cracking contribution; firstly, major fragmentation product was distinguished from the background signal, and secondly, assigning possible products at as few species as possible. In the case of a mass fragment resulting from multiple species other mass fragments were checked for consistency. The method used for TPD spectra analysis consists of:

(i) separation of the desorption peaks and categorizing into common domains of temperature

(ii) analysis of the fragmentation of each product independently

(iii) accounting all likely signals and start from the most intense fragment for each product and subtracting the corresponding amounts of each fragmentation pattern until the majority of the signals were accounted for.

Assuming constant heating rate, the relative yield Y_i of each species can be determined as a fraction of the entire sum of products

$$Y_i = \frac{PA_i \times CF_i}{\sum PA_i \times CF_i} \tag{1}$$

where PA_i is the area under the peak and CF_i is the correction factor.

The mass fragment desorption spectra are corrected to reflect rate of reaction and to allow relative product yields to be obtained. The correction factor, CF, relative to CO for product desorption is given as

$$CF_i = \frac{1}{I_x \times F_m} \sum_j \frac{F_m}{G_m \times T_m} \tag{2}$$

where I_x is the ionization efficiency, F_m is the mass fragment yield, G_m is the electron multiplier gain and T_m is quadrupole transmission. This allows correction for relative differences in ionization efficiency, mass fragment yield, electron multiplier gain and quadrupole transmission.

The ionization efficiency is primarily dependent on the number of electrons per molecule (n_e). A reasonable correction for the total ionization efficiency of a molecule relative to CO is given by

$$I_x = 0.6\left(\frac{n_e}{14}\right) + 0.4 \tag{3}$$

The gain of the electron multiplier is a function of ion mass, such that relative to CO, it can be approximated as

$$G_m = \left(\frac{28}{MW}\right)^{\frac{1}{2}} \tag{4}$$

Transmission of an ion through the quadrupole filter is also a function of ion mass and has been approximated by

$$T_m = 10^{\frac{30-MW}{155}} \quad MW > 30 \tag{5}$$

$$T_m = 1 \quad MW < 30 \tag{6}$$

Thus one has to calculate the CF for each fragment for each product following the outline procedure in order to perform quantitative analysis of the TPD data. The normalization of

desorption profiles were obtained by multiplying the product desorption spectrum by the correction factor.

For calculating raw area of the peak resulting from the desorption profile, Trapezoidal rule[9] was applied. Trapezoidal rule is an approximation for determining total area under numerous data points. The approximation can be expressed in

$$\int_a^b f(x)dx \approx T_n = \frac{\Delta x}{2}[f(x_0) + 2f(x_1) + 2f(x_2) + \dots$$
$$+ 2f(x_{n-2}) + f(x_n)] \tag{7}$$

where

$$\Delta x = \frac{b-a}{n} \tag{8}$$

$$x_i = a + i\Delta x \tag{9}$$

The area of the trapezoid that lies above the ith subinterval is

$$\Delta x \frac{f(x_i - 1) + f(x_i)}{2} = \frac{\Delta x}{2}[f(x_i - 1) + f(x_i)] \tag{10}$$

The sum of all these trapezoids results in the right side of equation (10).

Each time 50 mg and 31.2 mg of TiO$_2$ anatase and rutile catalysts, respectively, were loaded in the TPD reactor respectively. Based on the catalyst surface area, number density of fivefold coordinated Ti atoms on a TiO$_2$ surface loaded in the TPD reactor and assuming that all ethanol molecules in 1.0 µL of ethanol dosed are adsorbed, the surface coverage is estimated to be ~0.4.[33]

Results

X-ray diffraction revealed the broad features typical for the nano-crystalline TiO$_2$ polymorphs as well as for Au particles as shown in Fig. 1. The XRD pattern of micron size Au particles was used as a reference to monitor Au deposition on TiO$_2$. In the case of Au/TiO$_2$ anatase catalysts the overlap of Au (111) reflection with the TiO$_2$ anatase (004) reflection cannot be easily used to monitor Au particles but the Au (200) reflection region was clear. Au metal could be seen for the 4 and 8 wt-% due to its high loading on the support. However, in the case of Au supported rutile TiO$_2$ catalysts, both Au (111) and (200) reflections were clearly observed and increased in intensity with increasing Au loading. The sharpness of the Au peaks in the case of rutile catalysts as compared to those of anatase catalysts indicated the presence of comparatively larger Au particles in the former case in line with the results obtained by TEM studies.

Transmission electron microscopy images of 8 wt-%Au/TiO$_2$ anatase nanoparticles (Fig. 2a and b) and rutile nanofibers (Fig. 2c and d) confirm that deposition with urea method produces small gold particles in good contact with the TiO$_2$ support. It can be seen by comparing Fig. 2a to 2c that Au particles over rutile TiO$_2$ are much larger than those on anatase TiO$_2$. Figure 2c also indicates that, on 8 wt-%Au/TiO$_2$ rutile nanofibers, gold particles have wide size distribution range. The Au particle size distribution range and average size as a function of Au loading and phase type is summarized in Table 1. Detailed particle size analysis using

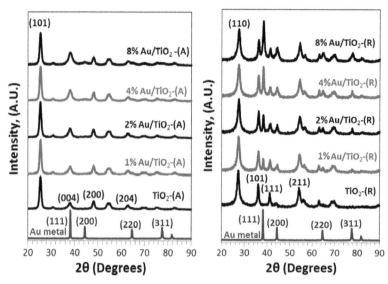

Figure 1 X-ray diffraction patterns for TiO₂ anatase nanoparticles (A) and rutile nanofibers (R) with different Au loadings

TEM images indicated that the particle size on both supports was in the nanoscale range; where Au particles were larger and have wider size distribution range on rutile support as compared to the corresponding anatase support.[20] The dependence of Au particle size on the nature of the support has been noted previously and is attributed to differences in metal/support interaction.[49] We have observed previously that Au particles over the rutile phase are larger than those on the anatase phase.[20] The reason can be attributed to the difference in the work function of both materials where the anatase is about $0.2\,eV$[52] higher making it easier for an incoming Au ion to interact with the surface than with another already formed Au cluster. This interpretation while useful needs to be taken as a guide or a possibility as a difference of $0.2\,eV$ is within errors and changes in crystallinity, crystallographic orientation and the presence of other adventitious dopants may affect the measurements. Yet we have constantly observed that Au particles on anatase are smaller than those on rutile. In addition, Au particle size depends on the preparation method, deposition precipitation method, using urea as basifying agent, is the most commonly used preparation method for Au catalysts as it

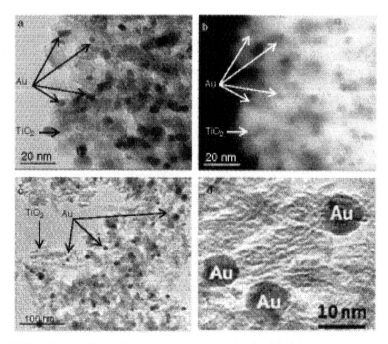

Figure 2 Images (TEM) of 8 wt-%Au/TiO₂ anatase nanoparticles in *a* bright field and *b* dark field modes to show presence of Au particles. TEM images *c* and *d* correspond to 8 wt-%Au/TiO₂ rutile nanofibers at two different magnifications

Table 1   Particle size analysis (TEM) of Au/TiO$_2$ catalysts (based on 100 particles)

TiO$_2$ phase	Au/wt-%	Range/nm	Mean particle size/nm
Anatase	1	2–14	5.4 ± 3.0
	2	2–18	8.8 ± 3.0
	4	6–22	12.8 ± 3.6
	8	4–26	15.9 ± 4.5
Rutile	1	14–32	22.6 ± 3.8
	2	14–36	23.8 ± 4.4
	4	16–40	28.5 ± 3.1
	8	18–42	30.0 ± 5.4

produces relatively smaller Au particles with uniform distribution.[53–55] Progressive decomposition of urea above 60°C allows the slow precipitation of gold (III) species and avoids abrupt increase of pH to avoid precipitation in solution.[56–58] Au ions in the freshly prepared Au/TiO$_2$ samples are converted to Au^0 after calcination above 300°C in air due to the instability of Au$_2$O$_3$ ($\Delta H_f = +19.3\,\text{kJ}\,\text{mol}^{-1}$) present on TiO$_2$ surface.[56–61]

Effect of prior H$_2$ treatment

All ethanol TPD experiments were performed on catalysts after H$_2$ treatment was carried out at 673 K overnight at a flow rate of about 10 mL min^{-1} at one atmospheric pressure. We have conducted a systematic study for ethanol TPD over TiO$_2$ that was prior oxidized and reduced and found some differences in particular related to the reaction yield (manuscript in preparation). However, because the objective of the present work is to probe into the role of Au we have opted to reduce all catalysts prior to reaction in order to prevent any possible changes in the oxidation of Au due to the O$_2$ annealing treatment. We have also previously studied, albeit briefly, ethanol reactions (dehydration and dehydrogenation) which occur on these catalysts.[33] Addition of Au changes the selectivity of catalysts form dehydration to dehydrogenation. The dehydrogenation reaction is further responsible for the production of other coupling products. In this study, we are interested in probing the reaction mechanism and possible sites identification responsible for enhancing the dehydrogenation reaction.

TiO$_{2-x}$ anatase and rutile catalysts

Temperature programmed desorption products profile following ethanol adsorption at 300 K on H$_2$ treated TiO$_2$ anatase and rutile is shown in Fig. 3. For both catalysts, the majority of the reaction products (>80%) desorb at high temperature (>600 K). For TiO$_2$ anatase catalyst, adsorbed ethanol starts to desorb at a temperature of 380 K until it starts converting into other products above 600 K. The desorption profile can be de-convoluted to two peaks; a small peak at about 460–480 K that can be attributed to the desorption of undissociatively adsorbed ethanol molecules as well as recombination of ethoxides adsorbed on non-defected surfaces. This peak, with a negligible carbon yield, is followed by a larger desorption peak (3.6%) at about 620 K attributed to re-combinative desorption of ethoxide and hydrogen atom of hydroxyl group, most likely on defects sites that further stabilize ethoxides due to higher binding

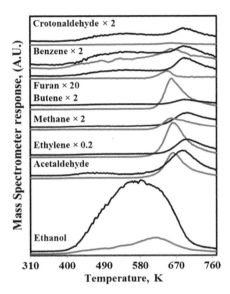

Figure 3   Comparison of TPD profiles of different desorption products after ethanol adsorption at room temperature on H$_2$ reduced bare anatase (red) and rutile (black) TiO$_2$ catalysts

energy (0.85 eV for stoichiometric and 2.27 eV for defect sites on TiO$_2$ (110) surface),[62] giving ethanol molecules.[63–65] Ethylene with carbon selectivity of 71.7% is the main product desorbed at this high temperature domain. Acetaldehyde (4.2%), butane (10%), and benzene (8.8%) are the other noticeable products desorbed. The carbon selectivities (%) of all products desorbed are given in Table 2. In contrast to anatase catalysts, for rutile nanofibers, ethanol is seen to desorb in a one large desorption peak in the temperature domain 380–700 K and accounted for 37% of the total product desorbed. This desorption originates from multiple sites, difficult to de-convolute involving contributions from perfect, defected and probably undercoordinated Ti cations. It may also involve a larger contribution of lateral interaction of adsorbates stabilized via hydrogen bonding. This indicates that rutile phase is less active for ethanol conversion into other products as compared to TiO$_2$ anatase where ethanol selectivity (amount of unreacted ethanol) is eight times lower. For both catalysts; ethylene is the major product produced by the β-hydrogen elimination of ethoxides, adsorbed either on perfect sites (producing water) or on already present defect sites as well as additional defects created during TPD (in this case regenerating surface oxygen atoms). For both catalysts, a smaller fraction of ethoxides gave acetaldehyde by dehydrogenation due to the removal of a hydride from the α-carbon of ethoxide. This hydride (H$^{\delta-}$, i.e. a negatively charged H) recombines with the hydrogen ion (H$^{\delta+}$) of the hydroxide to form H$_2$ while the remaining ethoxide is converted to acetaldehyde. A fraction of this acetaldehyde is desorbed while the other undergoes further coupling reactions forming butene, furan, crotonaldehyde and benzene. We have previously studied the production of these products from ethanol as a reactant on other oxides. See references[48,66] for butane,[47,48] furan,[67–71] crotonaldehyde[69,72] and benzene.[33–35]

Table 2 Carbon % yield and selectivity from ethanol-TPD on TiO_2 anatase and rutile (bold) catalysts, after overnigh reduction at 723 K with H_2 (relative errors are within 5%)*

Product	Peak temperature/K	Carbon yield/%	Carbon selectivity/%
Ethanol	380–550, 615/**(420–700)**	0.2, 3.6/**(37.0)**	...
Acetaldehyde	660/**(420–580, 682)**	4.0/**(1.5, 6.3)**	4.2/**(1.6, 7.1)**
Ethylene	660/**(700)**	71.7/**(34.4)**	74.5/**(38.2)**
Butene	660/**(420–580, 690)**	9.6/**(0.6, 2.8)**	10.0/**(1.5, 6.1)**
Methane	660/**(700)**	1.8/**(2.8)**	1.9/**(1.5)**
Crotonaldehyde	385–605, 665/**(420–580, 690)**	0.02, 0.4/**(1.0, 0.3)**	0.02, 0.42/**(2.2, 0.7)**
CO_2	610/**(680)**	0.1/**(0.9)**	0.1/**(0.5)**
Benzene	365–580, 650/**(420–580, 690)**	0.6, 7.9/**(3.8, 7.7)**	0.6, 8.2/**(12.8, 25.5)**
Furan	385–580, 650/**(420–580, 690)**	0.02, 0.1/**(0.4, 0.5)**	0.02, 0.1/**(1.2, 1.1)**
Total	LT, HT	0.84, 95.5/**(44.3, 55.7)**	0.64, 99.3/**(19.3, 80.7**

*LT and HT indicate total carbon % yield at low temperature and high temperature domains respectively. The carbon yield involve the corrected peak area of a desorbing product times its number of carbon. The carbon selectivity is the same taking away th reactant (ethanol in this case).

TiO$_2$ rutile nanofibers seem to be more selective towards dehydrogenation indicated by increased selectivity (from 4.2 to 8.7%) of acetaldehyde and most of the coupling products. Crotonaldehyde (1.0%, 0.3%), benzene (3.8%, 7.7%), butene (0.6%, 2.8%) and furan (0.4%, 0.5%) are the other products in parallel with acetaldehyde desorption from this rutile TiO$_2$. Also, in the high temperature desorption domain, except for ethanol all other products desorb at higher temperature for rutile catalyst as compared to anatase catalysts byabout 40 K. A small amount of methane with overall carbon selectivity of 1.9 and 1.5% for anatase and rutile catalysts (respectively) is desorbed at high temperature. CO$_2$ was seen to desorb in trace quantities at higher temperatures. This latter might originate from decomposition of surface carbonates.[21]

Au/TiO$_{2-x}$ anatase and rutile catalysts

We have conducted ethanol-TPD over Au/TiO$_2$ (anatase) and Au/TiO$_2$ (rutile) with Au% = 1, 2, 4 and 8 wt-%. Figure 4 presents TPD profile of different desorption products after ethanol adsorption at room temperature on H$_2$ treated 8 wt-%Au supported anatase and rutile catalysts, as they represented the largest difference; more details on the effect of Au loading are given further in the same section. A comparison of products desorption temperature and selectivity is given in Table 3. The presence of Au for the rutile and anatase catalysts, gradually affected the TPD products desorption temperature and distribution. Un-reacted ethanol ($m/z = 31$) started to desorb at 380 K very similar to TiO$_2$ nanoparticles (alone); however, conversion to other products started to occur at 90 K lower than in case of TiO$_2$ alone. The overall desorption of unreacted ethanol (20.3%) was found to be ca. 20% less than in case of TiO$_2$ alone.

Carbon selectivity for acetaldehyde, crotonaldehyde and benzene increased with Au loading; the increase in benzene yield with Au loading was lower in the case of rutile catalysts compared to that for anatase. Benzene has been postulated to be the result of a series of reactions from acetaldehyde via crotonaldehyde formation near Au sites. The decrease in benzene yield in the case of the rutile might be related to pore size. The larger pore size in the case of rutile fibers (10 nm) as compared to anatase nanoparticles (4 nm), determined from BET measurements (Table 4), may decrease the probability of bi-molecular reactions needed for the coupling.

The majority of the products desorbed at temperature above 530 K in two desorption domains in contrast to bar TiO$_2$ where only one desorption domain was observed. Th high temperature desorption domain can be de-convolute into two desorption peaks at about 600 and 650 K indicatin that Au loading shifts the desorption when compared t TiO$_2$ alone (650 K) to lower temperature, probably becaus ethoxides present in the proximity of Au react differently t give desorption products at 600 K. The carbon selectivity fo products desorbing at 600 K was found equal to 48.6% whil that at 650 K is equal to 39.0%. Carbon yields and carbo selectivities for individual products at different desorptio temperatures are summarized in Table 3. The middle ove high temperature product ratios (MT/HT) of Au supporte anatase is about six times higher than that on rutile phas indicating that the influence of Au on desorption products t be more pronounced in the case of the former. This ca

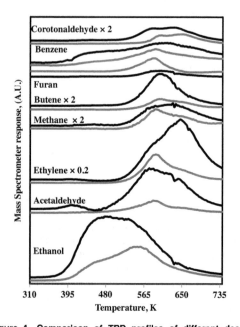

Figure 4 Comparison of TPD profiles of different deso rption products after ethanol adsorption at room tempera ture on H$_2$ reduced 8 wt-%Au supported anatase (red) an rutile (black) TiO$_2$ catalysts

Table 3 Carbon % yield and selectivity from ethanol-TPD on 8 wt-%Au/TiO$_2$ nanoparticles anatase and rutile (bold), after overnight reduction at 723 K with H$_2$ (relative errors are within 5%; more information is given in Table 2)*

Product	Peak temperature/K	Carbon yield/%	Carbon selectivity/%	Ratio MT/HT
Ethanol	355–500, 540 (**420–650**)	3.9, 8.3 (**20.3**)	...	...
Acetaldehyde	585, 640 (**390–530, 575, 648**)	3.9, 1.6 (**0.2, 7.9, 5.3**)	4.4, 1.8 (**0.2, 6.7, 4.5**)	2.4 (**1.5**)
Ethylene	590, 640 (**601, 648**)	7.0, 4.2 (**10.7, 26.1**)	8.0, 4.8 (**9.1, 22.0**)	1.7 (**0.4**)
Butene	600, 640 (**605**)	2.0, 1.7 (**4.4**)	2.3, 2.0 (**7.5**)	1.2
Methane	590, 640 (**601, 648**)	0.5, 0.5 (**0.7, 1.0**)	0.5, 0.5 (**0.3, 0.4**)	1 (**0.9**)
Crotonaldehyde	600, 640 (**601, 648**)	1.3, 1.8 (**2.2, 2.5**)	1.5, 2.0 (**4.0, 4.4**)	0.7 (**0.9**)
CO$_2$	650 (**540, 648**)	0.5 (**0.6, 1.7**)	0.6 (**0.2, 0.7**)	(**0.3**)
Benzene	355–477, 585 (**390–530, 601, 648**)	10.4, 50.3 (**4.8, 7.1, 2.8**)	11.8, 57.3 (**12.2, 18.0, 7.0**)	0.2 (**2.6**)
Furan	585 (**605**)	2.3 (**1.7**)	2.6 (**2.8**)	...
Total	LT, MT, HT	14.3, 76, 9.8 (**25.3, 29.2, 45.5**)	11.8, 76.6, 11.7 (**12.4, 48.6, 39.0**)	6.5 (**1.2**)

*LT, MT and HT indicate total % carbon yield at low, middle, and high temperatures, respectively.

partly be linked to the availability of more reaction sites in the proximity to Au in the case of anatase nanoparticles. A most likely explanation is particle dispersion: Au particle size is smaller in the case of anatase as compared to Au particle size in case of rutile with similar Au loading thus providing more reaction sites in the proximity of Au in the former case. However, the extent of reduction might be another reason: Au particles on the anatase may result in higher formation of oxygen vacancies than on rutile. These oxygen vacancies can increase the number of ethoxide species at the interface making them more poised to react. The work function of anatase phase is about 0.2 eV higher as compared to rutile phase work function (4.2 eV) indicating the easier reduction of the former; resulting in higher oxygen vacancies.[56,73,74]

Dehydration vs dehydrogenation on anatase catalysts

Figure 5 presents the desorption profiles of ethylene (Fig. 5a) and acetaldehyde (Fig. 5b) from Au/TiO$_2$ catalysts at the indicated Au loadings as a function of temperature. Two points can be extracted from Fig. 1):

(i) there is an overall decreasing trend in the desorption temperature as a function of Au loading
(ii) there is an abrupt drop in the rate of ethylene desorption from pure TiO$_2$ with increasing Au wt-%.

However, there is a gradual decrease in rate of acetaldehyde desorption with increasing Au loading. This leads to an increase in the acetaldehyde to ethylene ratio with increasing Au loading. Lowering of the desorption temperature is similar for both products with a maximum lowering of up to about 60 K in case of 8 wt-%Au/TiO$_2$ as compared to pure TiO$_2$.

Dehydration vs dehydrogenation on rutile catalysts

Figure 6 presents the desorption profiles of ethylene and acetaldehyde from Au/TiO$_2$ rutile nanofibers at the indicated Au loadings as a function of temperature. Figure 6a shows the effect of Au loading on ethylene desorption. The amount of desorbed ethylene decreases very slightly on rutile nanofibers as compared to anatase nanoparticles with increasing Au loading. However, an increase in Au loading has the following two pronounced effects:

(i) the appearance of a new desorption channel at lower temperature. This channel is more pronounced as a shoulder of ethylene desorption peak in case of pure 8 wt-%Au loading
(ii) a shift in the desorption of ethylene to lower temperature with increasing Au.

Figure 6b shows the effect of Au loading on acetaldehyde desorption. The amount of acetaldehyde increases with increasing Au loading. This increase is mainly in the form of a new desorption channel at lower temperature. This new channel appeared as a shoulder of acetaldehyde desorption peak in the case of pure 4% Au loading and becomes the main desorption channel in the case of 8% loading.

Discussion

As indicated in Fig. 7, the presence of Au on TiO$_2$ decreased the dehydration to ethylene reaction and increased the dehydrogenation to acetaldehyde. It has also increased the selectivity to other coupling products due to the further reactions of acetaldehyde. From Fig. 5 one can infer that there is actually, initially, an increase in acetaldehyde formation due to the presence of Au but further reactions to the other coupling products results in decreasing its overall

Table 4 BET surface areas, pore volume and pore radius for Au catalysts series

Au loading/wt- %	TiO$_2$ support	BET surface area/m^2 g^{-1}	Cumulative pore volume/cm^3 g^{-1}	Average pore radius/nm
0	Anatase	107	0.26	4.0
2	Anatase	105	0.26	4.0
4	Anatase	101	0.26	4.4
8	Anatase	104	0.24	4.0
0	Rutile	153	0.50	7.2
2	Rutile	172	0.72	9.5
4	Rutile	170	0.68	9.0
8	Rutile	168	0.74	10.0

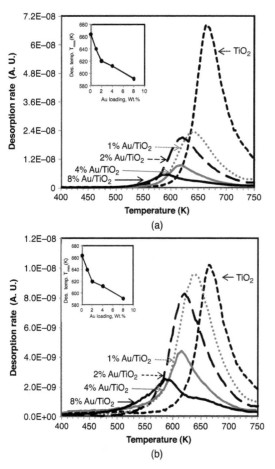

Figure 5 Temperature programmed desorption profiles of a ethylene and b acetaldehyde after ethanol adsorption on Au/TiO$_2$ anatase catalyst as function of Au loading

selectivity (the ratio acetaldehyde to ethylene still increases with increasing Au loading on both TiO$_2$).

Acidic oxides such as Al$_2$O$_3$ are found to make considerable amounts of ethylene while on the other hand, basic oxides such as CeO$_2$ have the opposite effect.[8,39,46] TiO$_2$ also gives high yield of ethylene. In this study, it has been seen that ethylene desorption rate is suppressed with increased Au loading. As the dehydration reaction on TiO$_2$ is thought to occur at oxygen defect sites, one possible explanation would be that Au particles block these as initially indicated by Matthey and co-workers.[59] While studying the adsorption of Au nanoparticle on TiO$_2$(110) single crystal they found that Au clusters preferred to adsorb on step edges and oxygen bridging vacancies. It is possible to make an estimate of the number of sites blocked by Au particles. Based on the catalyst surface area (107 m^2 g$_{Catal}^{-1}$) and number density of bridging oxygen atoms on a rutile TiO$_2$ surface (2 O atoms per 38.76 Å^2 = ~10^{18} O atoms per 1.0 m^2), the number of bridging O atoms available on the surface in 50 mg of TiO$_2$ loaded in the TPD reactor is equal to ~3 × 10^{19} atoms. We can assume that the number of oxygen defects sites prior to adsorption cannot reasonably exceed 30% (~1 × 10^{19} defects). From the average size of Au particles

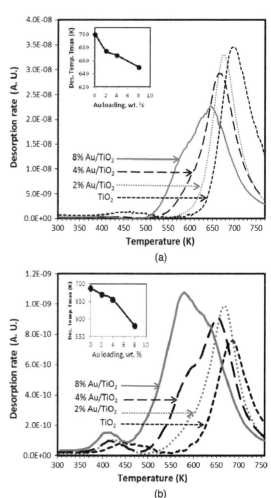

Figure 6 Temperature programmed desorption profiles of a ethylene and b acetaldehyde after ethanol adsorption on Au/TiO$_2$ rutile nanofibers as function of Au loading

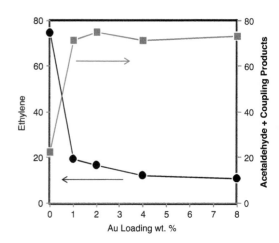

Figure 7 Comparison of carbon selectivity of different products on Au/TiO$_2$ anatase catalysts as function of Au loading

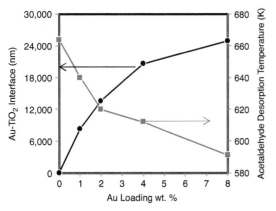

Figure 8 Comparison of product desorption temperature as function of Au/TiO$_2$ interface on amount of catalyst used. Interface was calculated using TEM analysis reported elsewhere on same catalysts.[20] Assuming that Au/TiO$_2$ interface for single metal particle will be about equal to circumference of metal particle, circumference of each individual particle among 100 metal particles was calculated and added to give total interface for each Au loading

(10 nm) present on 2 wt-%Au/rutile TiO$_2$ and the Au metal size (1.44 Å); the number of Au particles present on 50 mg of TiO$_2$ is equal to ~10^{14} Au nanoparticles. From these calculations, the defect coverage by Au particles is thus negligible. This indicates the involvement of other factors, i.e. electronic factors and creation of new reaction sites (dynamics) favoring the acetaldehyde formation which is further converted to coupling products. It appears clear from this work that the presence of Au particles in the case of Au/TiO$_2$ shifted the reaction selectivity from ethylene (dehydration) to acetaldehyde (dehydrogenation); the latter further reacts by condensation reactions leading ultimately to benzene (more details on this reaction can be found in Ref. 34). This means that the abstraction of the H atom in the alpha position from the C—O (as a hydride) is favored in the presence of Au compared to that of the H atom in the beta position of the C—O (as a proton) in the absence of Au. As such, this is not surprising as Au atoms in their cluster forms or as in organometallic compounds make strong hydrides.[60–62] Moreover, the binding energy for hydrogen has been computed (using DFT (GGA) including relativistic effective core potential) and found to increase with cluster size.[62] In addition, organometallic compounds of gold (I) hydride are found active for the dehydrogenation of alcohols.[61] It is also to be noted that the inclusion of relativistic effect is important in computing the acidity and basicity of Au atoms as it has been recently pointed out.[63] Taking all these together it is thus possible that dehydrogenation reaction occurs on Au/TiO$_2$ interface where ethanol is adsorbed as ethoxide bound to surface Ti atoms through its oxygen. The interaction leads upon activation to the hydride subtraction and the release of acetaldehyde. Acetaldehyde may further react to give coupling products. The results thus indicate that dehydrogenation reaction can, to some extent, be controlled by the Au metal particle size (and density) and the support porosity. Finally, it is worth noting that one can obtain a relationship between the interface length of Au-TiO$_2$ and the

reaction kinetics. Figure 8 presents an inverse relationship between the interface length of Au on top of TiO$_2$ (anatase) and the desorption temperature of acetaldehyde. The relationship may serve as a simple way to view the effect of Au metals on the acceleration of the reaction (reduction in desorption temperature) with the assumption that the desorption of these compounds is reaction limited. One possible interpretation can be related to molecular diffusion along the surface of TiO$_2$. The more the number of Au particles the more possible a diffusing particle (ethanol or ethoxide) would find itself at the interface Au-TiO$_2$ the higher the probability the dehydrogenation reaction would occur. Because at high temperatures acetaldehyde is not stable on the surface it will either desorb or it will further react to C4 and C6 products. This interpretation may explain the decrease in the desorption temperature with increasing the number of Au particles on the surface. A similar explanation has been given for the reaction of ethanol over Pd/TiO$_2$ when compared to TiO$_2$ alone.[49]

Conclusion

The presence of Au particles in the case of Au/TiO$_2$ lowered the overall desorption temperature by up to 60 K in addition to shifting the reaction selectivity from ethylene (dehydration) to acetaldehyde (dehydrogenation); the latter further reacted to give coupling products (mainly C4 hydrocarbons and benzene). This indicates that the abstraction of the H atom in the alpha position from the C—O (as a hydride) is favored in the presence of Au compared to that of the H atom in the beta position of the C—O (as a proton) in the absence of Au. A clear trend is also seen between the decrease in the reaction temperature of the desorbing products and the particle loading. This observation provides a simple method to track Au particles dispersion over TiO$_2$ using ethanol as a probe molecule.

Acknowledgements

The authors acknowledge Aberdeen City Council for a startup fund for the Aberdeen Chair in Energy Futures. M. A Nadeem acknowledges the Higher Education Commission of Pakistan for the award of a PhD scholarship.

References

1. M. Haruta, N. Yamada, T. Kobayashi and S. Iijima: *J. Catal.*, 1989, **115**, (2), 301–309.
2. M. Haruta: *Catal. Today*, 1997, **36**, (1), 153–166.
3. M. S. Chen and D. W. Goodman: *Catal. Today*, 2006, **111**, (1–2), 22–33.
4. M. S. Chen and D. W. Goodman: *Science*, 2004, **306**, (5694), 252–255.
5. A. S. K. Hashmi and G. J. Hutchings: *Angew. Chem. Int. Ed.*, 2006, **45**, (47), 7896–7936.
6. Q. Fu, H. Saltsburg and M. Flytzani-Stephanopoulos: *Science*, 2003, **301**, (5635), 935–938.
7. A. Corma and H. Garcia: *Chem. Soc. Rev.*, 2008, **37**, (9), 2096–2126.
8. M. Haruta: *Chem. Rec.*, 2003, **3**, (2), 75–87.
9. C. H. Christensen, B. Jørgensen, J. Rass-Hansen, K. Egeblad, R. Madsen, S. K. Klitgaard, S. M. Hansen, M. R. Hansen, H. C. Andersen and A. Riisager: *Angew. Chem. Int. Ed.*, 2006, **45**, (28), 4648–4651.
10. M. Haruta, S. Tsubota, T. Kobayashi, H. Kageyama, M. J. Genet and B. Delmon: *J. Catal.*, 1993, **144**, (1), 175–192.
11. P. Y. Sheng, G. A. Bowmaker and H. Idriss: *Appl. Catal. A: Gen.*, 2004, **261**, (2), 171–181.

12. X. Lai and D. W. Goodman: *J. Mol. Catal. A: Chem.*, 2000, **162**, (1–2), 33–50.
13. P. Claus, A. Bruckner, C. Mohr and H. Hofmeister: *J. Am. Chem. Soc.*, 2000, **122**, (46), 11430–11439.
14. G. J. Hutchings, M. S. Hall, A. F. Carley, P. Landon, B. E. Solsona, C. J. Kiely, A. Herzing, M. Makkee, J. A. Moulijn, A. Overweg, J. C. Fierro-Gonzalez, J. Guzman and B. C. Gates: *J. Catal.*, 2006, **242**, (1), 71–81.
15. L. M. Liu, B. McAllister, H. Q. Ye and P. Hu: *J. Am. Chem. Soc.*, 2006, **128**, (12), 4017–4022.
16. Z. -P. Liu, P. Hu and A. Alavi: *J. Am. Chem. Soc.*, 2002, **124**, (49), 14770–14779.
17. J. A. van Bokhoven, C. Louis, J. T. Miller, M. Tromp, O. V. Safonova and P. Glatzel: *Angew. Chem. Int. Ed.*, 2006, **45**, (28), 4651–4654.
18. D. Ricci, A. Bongiorno, G. Pacchioni and U. Landman: *Phys. Rev. Lett.*, 2006, **97**, (3).
19. T. Fujita, P. Guan, K. McKenna, X. Lang, A. Hirata, L. Zhang, T. Tokunaga, S. Arai, Y. Yamamoto, N. Tanaka, Y. Ishikawa, N. Asao, Y. Yamamoto, J. Erlebacher and M. Chen: *Nat. Mater.*, 2012, **11**, (9), 775–780.
20. M. Murdoch, G. Waterhouse, M. Nadeem, J. Metson, M. Keane, R. Howe, J. Llorca and H. Idriss: *Nat. Chem.*, 2011, **3**, (6), 489–492.
21. M. A. Nadeem, M. Murdoch, G. I. N. Waterhouse, J. B. Metson, M. A. Keane, J. Llorca and H. Idriss: *J. Photochem. Photobiol.*, 2010, **216**, (2–3), 250–255.
22. G. I. N. Waterhouse, A. K. Wahab, M. Al-Oufi, V. Jovic, D. H. Anjum, D. Sun-Waterhouse, J. Llorca and H. Idriss: *Sci. Rep.*, 2013, **3**, 2849.
23. G. I. N. Waterhouse, M. Murdoch, J. Llorca and H. Idriss: *Int. J. Nanotechnol.*, 2012, **9**, (1), 113–120.
24. A. Wahab, S. Bashir, Y. Al-Salik and H. Idriss: *Appl. Petrochem. Res.*, 2014, **4**, (1), 55–62.
25. S. Bashir, A. Wahab and H. Idriss: *Catal. Today*, 2014, **240**, Part B, 242–247.
26. Y. Z. Yang, C. H. Chang and H. Idriss: *Appl. Catal. B: Environ.*, 2006, **67**, (3–4), 217–222.
27. V. Jovic, P. H. Hsieh, W. T. Chen, D. Sun-Waterhouse, T. Sohnel and G. I. N. Waterhouse: *Int. J. Nanotechnol.*, 2014, **11**, (5), 686–694.
28. J. Yu, L. Qi and M. Jaroniec: *J. Phys. Chem. C*, 2010, **114C**, (30), 13118–13125.
29. Z. W. Seh, S. Liu, M. Low, S. Y. Zhang, Z. Liu, A. Mlayah and M. Y. Han: *Adv. Mater.*, 2012, **24**, (17), 2310–2314.
30. O. Rosseler, M. V. Shankar, M. K. -L. Du, L. Schmidlin, N. Keller and V. Keller: *J. Catal.*, 2010, **269**, (1), 179–190.
31. F. Gärtner, S. Losse, A. Boddien, M. M. Pohl, S. Denurra, H. Junge and M. Beller: *ChemSusChem*, 2012, **5**, (3), 530–533.
32. Y. Wang, D. Zhao, H. Ji, G. Liu, C. Chen, W. Ma, H. Zhu and J. Zhao: *J. Phys. Chem. C*, 2010, **114C**, (41), 17728–17733.
33. A. M. Nadeem, G. I. N. Waterhouse and H. Idriss: *Catal. Today*, 2012, **182**, (1), 16–24.
34. P. Y. Sheng, W. W. Chiu, A. Yee, S. J. Morrison and H. Idriss: *Catal. Today*, 2007, **129**, (3–4), 313–321.
35. A. Yee, S. J. Morrison and H. Idriss: *Catal. Today*, 2000, **63**, (2–4), 327–335.
36. H. Idriss and M. A. Barteau: 'Active sites on oxides: From single crystals to catalysts', in 'Advances in catalysis', (eds. H. K. Bruce and C. Gates), 261–331; 2000, Amsterdam, Academic Press.
37. H. Idriss, K. S. Kim and M. A. Barteau: *J. Catal.*, 1993, **139**, (1), 119–133.
38. A. Yee, S. J. Morrison and H. Idriss: *J. Catal.*, 2000, **191**, (1), 30–45.
39. H. Idriss and E. G. Seebauer: *Catal. Lett.*, 2000, **66**, (3), 139–145.
40. C. Diagne, H. Idriss and A. Kiennemann: *Catal. Commun.*, 2002, **3**, (12), 565–571.
41. P. Y. Sheng, A. Yee, G. A. Bowmaker and H. Idriss: *J. Catal.*, 2002, **208**, (2), 393–403.
42. V. Fierro, V. Klouz, O. Akdim and C. Mirodatos: *Catal. Today*, 2002, **75**, (1–4), 141–144.
43. J. P. Breen, R. Burch and H. M. Coleman: *Appl. Catal. B. Environ.*, 2002, **39**, (1), 65–74.
44. S. Cavallaro and S. Freni: *Int. J. Hydrog. Energy*, 1996, **21**, (6), 465–469.
45. A. M. Nadeem, J. M. R. Muir, K. A. Connelly, B. T. Adamson, B. J. Metson and H. Idriss: *Phys. Chem. Chem. Phys.*, 2011, **13**, (17), 7637–7643.
46. H. Idriss and E. G. Seebauer: *J. Mol. Catal. A: Chem.*, 2000, **152**, (1–2), 201–212.
47. S. V. Chong, T. R. Griffiths and H. Idriss: *Surf. Sci.*, 2000, **444**, (1–3), 187–198.
48. H. Madhavaram and H. Idriss: *J. Catal.*, 2004, **224**, (2), 358–369.
49. F. Cárdenas-Lizana, S. Gómez-Quero, H. Idriss and M. A. Keane: *J. Catal.*, 2009, **268**, (2), 223–234.
50. S. Porsgaard, P. Jiang, F. Borondics, S. Wendt, Z. Liu, H. Bluhm, F. Besenbacher and M. Salmeron: *Angew. Chem.*, 2011, **123**, (10), 2314–2317.
51. E. I. Ko, J. B. Benziger and R. J. Madix: *J. Catal.*, 1980, **62**, (2), 264–274.
52. G. Xiong, R. Shao, T. C. Droubay, A. G. Joly, K. M. Beck, S. A. Chambers and W. P. Hess: *Adv. Funct. Mater.*, 2007, **17**, (13), 2133–2138.
53. A. Primo, A. Corma and H. García: *Phys. Chem. Chem. Phys.*, 2011, **13**, (3), 886–910.
54. G. R. Bamwenda, S. Tsubota, T. Kobayashi and M. Haruta: *J. Photochem. Photobiol. A: Chem.*, 1994, **77**, (1), 59–67.
55. S. Oros-Ruiz, R. Zanella, R. López, A. Hernández-Gordillo and R. Gómez: *J. Hazard. Mater.*, 2013, **1**, (Part 1), 2–10.
56. R. Zanella, L. Delannoy and C. Louis: *Appl. Catal. A: Gen.*, 2005, **291**, (1), 62–72.
57. R. Zanella, S. Giorgio, C. -H. Shin, C. R. Henry and C. Louis: *J. Catal.*, 2004, **222**, (2), 357–367.
58. A. Sandoval, A. Aguilar, C. Louis, A. Traverse and R. Zanella: *J. Catal.*, 2011, **281**, (1), 40–49.
59. A. Zwijnenburg, A. Goossens, W. G. Sloof, M. W. Crajé, A. M. van der Kraan, L. Jos de Jongh, M. Makkee and J. A. Moulijn: *J. Phys. Chem. B*, 2002, **106B**, (38), 9853–9862.
60. G. C. Bond: *Catal. Today*, 2002, **72**, (1), 5–9.
61. R. Zanella, S. Giorgio, C. R. Henry and C. Louis: *J. Phys. Chem. B*, 2002, **106B**, (31), 7634–7642.
62. J. N. Muir, Y. Choi and H. Idriss: *Phys. Chem. Chem. Phys.*, 2012, **14**, (34), 11910–11919.
63. L. Gamble, L. S. Jung and C. T. Campbell: *Surf. Sci.*, 1996, **348**, (1–2), 1–16.
64. J. Muir, Y. Choi and H. Idriss: *Phys. Chem. Chem. Phys.*, 2012, **14**, (34), 11910–11919.
65. E. Farfan-Arribas and R. J. Madix: *J. Phys. Chem. B*, 2002, **106B**, (41), 10680–10692.
66. H. Madhavaram and H. Idriss: *J. Catal.*, 1999, **184**, (2), 553–556.
67. H. Idriss and M. Barteau: *Catal. Lett.*, 1996, **40**, (3–4), 147–153.
68. J. E. Rekoske and M. A. Barteau: *Langmuir*, 1999, **15**, (6), 2061–2070.
69. S. Luo and J. L. Falconer: *Catal. Lett.*, 1999, **57**, (3), 89–93.
70. H. Idriss, C. Diagne, J. Hindermann, A. Kiennemann and M. Barteau: *J. Catal.*, 1995, **155**, (2), 219–237.
71. J. Raskó, T. Kecskés and J. Kiss: *Appl. Catal. A: Gen.*, 2005, **287**, (2), 244–251.
72. H. Nakabayashi: *Bull. Chem. Soc. Jpn*, 1992, **65**, (3), 914–916.
73. A. Imanishi, E. Tsuji and Y. Nakato: *J. Phys. Chem. C*, 2007, **111C**, (5), 2128–2132.
74. Z. Zhao, Z. Li and Z. Zou: *J. Phys.: Condens. Matter*, 2010, **22**, (17), 5008.

16

Catalysis in Diesel engine NO$_x$ aftertreatment

Marco Piumetti, Samir Bensaid, Debora Fino, and Nunzio Russo*

Department of Applied Science and Technology, Politecnico di Torino, Corso Duca degli Abruzzi, 24, 10129 Torino, Italy

Abstract The catalytic reduction of nitrogen oxides (NO$_x$) under lean-burn conditions represents an important target in catalysis research. The most relevant catalytic NO$_x$ abatement systems for Diesel engine vehicles are summarized in this short review, with focus on the main catalytic aspects and materials. Five aftertreatment technologies for Diesel NO$_x$ are reviewed: (i) direct catalytic decomposition; (ii) catalytic reduction; (iii) NO$_x$ traps; (iv) plasma-assisted abatement; and (v) NO$_x$ reduction combined with soot combustion. The different factors that can affect catalytic activity are addressed for each approach (e.g. promoting or poisoning elements, operating conditions, etc.). In the field of catalytic strategies, the simultaneous removal of soot and NO$_x$ using multifunctional catalysts, is at present one of the most interesting challenges for the automotive industry.

Keywords Environmental catalysis, Air pollution, Diesel engine, NO$_x$ abatement, Nitrogen oxides

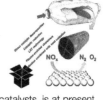

Introduction

Air pollution from mobile sources, such as cars and trucks, contributes to a great extent to air quality problems and induces health risks in rural, urban, and industrialized areas in both developed and developing countries. About 60 million cars are produced every year and over 700 million cars are used worldwide. Moreover, the vehicle population is expected to grow to almost 1300 million by the year 2030.[1]

Most vehicle transport relies on the combustion of gasoline and Diesel fuels, and hence the emission of carbon monoxide (CO), unburned hydrocarbons (HC), nitrogen oxides (NO$_x$), and particulates matter (PM) is of particular concern.[2-4] The incomplete combustion of fuels causes the emission of partial oxidation products, such as alcohols, aldehydes . As a result of the thermal cracking reactions that occur in the flame, especially for incomplete combustion, hydrogen, as well as different hydrocarbons from those present in the fuel, are formed and emitted. Therefore, the total conversion of engine-out emissions into CO$_2$, N$_2$, and H$_2$O using effective catalytic devices remains one of the most pressing challenges in the automotive industry.[5-8]

In the case of Diesel engines, the lean-burn conditions that are found in the combustion chamber lead to the following average composition of the emissions: CO$_2$ 2–12%, H$_2$O 2–12%, O$_2$ 3–17%, and N$_2$ balance. However, the features of the Diesel fuel itself, and of the Diesel engine operating conditions (air-to-fuel ratios greater than 22) lead to the formation of both gaseous (NO$_x$, CO, HC) and solid/liquid (PM) pollutants.[2] These Diesel engine emissions may originate from the incomplete

combustion of fuel, from operating conditions that favor the formation of particular pollutants, or from the oxidation of nitrogen-and sulfur-containing compounds present in the fuel which are not hydrocarbons.[3,4] A common engine management strategy to control NO$_x$ emissions is Exhaust Gas Recirculation (EGR). In this strategy, part of the exhaust (with O$_2$, N$_2$, H$_2$O and CO$_2$) is recycled back to the combustion chamber. In fact, since the heat capacity of CO$_2$ in the recirculated exhaust gas is about 20–25% higher than that of O$_2$ and N$_2$, the energy released from the fuel combustion results in a lower temperature rise, and hence a lower peak cycle temperature, with consequent lower NO$_x$ levels. However, this strategy alone is not sufficient to meet the recent NO$_x$ regulations throughout the world.

The application of strict measures to control Diesel engine emissions has been the main reason for the reductions in emissions in western European countries. The introduction of new vehicle technologies (e.g. cooled EGR) and stringent inspection systems related to Euro standards (Table 1) have led to a progressive reduction in road traffic emissions, such as NO$_x$ since 1990, despite the increase in fuel consumption.[9] Since NO$_x$ emission regulations have become more stringent over the last few years (Fig. 1), several catalytic DeNO$_x$ approaches have been investigated for lean-burn conditions, such as the direct decomposition of NO$_x$, selective catalytic reduction (SCR) using different reducing agents (e.g. ammonia/urea, hydrocarbons), and NO$_x$ storage-reduction.[10-16] On the other hand, modern three-way catalysts cannot reduce NO$_x$ in the presence of excess oxygen, because high levels of oxygen suppress the necessary reducing reactions. Consequently, there is a strong driving force to develop multifunctional catalysts capable of reducing NO$_x$ to N$_2$ and of oxidizing PM, HC, and CO

*Corresponding author, email nunzio.russo@polito.it

Table 1   European emission standards for passenger cars, g/km

Tier	Date	CO	HC	VOC	NO$_x$	HC + NO$_x$	PM
Diesel cars							
Euro 1[a]	July 1992	2.72 (3.16)	–	–	–	0.97 (1.13)	0.14 (0.18)
Euro 2	January 1996	1.0	–	–	–	0.7	0.08
Euro 3	January 2000	0.64	–	–	0.50	0.56	0.05
Euro 4	January 2005	0.50	–	–	0.25	0.30	0.025
Euro 5	September 2009	0.50	–	–	0.180	0.230	0.005
Euro 6	September 2014	0.50	–	–	0.080	0.170	0.005
Gasoline cars							
Euro 1[a]	July 1992	2.72 (3.16)	–	–	–	0.97 (1.13)	–
Euro 2	January 1996	2.2	–	–	–	0.5	–
Euro 3	January 2000	2.3	0.20	–	0.15	–	–
Euro 4	January 2005	1.0	0.10	–	0.08	–	–
Euro 5	September 2009	1.0	0.10	0.068	0.060	–	0.005[b]
Euro 6	September 2014	1.0	0.10	0.068	0.060	–	0.005[b]

[a]Values in brackets are conformity of production (COP) limits.

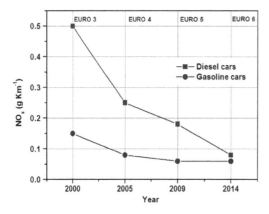

Figure 1   Trend of the European emission NO$_x$ limits for both Diesel and gasoline cars

to CO_2 and H_2O under lean conditions. In the present paper, the role of aftertreatment catalysis in Diesel NO$_x$ abatement is discussed comprehensively.

Direct catalytic NO decomposition

Engine-out NO$_x$ emissions mainly consist of NO (~90–95%) and, to a lesser extent, of NO$_2$. The main source of NO formation is the thermal (Zeldovich) mechanism. This mechanism takes place through a chain of high temperature reactions (greater than 1600 °C) and it is responsible for more than 90% of NO$_x$ emissions from road transport.[9] During flame combustion, the reaction between N$_2$ and O$_2$ is thermodynamically favored ($\Delta H_{298\,K} = 180.6$ kJ mol^{-1}), thus resulting in the formation of NO, according to the reaction:

$$N_2 + O_2 \rightarrow 2NO \tag{1}$$

Direct NO decomposition (the reverse of equation (1)) has received considerable attention in the field of environmental catalysis, since the overall process is thermodynamically favored below 1000 °C, and the use of reducing agents is not required.[17, 18] In the catalytic decomposition of NO, the exhaust containing NO is made to flow over a solid catalyst, where the NO compound is split into its elements (2NO → N$_2$ + O$_2$). The main concern is to find a material that is both active and

oxidation-resistant.[19] Therefore, since the early 1990s, several materials have been investigated for direct NO decomposition.

Noble metals

Direct NO decomposition is assumed to occur through four elementary steps[18]:

$$NO + {}^* \leftrightarrow NO^* \tag{2}$$

$$NO^* + {}^* \leftrightarrow N^* + O^* \tag{3}$$

$$N^* + N^* \leftrightarrow N_2 + 2^* \tag{4}$$

$$O^* + O^* \leftrightarrow O_2 + 2^* \tag{5}$$

where NO*, N*, and O* are the adsorbed species on the catalytic active sites (*). Overall, the NO decomposition reaction consists of two parts: (i) adsorption and dissociation of the NO species (equations (2) and (3)); (ii) recombination and the removal of N$_2$ and O$_2$ from the catalyst surface (equations (4) and (5)).

Many studies have been conducted regarding NO decomposition over Pt–Al$_2$O$_3$ catalysts.[17, 20, 21] Freund et al.[20] have investigated the role of different adsorption and reaction sites on a structurally well-defined Pd/Al$_2$O$_3$ catalyst. They found that atomic nitrogen and oxygen species adsorb preferentially at sites in the vicinity of edges and defects of solid surfaces. The presence of these atomic species critically controls the NO dissociation activity. In other words, the direct decomposition of NO was found to be dominated by particle edges, steps, defects, and (100) sites rather than by (111) facets. Similarly, Ge and Neurock[21] noted an exceptional low energy barrier for the dissociation of NO adsorbed on Pt (100) surfaces.

It has been shown that NO adsorption onto a solid surface is the kinetically controlling step, and the following kinetic equation has been proposed:

$$r = N\,k\,(NO)/(1 + K\,(O_2)) \tag{6}$$

where r is the reaction rate, N is the Avogadro number, k is the kinetic constant for NO adsorption, and K is the equilibrium constant for O$_2$ adsorption. This kinetic equation has been

Table 2 Catalytic activity of metal oxides for the decomposition of NO. Reaction conditions: P_{NO} = 2.0%; W/F = 0.5 g s cm^{-3}

Catalyst	Oxygen content (%)	Conversion of NO to N_2 (%) 500 °C	Conversion of NO to N_2 (%) 600 °C
Fe_2O_3	0	3.8	11
	5	0.2	1.8
Co_3O_4	0	25	77
	5	2.0	39
NiO	0	3.5	15
	5	–	1.9
CuO	0	3.7	9.7
	5	0.7	1.8
CeO_2	0	0.4	0.1
	5	–	–
$Ag-Co_3O_4$	0	45	38
	5	18	25

Source: Adapted from Ref.[29]

used to describe the catalytic behavior of several noble metal-containing catalysts. The main problem related to the use of noble metals for direct NO decomposition is likely that such materials are easily oxidizable and therefore not very active in the presence of oxygen. In fact, the high K value takes into account the inhibiting effect of oxygen on the kinetics of this reaction.[22]

The results obtained by Hamada et al.[23] with Pt, Rh, Pd, and Ru on Al_2O_3 (noble metal content = 0.5 wt.%) had shown that the NO decomposition activity increased with a decrease in the affinity of the metals toward oxygen. Thus, Pt is the most active metal, and this is followed by Rh, Pd, and Ru. Ogata et al.[24] observed that Pd exhibits more activity when it is dispersed on oxides containing Mg^{2+} ions (e.g. $MgAl_2O_4$, MgO, $Mg_2Si_3O_8$, $MgZrO_3$) than when it is supported on Al_2O_3. The Pd–Mg interactions significantly increase the catalytic activity, since the two components are in intimate contact with each other on an atomic scale.[25] Thus, interactions among active sites may lead to beneficial effects (synergies) for catalysis.

On the other hand, Wu et al.[26] have shown the beneficial effect of small amounts of Au or Ag on the catalytic activity of Pd/Al_2O_3 catalysts (namely 0.5% Pd - 0.03% Au/Al_2O_3 and 0.5% Pd - 0.06% Ag/Al_2O_3) obtained by co-precipitation, due to an easier reducibility of the active phase.

Frank et al.[27, 28] have observed good performances for Pt–Mo-based materials. Specifically, the catalyst with the best activity exhibited Pt–Mo–Co components, thus suggesting cooperation phenomena among several active centers.[27]. The latter catalyst decomposed ca. 60% NO at 150 °C (in the absence of CO). However, when CO was increased to 0.6%, the maximum NO_x conversion decreased to ca. 40% at 220 °C. The authors explained this catalytic behavior by showing that NO_x abatement begins when CO is fully oxidized and they argued that CO does not participate in NO_x reduction.

Metal oxides

Several metal oxides have been studied as catalysts for direct NO decomposition, as can be seen in Table 2.[29] Among these, cobalt oxide (Co_3O_4) is one of the most active compounds. The high activity of Co_3O_4 seems to be related to the relatively weak Co–O bonds, which lead to an easy desorption of lattice oxygen (β-species), especially at low temperature. Co_3O_4 shows a spinel structure with Co^{2+} and Co^{3+} cations. During the direct NO decomposition reaction, Co^{2+} may partially oxidize

to Co^{3+} and form a Co_2O_3-like phase on the surface. The latter is not stable and decomposes to Co_3O_4, thus favoring NO dissociation.[30].

Iwamoto et al.[31] have found that promoting Co_3O_4 catalysts with small amounts of Ag results in a significant increase in the catalytic activity. Moreover, the decrease in the NO decomposition activity with an increase in the O_2 concentration was less marked when Co_3O_4 was doped with Ag. With the addition of either Ag or Na, the activity in fact enhances primarily due to the excess electron density on the surface.[30]

The NO decomposition activity of metal oxides is closely related to the bond strength between the metal and the oxygen in the lattice.[17] NO usually dissociates on the metal oxides, although NO dissociation depends on several factors, including the temperature, surface coverage, crystal planes, and surface defects.[17, 20, 30] The catalytic activity of metal oxides towards the NO decomposition is usually lower than that of noble metals.[29, 30] Therefore, the operating temperature must be kept at quite high values (up to 1000 °C), which can lead to catalyst sintering, and it is of no interest for Diesel exhaust applications. As a whole, the inhibiting effect of oxygen is lower for metal oxides than for noble metals.[18, 29, 30]. Moreover, the oxygen "self-poisoning effect" towards NO decomposition can be much lower on ceria-based materials, as a consequence of oxygen spillover phenomena.[32, 33] Zhang et al.[34] doped La_2O_3 catalysts with Sr (namely 4% Sr/La_2O_3), thus promoting NO decomposition activity under different operating conditions. Other authors[35] have performed studies on CeO_2–ZrO_2 mixed oxides, which exhibit interesting activities in NO_x decomposition.

Perovskites

Perovskite-type catalysts have been widely investigated for the direct decomposition of NO, as reported in Table 3.[29] Mixed oxides with an ABO_3 chemical composition, in which the A cation is either a rare earth or an element of the II group (e.g. La, Sr, Ba, Y, etc.) and B is a transition 3d metal, belong to this category of catalysts. The substitution of A or B ions with other metals creates oxygen defects in the lattice, which are generally related to the catalytic activity, although the nature of these interactions has not yet been fully understood.[36] However, it is known that oxygen-deficient perovskites adsorb large amounts of oxygen, and that the nature and reactivity of the adsorbed oxygen, which are more weakly bonded with metal cations, are quite different from the oxygen in the

Table 3 Catalytic activity of perovskite-type oxides for the decomposition of NO. Reaction conditions: P_{NO} = 2.0% W/F = 0.5 g s cm^{-3}

Catalyst	Oxygen content (%)	Conversion of NO to N$_2$ (%) 500 °C	Conversion of NO to N$_2$ (%) 600 °C
SrFeO$_{3-x}$	0	0	0.12
	5	–	–
LaCoO$_3$	0	1.5	3.8
	5	–	0.44
La$_{0.8}$Sr$_{0.2}$CoO$_3$	0	1.6	6.3
	5	–	–
YBa$_2$Cu$_3$O$_{7-x}$	0	–	0.49
	5	–	–

Source: Adapted from Ref.[29]

lattice.[17] The order of the desorption temperature for oxygen is correlated to the NO decomposition activity: the weaker the adsorption of oxygen on the catalyst surface, the greater the mobility of the oxygen, and hence the greater the activity toward NO decomposition.[37] This suggests that surface oxygen species may act as catalytic active sites.[38] On the other hand, oxygen-deficient compounds, such as the YBa$_2$Cu$_3$O$_{7-x}$ perovskite, have been proved to be active for NO decomposition,[39] especially when supported on MgO. In perovskites, the complexity of surface defects (e.g. oxygen vacancies) is at least one order to magnitude higher than that of binary metal oxides and metals, because of the presence of ions and cations, which can assume a variety of charged states.

The effects of dopants in BaMnO$_3$ perovskites on the direct NO decomposition activity have been investigated by Iwakuni et al.[40] The authors observed that the activity of NO decomposition increased by an order of Mg > Zr > Fe > Ni > Sn > Ta for the Mn-site dopant, and La > Pr for the Ba site.

Teraoka et al.[41] considered the effect of the preparation method and catalyst composition on the activity toward NO decomposition for several perovskites with a general formula La$_x$Sr$_{1-x}$XO$_3$ (X = Co, Cr, Mn, Fe, Ni and Cu).

Kim et al.[42] observed that under realistic conditions, La$_{1-x}$Sr$_x$CoO$_3$ catalysts exhibit higher NO-to-NO$_2$ conversions than commercial Pt-based catalysts.

Complex perovskite-type catalysts have been prepared by Monceaux et al.[43] The authors have found that the substitution of a small quantity of Pt for Mn or Co makes it possible to prevent sulfur poisoning and to increase the catalytic performances of these materials toward several oxidation reactions. Perovskites exhibit high thermal stability, compared to metal oxides, and do not suffer to any great extent from the oxygen "self-poisoning effect." Since they can easily desorb oxygen, these materials should be good candidates for purifying exhaust gas from Diesel engines, particularly for the removal of soot and for NO abatement. The most significant disadvantage of NO decomposition on perovskite catalysts is the high reaction temperature required to achieve a high NO decomposition activity. Other significant drawbacks are their low surface areas and pore volumes. The ceramic solid–solid reaction and co-precipitation methods, which are commonly used for the synthesis of perovskite-type materials, involve a high reaction temperature (>900 °C) and hence yield perovskite-type oxides with low surface areas (<2 m^2 g^{-1}), due to their sintering.[44] Research should therefore be addressed to the development of new preparation techniques, aimed at obtaining perovskite structures with better textural properties.

Zeolites

Several types of metal-ion-exchanged zeolites, including X- and Y-faujasite, mordenite, ferrierite, and ZSM-5-type zeolites with different Si/Al molar ratios, have been studied for direct NO decomposition.[45, 46] Among zeolite-based catalysts, the Cu/ZSM-5 catalyst has been paid much attention, due to its superior activity and N$_2$ selectivity in a wide temperature range.[47] However, the Cu/ZSM-5 has been found to deactivate readily during high temperature hydrothermal treatment.[48] It has been reported that the primary causes for the thermal deactivation, such as agglomeration of the active metal, the migration of the reaction active sites, and the dealumination of zeolite support, can be obviously suppressed by the introduction of a second metal.[49, 50] In particular, the addition of Ce greatly improves the hydrothermal stability of the Cu/ZSM-5 catalysts, since Ce species stabilize the dispersion of CuO and suppress the bulk-type CuO crystallites formation.[51, 52]

As a whole, the NO decomposition over Cu-ZSM-5 proceeds via a redox-type mechanism.[30] In other words, Cu$^+$ ions created through a thermal pre-treatment can be oxidized to Cu^{2+} by gaseous oxygen. Thus, Cu^{2+} acts as an adsorption center for NO, as follows:

$$Cu^{2+} + NO \leftrightarrow Cu^+NO^+ \qquad (7)$$

NO molecules can be chemisorbed as NO$^+$, NO$^-$, and NO$_2^-$ species on the zeolite surface. A fraction of the Cu^{2+} ions is reduced to Cu$^+$ through the desorption of O$_2$, whereas catalyst re-oxidation with NO restores Cu^{2+} sites, thus forming N$_2$.

The catalytic behavior of Cu-ZSM-5 is negatively affected by the relatively high oxygen concentration in the feed, thus limiting their use for applications in Diesel exhausts. For instance, a maximum conversion rate of 6% has been obtained at 500 °C for a feed similar to Diesel exhausts (1000 ppm NO and 10 vol.% O$_2$), which is far too low to be acceptable.[29] Moreover, this material is not stable in water vapor conditions for long periods of time and it has proved to be sensitive to SO$_2$ poisoning.[29] As a whole, the presence of water vapor has an inhibition (although reversible) effect on the decomposition of NO, whereas SO$_2$ poisons the catalyst surface.

On the other hand, Weisweiler et al.[53] have observed that Pt-ZSM-5 catalysts may/are able to adsorb NO under controlled dynamic conditions (simulation of a driving cycle), although real NO abatement cannot be attained at low temperatures (below 180 °C).

Researchers are usually critical about the use of zeolite-type catalysts for direct NO decomposition, since they have shown low hydrothermal stability and low SO$_2$ resistance.[54, 55] On the

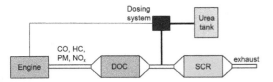

Figure 2 Schematization of DOC and SCR systems in Diesel vehicles
Source: Adapted from Ref.[10]

other hand, since the operating temperatures of these catalysts are usually higher than those required for the catalytic reduction of NO, the decomposition route, even though very attractive, is actually not very interesting for application in aftertreatment exhaust devices.

Selective catalytic reduction (SCR) of NO

Since the ideal solution of direct catalytic NO decomposition has not been successful for the control of Diesel engine emissions, researchers have begun to investigate alternative approaches, such as the selective catalytic reduction (SCR) of NO.

The catalytic reduction of NO has been studied using a number of reductants, of which ammonia, urea, and hydrocarbons are the most frequently reported.[4, 16] Catalytic reduction with ammonia (or urea) is usually referred to as NH_3-SCR, while reduction with hydrocarbons is often referred to HC-SCR.

Several catalytic materials have been developed for the SCR of NO_x since the early 1970s.[56] The first generation of commercial SCR catalysts for mobile applications was monoliths containing $V_2O_5/WO_3/TiO_2$, which were similar to the V_2O_5-based catalysts used for NH_3-SCR in stationary applications.[2] However, the stringent legislations on NO_x emissions, the necessity of new materials to extend the temperature operation window, and the toxicity of vanadium have driven the research toward the development of more effective catalysts.[30]

Use of ammonia/urea as a reductant

The reduction of NO_x using ammonia is a widely commercialized technology for large stationary combustion plants (e.g. power plants, heaters and boilers in the process industry). In Japan, USA, and Europe, large-scale applications of SCR have been introduced over the last few decades.[57–59] Ammonia is commonly used as a reductant agent in large commercial scale SCR. NH_3 is supplied to the SCR process using a gaseous solution (anhydrous form), an aqueous solution, or a solution of urea. The choice depends on the economic and safety issues involved in the handling of the preferred solution, that is, anhydrous ammonia. As a whole, a high efficiency of NO_x removal can be obtained with the NH_3-SCR process (namely 70–98%).[3] The uniqueness of this reaction with NH_3 is that it can occur in the presence of excess O_2. Thus, this technology has received a great deal of attention for Diesel-engine vehicles.

The overall NH_3-SCR reaction is[60]:

$$4NO + 4NH_3 + O_2 \leftrightarrow 4N_2 + 6H_2O \quad \text{``Standard SCR reaction''} \tag{8}$$

The role of oxygen is to donate one electron to the redox process.[17] Thus, oxygen enhances the rate of the $NO–NH_3$ reaction.[57] A complex reaction network can be observed on the catalyst surface. The main prevailing reactions are[60]:

$$6NO + 4NH_3 \leftrightarrow 5N_2 + 6H_2O \tag{9}$$

$$4NO + 4NH_3 + O_2 \leftrightarrow 4N_2 + 6H_2O \tag{10}$$

$$6NO_2 + 8NH_3 \leftrightarrow 7N_2 + 12H_2O \quad \text{``NO_2-SCR reaction''} \tag{11}$$

$$2NO_2 + 4NH_3 + O_2 \leftrightarrow 3N_2 + 6H_2O \tag{12}$$

$$NO + NO_2 + 2NH_3 \leftrightarrow 2N_2 + 3H_2O \quad \text{``Fast SCR reaction''} \tag{13}$$

These reactions are inhibited by water, which can be present in the exhaust gases. However, other reaction pathways may occur and, as a result, undesired products can be formed. These reactions may include a partial reduction of NO_x, which leads to N_2O (equations (14)–(16)), or the direct oxidation of NH_3, which forms NO (equations (17)–(18))[60]:

$$8NO_2 + 6NH_3 \leftrightarrow 7N_2O + 9H_2O \tag{14}$$

$$4NO_2 + 4NH_3 + O_2 \leftrightarrow 4N_2O + 6H_2O \tag{15}$$

$$2NH_3 + 2O_2 \leftrightarrow N_2O + 3H_2O \tag{16}$$

$$4NH_3 + 3O_2 \leftrightarrow 2N_2 + 6H_2O \tag{17}$$

$$4NH_3 + 5O_2 \leftrightarrow 4NO + 6H_2O \tag{18}$$

Particular temperature conditions (100–200 °C) may lead to the formation of NH_4NO_3, which is explosive and deposits in the cavities of the catalytic material, causing temporary deactivation.[61] One possible way of reducing the ammonium nitrate or other byproducts is to tailor the reductant injection with different amounts rather than stoichiometric with respect to NO_x[60]:

$$2NH_3 + 2NO_2 + H_2O \leftrightarrow NH_4NO_3 + NH_4NO_2 \tag{19}$$

The molar ratio of ammonia to NO_x is set below one (sub-stoichiometric) to minimize ammonia slip. The typical SCR process operates using/with an oxidation catalyst (e.g. V_2O_5–WO_3/TiO_2) downstream from the SCR, which prevents the unreacted ammonia from leaving the reactor. The oxidation catalyst may also favor the oxidation of CO and HC emissions.[60] On the other hand, an increase in N_2O and NO in the exhaust gases may occur due to the oxidation of ammonia.[63] Similarly, the catalyst may enhance SO oxidation and hence cause an increase in sulfate emissions. Since ammonia poses health and practical problems (NH_3 is a toxic gas that has to be stored under

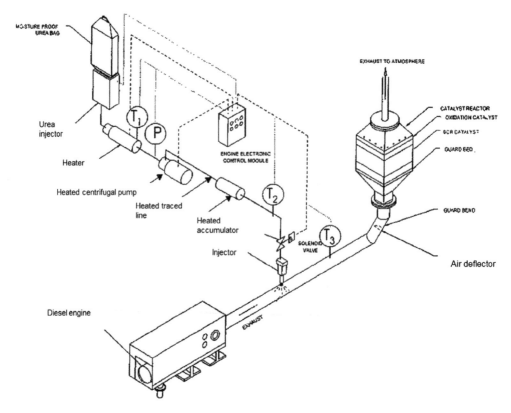

Figure 3 Experimental plant developed by Tarabulski et al. to test solid urea as a reducing agent for NO_x abatement[65]

pressure), an alternative source of NH_3 has been developed, in the form of urea, especially for non-stationary Diesel engines.[14, 30] Urea, which is a solid that is highly soluble in water, can be injected, as an aqueous solution, into the exhaust gases, where it decomposes, according to equation (20), at about 200 °C:

$$H_2N\text{-}CO\text{-}NH_2 + H_2O \rightarrow 2NH_3 + CO_2 \qquad (20)$$

Equation (20) is the result of the following two steps[62]:
(i) Thermal decomposition (hydrolysis)

$$NH_2\text{-}CO\text{-}NH_2 + H_2O \rightarrow NH_3 + HNCO \qquad (21)$$

(ii) Isocyanic acid reaction with water

$$HNCO + H_2O \rightarrow NH_3 + CO_2 \qquad (22)$$

The layout of an SCR process for mobile Diesel engines, fueled with urea, is generally structured as an open-loop control, namely the amount of injected urea follows a pre-determined route of the NO_x emissions as a function of the engine operating conditions. This technology has been shown to yield above 80% conversion of the engine-out NO_x.[2, 14] The urea solution is injected into the exhaust line upstream from the SCR catalyst. The atomization allows the solution that has been tailored to obtain a good mixing with the exhaust gasses to evaporate quickly, a process that can be assisted through the use of static mixers. A uniform distribution of the flow in the catalytic converter is necessary to reach high conversion efficiencies.[64] The SCR is usually placed after the Diesel oxidation catalyst (DOC), which is used to oxidize CO, HC, and part of the NO (Fig. 2).

The DOC oxidizes the NO to NO_2 and this compound is more reactive and extends the operating temperature window for the SCR process. In this way, the catalyst can take advantage of the "fast SCR" (equation (13)) to significantly enhance the $DeNO_x$ efficiency at low temperatures.[10, 14] The SCR catalyst can be fouled and deactivated due to the deposition of ammonium sulfate and disulfate, resulting from the oxidation SO_2, with the subsequent formation of H_2SO_4 in the DOC, and the reaction with NH_3 in the SCR. The SCR deactivation occurs at temperatures below 250 °C; hence, at low temperatures (150–250 °C), the urea injection can be interrupted to prevent SCR catalyst deactivation. Urea has mainly been selected as the best ammonia source, due to its low toxicity, safety, availability, and low cost. However, 32.5% urea solutions have freezing temperatures of −11 °C, which is not acceptable for winter conditions in cold climates. Thus, the use of ammonium formate (HCO_2NH_4) has been proposed for SCR applications in cold climates (a 40% aqueous solution of HCO_2NH_4 has a freezing point of −35 °C), but it has a lower NH_3 content than urea. An alternative reductant supply method is to use solid urea rather than aqueous solutions.

An interesting approach has been introduced by Tarabulski et al.[65] in which urea, or another reducing agent, is employed in the SCR process in a solid state (Fig. 3). Aqueous solutions of urea or other reagents are not required in the Tarabulski process. The solid reagent is fed to a gas generator that produces a reactant gas through heating; the latter gas is rich in NH_3 and can therefore be added to the exhaust gas on an as-needed

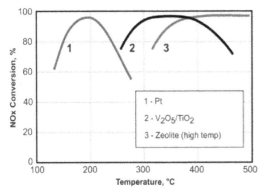

Figure 4   Operating temperature windows for different NH₃-SCR catalysts
Source: Reprinted with permission from Ref.[66]

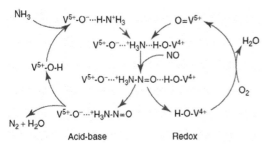

Figure 5   Catalytic cycle of the SCR reaction over a V₂O₅/TiO₂ catalyst
Source: Copyright Wiley-VCH Verlag GmbH & Co. KGaA. Reproduced with permission from Ref.[3]

basis for NO$_x$ abatement. Using urea or another solid reducing agent can cause nozzle plugging and fouling of the catalyst. This technology offers several advantages, including the realization of significant savings in energy, which would otherwise be necessary to vaporize the water, and savings on the cost of antifreeze additives. The temperature that must be reached for urea gasification is about 400 °C and, to reduce the vessel volume, it is possible to use solid catalysts, such as platinum, palladium, oxides of vanadium, titanium, and chromium. However, the SCR systems that have been proposed for dosing solid urea appear more complex than those that utilize urea water solutions, which are now the most common applications for Diesel engines.

Platinum, vanadium oxide and zeolites

Two important features of an SCR catalyst are that the material is active as an oxidation catalyst, and that materials which are effective in partial oxidation, when supported on TiO$_2$ (namely anatase), are usually good SCR catalysts.[3] Such materials can be based on V$_2$O$_5$–WO$_3$ on TiO$_2$ or V$_2$O$_5$–MoO$_3$ on TiO$_2$, although other materials (e.g. zeolites) have been considered. The main reason for such dominance is that they offer excellent performances, yet at the same time they are very tolerant toward poisons in the flue gasses.[2–4] The anatase form of TiO$_2$ is the preferred support, mainly because SO$_2$ poisoning has a lower influence on the TiO$_2$ surface.[68]

The first SCR technology that was developed was based on a Pt-containing catalyst. However, NO$_x$ reduction over the Pt surface is only effective at temperatures below 250 °C (Fig 4).[2] In fact, at temperatures between 225 and 250 °C, the oxidation of NH$_3$ to NO and H$_2$O (equation (18)) becomes dominant and, as a result, poor selectivity toward N$_2$ can be achieved/observed. On the other hand, low temperatures (150–200 °C) may lead to the above-mentioned NH$_4$NO$_3$ formation, which entails a very narrow range of available working conditions.[2] Moreover, Pt catalyzes the reduction of NO$_x$ to N$_2$O, which is a powerful greenhouse gas. Pt has the benefit of generally being insensitive to SO$_2$ and possesses good thermal stability, but it may favor the formation of SO$_3$.[2]

Vanadium oxide catalysts act well in a wider and upper temperature range, from 260 °C up to 450 °C, with the best SCR performances taking place between 300 and 400 °C.[67] This range is optimal for both light-duty (lower limit) and high duty (upper limit) applications. Catalytic materials, such as V$_2$O$_5$/TiO$_2$ or V$_2$O$_5$–WO$_3$/TiO$_2$, are capable of NO$_x$ reduction in excess of 90%, and they are most probably the best candidates to meet severe NO$_x$ reduction goals.[14] Despite discrepancies concerning the detailed nature of the active centers, there is general consensus in experimental analyses that the SCR reaction involves both Brønsted (V–OH) and Lewis (V=O) acidic sites, and hence the presence of water has significant effects on the SCR process.[14] The terminal V^{5+}=O groups of V$_2$O$_5$ appear to be essential in carrying out this reaction, since they are the energetically favored sites and, in addition, are accessible for the formation of Brønsted acidic sites. Thus, ammonia can readily adsorb on V^{4+}–OH species to form an NH$_4$$^+$ intermediate, which subsequently reacts with co-adsorbed NO to form the adsorbed (NH$_3$–NHO)$^+$ intermediate. The N–H bond in the ammonium intermediate is broken as it reacts with NO (from the gas-phase). The proton which transfers from the NH$_4$$^+$ to the V$_2$O$_5$ surface during this initial step is subsequently transferred back to the NO molecule from the V$_2$O$_5$ (redox cycle). Ultimately, V^{5+}=O groups can be restored via proton transfer. This produces gas-phase NH$_2$NO, which can be converted into N$_2$ and H$_2$O through subsequent reactions over V$_2$O$_5$.[3, 17]

Several studies have confirmed an Eley–Rideal-type mechanism (ER) for the SCR reaction, where ammonia adsorbs onto the V$_2$O$_5$ surface and NO molecules react from the gas-phase (or as weakly adsorbed species) on the solid surface. Although ER is the prevalent mechanism in most of the operating conditions, at low temperatures (<200 °C), the reaction seems better described as a Langmuir–Hinshelwood-type mechanism (LH), thus suggesting that the SCR process takes place between adsorbed NH$_3$ and NO species on the solid surface.[3] Moreover, Dumesic et al.[69] proposed an SCR catalytic cycle, which consists of two cycles interacting with each other (namely acid-base and redox cycles), thus confirming the complexity of this catalytic system. V-species can in fact act simultaneously as Lewis/Brønsted acidic and redox centers (Fig. 5).

Over the last few years, reports on health issues concerning vanadium emissions from SCR catalysts in mobile applications,[70] V$_2$O$_5$ having been classified as possibly carcinogenic to humans (group 2B),[71] have limited its further exploitation for mobile systems.

Therefore, zeolites containing transition metal ions (Cu, Fe, Cr, Mn, Ni, Ce, etc.) have been investigated extensively for mobile applications.[14] Zeolite-type catalysts have been considered for special applications, in particular for low-sulfur fuels. Unfortunately, however, zeolites are relatively expensive and thus are not suitable for extruded monoliths, although they are suitable for coated monoliths.[3]

Zeolitic materials have a relatively wide temperature range of application (300–450 °C), and the most frequently studied material for the SCR reaction is Cu-exchanged ZSM-5.[14, 15, 29, 30, 50] This catalyst does not oxidize NH_3 to NO_x at high temperatures, and the upper temperature limit for the SCR process depends on its structural stability.[72]

Low-temperature zeolites have also been synthesized, with the aim of widening the temperature range at which they are operative, and moderate NO_x to N_2 conversion efficiencies of between 200–400 °C have been reached.[73] Metal-exchanged zeolites have been found to adsorb considerable amounts of NH_3 under certain operating conditions, and this results in rather slow responses to temperature variations or changes in NO_x concentration. The NH_3 adsorption capacity of catalysts depends to a great extent on the temperature; unwanted ammonia slip occurs for the increases of fast temperature.[61]

Moreover, it has been observed that the SCR reaction shows considerable sensitivity to the nature of the support, and hence comparative studies on different zeolite supports (ZSM-5, A, beta, FAU, ferrierite, CHA, Linde type L) have been performed. From these studies, it emerged that Fe- or Cu-exchanged ZSM-5 zeolite showed good activity and selectivity for N_2 production.[14] However, they demonstrate a lack of hydrothermal stability at temperatures above 700 °C. In particular, Cu-CHA catalysts have become the state of the art in NH_3-SCR catalysis for Diesel vehicles, due to their excellent low temperature activity and high hydrothermal stability,[74, 75] which makes it a valuable candidate in the functionalization of DPFs with NH_3-SCR catalysts, in order to increase the operating temperatures of the SCR catalysts by being close-coupled with the DOC instead of in underfloor position, in the so-called "SCR on filter" concept. Indeed, compared to ZSM-5 or beta-type zeolites, the chabazite zeolite contains small-sized pores and can coordinate isolated Cu^{2+} species, which are more resistant to hydrothermal aging.[72]

Many other NH_3-SCR catalysts have been investigated over the years. Among these, it is worth mentioning Fe_2O_3, Fe-containing mixed oxides, and Fe-exchanged materials that may show good SCR performances.[30, 73] Similarly, Mn-oxides have received much/ a great deal of interest for the SCR reaction because of their low-temperature activity, although they exhibit low selectivity for N_2 production. In particular, $MnCr_2O_4$ spinel-Oxide Catalysts have shown promising catalytic results, as they can reach an NO conversion of 96% and selectivity to N_2 of 97% at 125 °C.[76]

Use of hydrocarbons as reductants

An alternative to the use of ammonia as a reductant agent is the employment of hydrocarbons, namely the HC-SCR process. The latter is also known as $DeNO_x$ or lean-NO_x process.

According to this approach, the hydrocarbons can be oxidized by the oxygen present in NO, as follows:

$$NO + \text{hydrocarbon} + O_2 \rightarrow N_2 + CO_2 + H_2O \quad (23)$$

The reaction of equation (23), which leads to N_2, CO_2, and H_2O, is not the only path involved in the NO reduction, since undesired products, such as N_2O, can also be obtained.

It is thought that the HC-SCR reaction proceeds competitively with the combustion reaction of hydrocarbons, and the selectivity to N_2 determines the feasibility of the HC-SCR process. The selectivity can be defined as the ratio between the amount of hydrocarbon oxidized by NO, with subsequent N_2 formation, and the total reacted hydrocarbon. Thus, the HC-promoted SCR reaction of the NO reduction and the simple HC combustion are[60]:

$$C_XH_Y + (2X + 1/2Y)NO \leftrightarrow (X + 1/4Y)N_2 + 1/2YH_2O \quad (24)$$

$$C_XH_Y + (X + 1/4Y)O_2 \leftrightarrow XCO_2 + 1/2YH_2O \quad (25)$$

Consequently, catalyst selectivity is the key parameter that has to be optimized using suitable catalytic materials, as well as a suitable reducing agent, HC/NO_x ratio, temperature range, and so on.

As highlighted by/in Zelenca et al.,[77] the research should be conducted in three main directions:

- Choice, characterization, and improvement of suitable catalytic materials, which should be industrially available;
- Determination of the most efficient hydrocarbon through the injection of different hydrocarbon compounds into the synthetic gas ahead of the catalyst to increase NO_x conversion rates;
- Optimization of the different parameters that affect the catalyst performance, in order to improve its use with vehicles.

An important parameter for NO_x abatement is the HC to NO_x ratio; generally, a two to fourfold surplus of hydrocarbons (expressed as ppm-HC) relative to the NO_x concentration (ppm-NO_x) is necessary to reach ca. 80% NO_x conversion.[78–80] As such a surplus of hydrocarbons is not usually present in Diesel exhaust gases, hydrocarbons or Diesel fuel would have to be added to the Diesel exhaust gases.[81–83] There are two main possibilities of HC enrichment, that is, low or high pressure injection of Diesel fuel ahead of the catalyst ("active $DeNO_x$") and utilization of unburned HC emissions directly from the engine exhaust gas ("passive $DeNO_x$"). In "active $DeNO_x$" systems, the increase in the HC amount, and the control of their concentration to optimize the SCR catalyst can be realized by two means: the injection of hydrocarbons, preferably Diesel fuel, into the exhaust system upstream of/ from the catalyst, or late in-cylinder injection in a common rail fuel system. The DOC system, which has a classical CO/HC emission reduction function, can be positioned downstream from the $DeNO_x$ system, in order to also undertake the role of preventing possible HC emissions.

On the other hand, "passive $DeNO_x$" should be a simpler and cheaper option, since no additional injection equipment would be necessary. However, since the HC concentration in the exhausts is dependent on the engine points and it is

Table 4 Catalytic performances of several materials towards the HC-SCR reaction

Catalyst	Reductant	Catalytic performances	Note
Cu/ZSM-5	C_3H_6	Maximum conversion: 33% (370 °C) Conversion higher than 20% (320–450 °C)	No production of N_2O
Co/Al_2O_3	CH_3OH	Maximum conversion: 40% (370 °C) Conversion higher than 20% (>290 °C)	No production of N_2O but CH_3OH emissions
Ag/Al_2O_3	C_2H_5OH	Maximum conversion: 31% (460 °C) Conversion higher than 20% (>370 °C)	N_2O and aldehydes produced from 370 °C
Pt/Al_2O_3	C_3H_6	Maximum conversion: 41% (215 °C) Conversion higher than 20% (200–270 °C)	High N_2O production
Pt/ZSM-5	C_3H_6	Maximum conversion: 33% (240 °C) Conversion higher than 20% (220–270 °C)	High N_2O production from 240 °C

Source: Adapted from Ref.[85]

somewhat limited, low conversion efficiencies can be reached. In order to achieve the necessary HC concentrations, engine modifications should be undertaken, in order to obtain higher hydrocarbon emissions. As a result, active DeNO$_x$ systems offer higher NO$_x$ conversion efficiency, but at the cost of increased system complexity and a fuel economy penalty.

A wide variety of metal oxides, alumina-based catalysts, zeolites, and perovskites have so far been tested for this promising technology.[84] The first catalyst screening, which shows the catalytic behavior of several metals (Pt, Au, Cu, Co and Ag) supported on Al_2O_3 and ZSM-5, was published by Obuchi et al.[85] Then, Ag/alumina have been found to be promising catalysts for the selective catalytic reduction of NO$_x$ to N_2 by hydrocarbons in laboratory tests as well as in full-scale diesel engine operation.[86]

The addition of small amounts of H_2 can promote the HC-SCR activity of Ag-based catalysts for low temperature NO$_x$ reduction. Over the years, several explanations have been proposed for this beneficial "H_2-effect", including the enhancement of the partial oxidation of the reducing agent, the formation of reactive N species (NCO-like groups or gas phase radicals) from the reducing agent, the easier formation of active cationic Ag clusters, and the destabilization of surface nitrates blocking active Ag sites. Hence, H_2 promotes the HC-SCR activity of Ag-based catalysts through multiple roles, involving morphological, chemical, and kinetic changes.[87, 88]

Appropriate reductants, such as C_3H_6, CH_4, CH_3OH and C_2H_5OH (which are efficient reductant agents), were used. The catalytic performances for each catalyst are summarized in Table 4. The most promising materials are based on platinum, although the high emission of N_2O is a difficult problem to solve. For this reason, the most suitable catalyst seems to be the Co/Al_2O_3 one, with the addition of an oxidation catalyst downstream to prevent the leakage of non-reacted or incompletely oxidized reductants. De Soete[89] has published an interesting work that shows the reduction rate expressions of NO to N_2 in the HC-SCR reaction over Cu/ZSM-5. The author has found that C_2H_4 and C_6H_{14}, when used as reductants, show different catalytic behavior: the reaction order in C_2H_4 is negative, whereas it is positive for C_6H_{14}. In fact, ethylene is less active in NO reduction than n-hexane, although HC oxidation is lower.

Many studies[82, 89] have shown that hydrocarbons do not participate directly in the reduction of NO, but they are first partially oxidized to active intermediates (e.g. aldehydes) which then can react with NO to form N_2 and O_2. On the other hand, Bell et al.[90–93] reported that, for HC-SCR over Co-, Mn-,

Fe- and Pd-ZSM-5, the highly active species are CN groups which react with NO_2 to form N_2 and CO_2.

Zeolite-type catalysts

As a whole, the HC-SCR activities of zeolite-type catalysts are better than those of metal oxides, such as alumina. The crystalline structure seems to contribute to the high activity of the zeolites. Thus, several zeolite-related compounds, such as metallo-silicates and silicoaluminophosphates, have been reported to be active for the HC-SCR reaction.[14, 30, 46]

Many ion-exchanged zeolites, such as Cu-, Fe-, Pt-, Co-, Ga-, Ce- and H-exchanged zeolites, have been found to be active for this reaction. In particular, Cu- and Co-ZSM-5 have received a great deal of attention over the last few decades.[2, 3, 11, 17, 46] Several studies have revealed/pointed out the key role of reductant agents toward selectivity to N_2. It has been observed, for Cu-ZSM-5, that some hydrocarbons (namely C_2H_4, C_3H_8, C_4H_8, and alcohols) behave as selective reductants, while other reductants (i.e. H_2, CO and CH_4) are non-selective toward N_2 production.[94] Conversely, CH_4 has been shown to be a selective reductant over Co-ZSM-5 and Ga- or In-ZSM-5 catalysts.[95, 96] Erkfeldt et al.[97] established that a C–C bond in the reducing agent is required for lean NO$_x$ reduction over Cu-ZSM-5. The influence of the hydrocarbon concentration on the activity of a Cu-ZSM-5 catalyst has been investigated by Konno et al.,[81] who reported that the reduction rate increases with the hydrocarbon concentration up to an HC/NO molar ratio of ca. 8. The presence of some co-exchanged cations (Ca, Sr, Fe, Co and Ni) had the effect of expanding the temperature range over which a Cu-ZSM-5 can be active for the HC-SCR process.[42, 98] Interestingly, lanthanum co-exchange resulted in an improvement in the NO adsorption capacity of the zeolite, both in the absence and in the presence of water.[99]

Pt-ZSM-5 is more stable than Cu-ZSM-5: Iwamoto et al.[94] investigated the long-term stability of Pt-ZSM-5 under simulated and actual exhaust conditions, and found that its activity did not decrease in the presence of water vapor or SO_2 in the reactant stream. Moreover, they observed that the catalyst activity barely changes after 1000 h exposure to water vapor.[100] On the other hand, the catalytic performances of Cu-exchanged zeolites are affected by the presence of SO_2 (Fig. 6).

Falley et al.[101] have prepared a new catalytic material, based on a zeolite chosen from between β-Zeolite Y-Zeolite and ZSM-5, in which Cu, Co, and Fe have been incorporated as active species. They observed that a combination of these three metals tends to lower the temperature at which a Cu-containing

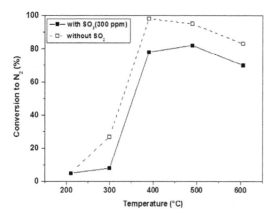

Figure 6 Conversion of NO + O$_2$ + C$_3$H$_6$ system into N$_2$, with and without SO$_2$. Catalyst Cu-MFI. Without SO$_2$: P$_{NO}$ = 500 ppm, P$_{O2}$ = 1.0%; P$_{C3H6}$ = 990 ppm, W/F = 0.1 g cm^{-3}; With SO$_2$) P$_{NO}$ = 530 ppm, P$_{O2}$ = 1.0%; P$_{C3H6}$ = 1000 ppm, W/F = 0.1 g cm^{-3}, P$_{SO2}$ = 300 ppm
Source: Adapted from Ref.[94]

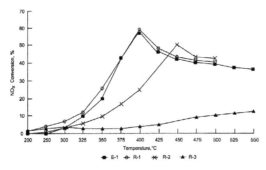

Figure 7 NO$_x$ conversion vs. temperature obtained using fresh ZSM-5-based catalytic material, with the addition of small amounts of Cu, Co and Fe: E-1 3.31% Cu, 2.27% Fe, 0.72% Co; R-1 3.22% Cu, 1.96% Fe; R-2 3.17% Cu, 3.13% Co; R-3 3.3% Co, 1.98% Fe
Source: Adapted from Ref.[94]

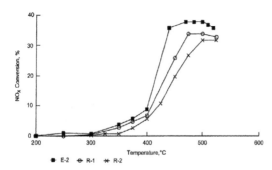

Figure 8 NO$_x$ conv. vs. T obtained using an aged (through exposure to a mixture of 10% steam in air for 5 h at 700 °C) ZSM-5-based catalytic mat. with addition of small amounts of Cu, Co and Fe: E-2 3.46% Cu, 2.03% Fe, 1.35% Co; R-1 3.22% Cu, 1.96% Fe; R-2 3.17% Cu, 3.13% Co
Source: Adapted from Ref.[101]

catalyst reaches its optimum NO$_x$ conversion rate (Fig. 7), particularly after aging (Fig. 8). This trimetallic-based system lends stability to the catalytic material, so that NO$_x$ conversion rates after an accelerated ageing cycle are higher than those of comparative materials.

Although Cu is the exchange metal chosen for most studies, many other metals have been tested, either alone or as co-cations: interesting tests were carried out using Gallium and Cerium. Gallium exchanged ZSM-5 was tested in SCR with 1000 ppmv of NO and C$_3$H$_6$ by Yogo et al.,[96] who found that this catalyst exhibits similar catalytic behavior to that of Cu-based systems.

Over the last few years, new multicomponent catalysts have been designed to expand the lean NO$_x$ reduction capacity of zeolite-type catalysts. For instance, Deeba et al.[102] reported interesting NO$_x$ conversions over new four-way catalysts, which were active over a temperature range of between 150 and 320 °C. These materials allow hydrocarbons to be activated and stored; these hydrocarbons then reduce NO$_x$ selectively. Moreover, synergistic phenomena (e.g. cooperative adsorption) in the HC-SCR reaction can be observed for zeolite-type catalysts when different hydrocarbons (e.g. CH$_4$ and C$_3$H$_8$) are used as reductant agents.[103, 104]

Supported noble metals
Due to the low hydrothermal stability of zeolitic materials, several researchers have addressed their efforts to the development of noble metal catalysts.[105–108] An interesting work on the catalyst choice for the HC-SCR process was conducted by Nakatsuji et al.,[105] and remarkable results were shown for the Ag/Al$_2$O$_3$ catalyst. This catalyst can be active at 300 °C and exhibit good catalytic stability in the presence of SO$_x$, which is further improved by the addition of WO$_3$, MoO$_3$, and Pt. The authors observed that catalyst activity can be improved by adding the aldehydes that form from the injected fuel in the exhaust gas. Thus, they tried to partially oxidize fuel in a catalytic oxidizing reactor before injecting it into flue gas. The results obtained using this system are very interesting: at 450 °C, the engine bench test exhibited an NO$_x$ conversion of 75%.

Several works have been published on platinum-group metals, mainly due to their relatively high surface stability.[107, 108] Bamwenda et al.[109] studied platinum-group metals (Pt, Pd, Rh and Ir) deposited on different supports (TiO$_2$, ZnO, ZrO$_2$ and Al$_2$O$_3$) for the SCR of NO in the presence of C$_3$H$_6$. They found that the alumina supported catalysts showed both the highest activity for NO conversion and the highest selectivity toward N$_2$ formation. The catalytic activities followed the Pt/Al$_2$O$_3$ > Rh/Al$_2$O$_3$ > Pd/Al$_2$O$_3$ > Ir/Al$_2$O$_3$ order, whereas Rh/Al$_2$O$_3$ was found to have the highest selectivity toward N$_2$ formation.[110]

Other studies have been carried out for Ag supported on Al$_2$O$_3$[105–108] or saponite.[108] Miyadera et al.[105] have shown that 2 wt.% Ag/Al$_2$O$_3$ can reduce NO with ethanol (a higher molar ratio of C$_2$H$_5$OH/NO$_x$ than 1.25), even in the presence of water vapor, although several byproducts may appear (e.g. N$_2$O, NH$_3$, CH$_3$CN and HCN). Cyanide is the dominating byproduct below 400 °C (the typical exhaust temperature of a Diesel engine), although its concentration decreases at higher temperatures.

The influence of the reducing agents (C_1–C_3 hydrocarbons) on the HC-SCR of NO in a Ga_2O_3–Al_2O_3 system has been studied by Miyahara et al.[110]: the most efficient NO abatement was achieved with C_2 hydrocarbons, whereas C_3 hydrocarbons were less selective for N_2 production. Thus, it has been proposed that only one carbon atom per molecule is used for/in the HC-SCR process, and the other carbons are consumed by the combustion reaction. However, under wet conditions, the NO conversion on the Ga_2O_3–Al_2O_3 catalyst decreased for all the hydrocarbons. The decrease in NO conversion caused by the presence of water is due to the preferential adsorption of water, which inhibits the HC adsorption on the catalyst surface (competitive adsorption).[110] Noble metals should be added to avoid the retarding effect of water vapor. For instance, Haneda et al.[109] have studied the catalytic behavior of an Indium supported catalyst (5 wt.% In/TiO_2–ZrO_2) for the SCR of NO with C_3H_6 with or without the presence of water: a decrease in NO conversion occurred under wet conditions. However, the authors observed that small amounts of Pd (0.005–0.02 wt.%) may improve the "resistance" of In/TiO_2–ZrO_2 against the presence of water (up to ca. 400 °C).

Highly dispersed Au catalysts have been studied for the HC-SCR of NO in recent years.[111] These catalysts are active in the reduction of NO with C_3H_6 in the presence of O_2 and moisture. The catalytic behavior of Au-species depends on the metal oxide support. In fact, an activity scale has been drawn up: α-Fe_2O_3 ~ ZnO < MgO ~ TiO_2 < Al_2O_3. The highest conversion to N_2 (around 70%) at 427 °C has been achieved over a 0.1–0.2 wt.% Au/Al_2O_3 catalyst.[111, 112] It is worth noting that the conversion of NO to N_2 over Au/Al_2O_3 was increased slightly by the presence of water. Thus, in order to improve the performances of the latter catalyst, Mn_2O_3 was then mixed with Au/Al_2O_3.[113] This mechanical mixture exhibited interesting results for the HC-SCR of NO in the 250 and 500 °C temperature range. This catalyst has been considered one of the most effective for the NO emission control of lean-burn gasoline and Diesel engines.[111–113]

NO$_x$ traps

NO$_x$ traps (or NO$_x$ adsorbers) constitute an interesting NO$_x$ control technology for gasoline direct injected engines and for Diesel engines. NO$_x$ traps are also referred to using different terms: Lean NO$_x$ traps (LNT), NO$_x$ adsorber catalysts (NAC), DeNO$_x$ traps (DNT), NO$_x$ storage catalysts (NSC), NO$_x$ storage/reduction (NSR) catalysts.

NO$_x$ traps, which are incorporated into the catalyst washcoat, chemically bind the NO$_x$ species and convert them into solid species (metal bonded nitrates). NO$_x$ accumulation is carried out during lean conditions, and proceeds until the adsorber capacity is saturated. Then, the NO$_x$ trap is regenerated, and the released NO$_x$ species are reduced to N_2 when a rich air-to-fuel mixture is injected.[114]

Since the rich mode of operation is not feasible for Diesel engines, periodic fuel injections are necessary. The amount of fuel and the periodicity of the injections, as well as the storage capacity of the materials, are the main parameters that need to be optimized to reduce the fuel penalty associated with this technology.

Among the main drawbacks, it should be pointed out that the supply of additional fuel to either the cylinder or directly to the exhaust pipe, causes PM, CO, and HC emissions, whose concentration must comply with the current legislation limits. Moreover, the regeneration step must be carried out as efficiently as possible in order to prevent it from having too much of an impact on the fuel economy of the vehicle. Finally, NO$_x$ adsorbers can be poisoned by sulfur compounds,[2, 115, 116] thus requiring both the use of low sulfur Diesel fuels and the development of efficient desulfation strategies. Therefore, low sulfur fuels favor NO$_x$ conversion levels and reduce the frequency of the desulfation step.

Operating principles

During operating conditions, the NO$_x$ reduction takes place according to a two-stage mechanism, as shown in Fig. 9. The NO$_x$ trap combines the effect of an oxidation catalyst (e.g. platinum), an adsorbent material (e.g. BaO), and a reduction catalyst (e.g. rhodium).

The exhaust is rich in NO (a thermodynamically stable species at high-temperatures), but traps are more effective toward NO_2 entrapment. Therefore, the step immediately before NO$_x$ adsorption should be the NO oxidation to NO_2 step. This operation is performed by means of an oxidation catalyst, such as a Pt-containing system, which is able to operate at the low exhaust temperatures of light-duty Diesel engines, as follows[60, 117]:

$$NO + 1/2 O_2 \leftrightarrow NO_2 \tag{26}$$

The NO_2 formed on the solid surface is trapped on an adsorbent (BaO) in the form of a nitrate (e.g. $Ba(NO_3)_2$), which is chemically stable at the operating conditions.

Thus, the following reaction steps can occur[10, 60]:

$$BaO + NO_2 \leftrightarrow BaO - NO_2 \tag{27}$$

$$BaONO_2 \leftrightarrow BaO_2 + NO \tag{28}$$

$$BaO_2 + 2NO_2 \leftrightarrow Ba(NO_3)_2 \tag{29}$$

The adsorption capacity of an NO$_x$ trap depends on the accessibility of the BaO sites, which must be regenerated when a certain NO$_x$ concentration is attained at the exit of the converter.

During the regeneration process, the oxygen concentration decays to almost zero, and reductant conditions for NO abatement are thus achieved.

The first regeneration step consists of the decomposition of the $Ba(NO_3)_2$ and the recovery of the BaO active phase (equation (30)). In this step, NO is released and hence a suitable amount of fuel has to be dosed in order to reduce the released NO.[60]

$$Ba(NO_3)_2 \leftrightarrow BaO + 2NO + 3 1/2 O_2 \tag{30}$$

NO reduction is carried out by means of a reducing catalyst, such as an Rh-based system, incorporated in the catalyst formulation.[118] This reduction step is similar to that which occurs in a conventional three-way converter for the treatment of the exhausts from gasoline fueled engines. When the engine is switched to a fuel-rich condition, HC, CO, and H_2 react with the

**Lean Conditions
(adsorption)**

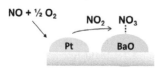

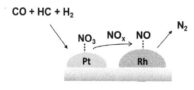

**Rich Conditions
(regeneration)**

Figure 9   The basic concept of NO$_x$ storage and reduction mechanisms
Source: Reprinted from Refs.[10,66]

NO species to form N$_2$, CO$_2$, and H$_2$O (equations (24), (31) and (32)).[60]

$$NO + CO \leftrightarrow 1/2N_2 + CO_2 \qquad (31)$$

$$NO + H_2 \leftrightarrow 1/2N_2 + H_2O \qquad (32)$$

The operating temperature of these NO$_x$ traps has a lower limit, which is determined by the Pt activity toward the NO oxidation, as well as the NO$_x$ release and reduction in the regeneration step; on the other hand, the upper limit is related to the stability of the NO$_3$ species, which undergo thermal decomposition at high temperatures, even under lean conditions.[119]

It is worth noting that several complex phenomena may appear on the surface of NO$_x$ traps, since they are multicomponent materials which have different functionalities. For instance, BaCO$_3$ and Ba(OH)$_2$ coexist with BaO on the catalyst surface.[120] Moreover, NO$_x$ release and reduction may not occur as consecutive steps, but, as a first step, it could appear via a direct nitrate reduction, without thermal decomposition of the adsorbed NO$_x$ species.[121] Therefore, it is very difficult to investigate the NO$_x$ storage/release mechanisms of these catalytic materials.

NO$_x$ adsorbents are sensitive to sulfur species: sulfur is a poison for Pt-sites and, in the form of SO$_3$, it is competitive with NO$_2$ in the formation of barium salts (e.g. BaSO$_4$), thus causing a loss of activity toward the adsorption of NO$_2$ (= competitive adsorption). BaSO$_4$ is more stable than the corresponding nitrate, and hence its decomposition occurs at higher temperatures.[119] Finally, the possibility of having parasitic reactions that lead to undesired products (e.g. NH$_3$, N$_2$O, H$_2$S) reduces the NO$_x$ trap efficacy, and a careful control of the secondary emissions is also necessary.

Multifunctional catalysts

NO$_x$ traps are multifunctional catalysts that display adsorbent, oxidation, and reduction functionalities. These materials consist of alkaline earth metals (Ba, Ca, Sr, Mg), alkali metals (K, Na, Li, Cs), and rare earth metals (La and Y).[121, 122] As a whole they appear in the form of binary oxides, mixed oxides (perovskites), and metal-containing zeolites. In particular, alkali metals, such as potassium, show the highest NO$_x$ conversion efficacy among these metal groups: the inclusion of K-, Na-, and Cs- based oxides in the adsorbent catalyst formulation increases the NO$_x$ reduction to between 350 and 600 °C.[12] Another remarkable advantage of alkali metals is their good resistance to sulfur poisoning: their inclusion in Ba-containing materials leads to a better performance; however, this in turn leads to a lower NO conversion, due to the hydrocarbons which makes the adsorbent catalyst more fuel demanding.[12] Finally, alkali-based materials exhibit low performances and give rise to leaching effects in the presence of water vapor.[12] It is therefore necessary to find a trade-off between the good NO$_x$ reduction activity at high temperatures of alkali metal oxides and their excessive mobility, in order to include them in Ba-based systems.

NO$_x$ traps for light-duty Diesel engines, which operate at lower temperatures, may eliminate the necessity of using alkali metals.

A washcoat is usually obtained using γ-Al$_2$O$_3$. Washcoats are often employed due to their high surface area (>100 m^2 g^{-1}) this allows high dispersion of the active sites. Ba and Al mixed oxides are designed to limit BaO sintering in the 700–800 °C temperature range.[123] Another washcoat component is TiO$_2$, whose acidity gives a lower sulfur affinity, although it also reduces the stability of nitrates.[126] CeO$_2$ can be considered a good washcoat component as it prevents Pt-sintering. However, its high oxygen storage capacity (OSC) can cause a higher fuel penalty during rich fuel conditions, since some of the hydrocarbons react with the released oxygen. The effects of BaO loading on the trapping capacity of NO$_2$ and on the overall conversion performance of NO$_x$ are summarized in Fig. 10.[127] In this work, the catalyst consisted of Pt (2.20 wt.%) and BaO (16.3 wt.%) over Al$_2$O$_3$, and it showed a mean NO conversion of 85% over cycles lasting 60 s, with injection of C$_3$H$_6$ every 10 s. The complete cycle involves a fuel penalty which can be kept under 4% with an overall NO conversion above 80%. The regeneration strategy is closely related to the nature of the reductant species. The regeneration step is constituted by a short pulse of a reducing agent, such H$_2$, CO, or HC, to convert the stored nitrates. The activity of a complex multicomponent catalyst toward NO$_x$ reduction is reported in the work by Takahashi et al.[128]: the catalyst consisted of an Al$_2$O$_3$-based washcoat of CeO$_2$–ZrO$_2$ oxygen storage materials, with Ba and K oxides as the NO$_x$ storage compounds and with Pt and Rh as the supported noble metals. Figure 11 reports a schematization of the adsorption capacity of the fresh/regenerated catalyst: the capacity of the reductant species to restore the BaO sites is considered. "Lean 1" is the inlet composition of the gas fed (namely at 250 °C), and represents a gas model for the exhausts; when the lean atmosphere is switched on, the outlet NO$_x$ concentration gradually increases with time, up to reach a constant level.

Shaded area A is related to the NO$_x$ amount stored in the catalyst, while shaded area B is related to the number of regenerated NO$_x$ storage sites on the catalyst. Takahashi et al.[128] then

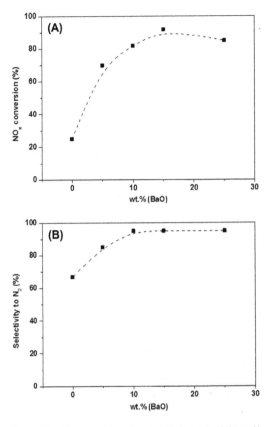

Figure 10 NO$_x$ conversion (Section A) and selectivity to N$_2$ (Section B) on BaO loading. NO = 500 ppm, O$_2$ = 5% (with C$_3$H$_6$ = 0.7% during regeneration), temperature = 375 °C and GHSV = 60,000 h^{-1}
Source: Adapted from Ref.[127]

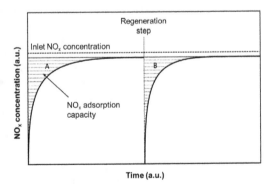

Figure 11 NO$_x$ concentration curves, during lean and rich periods, in the inlet and outlet using Lean1 and gases at 250 °C
Source: Adapted from Ref.[128]

established that the effectiveness of the reducing agents follows the following order: C$_3$H$_6$ < CO < H$_2$.

One extremely interesting attempt to improve the NO$_x$ absorbers activity is the one from Toyota,[129] which investigated the effect of an oscillating air: fuel ratio for the regeneration

of the trap, in the so-called Di-Air system. The fundamental finding of this phenomenon is that a high-frequency injection of hydrocarbons might considerably improve the NO$_x$ regeneration phase, at high temperatures, with a very moderate fuel penalty (<2%).

The combination of different catalytic functionalities could also be implemented to achieve synergies that increase the overall NO$_x$ abatement efficiency. A relevant example is the embodiment of an SCR catalytic functionality in a NO$_x$ trap. This system was presented by Ford in 2004, and widely studied by the group at the University of Houston (see the general concept in[130]). The concept is based on the fact that, in the presence of H$_2$ during the rich pulses of the NO$_x$ trap regeneration, NH$_3$ can be formed by direct reaction with the previously stored NO$_x$[131]; CO instead leads to NH$_3$ formation through intermediate isocyanate species, in water-assisted reaction. The water gas shift reaction makes both pathways to co-exist, due to the presence of CO$_2$ and H$_2$O in the exhausts.[131] The suitable proportion of the SCR and NO$_x$ trap catalysts in the overall formulation is determined by the amount of NH$_3$ generated in the NO$_x$ trap during its regeneration, which should be utilized by the SCR adsorption sites. Since it was found that only the neighboring SCR catalyst concurs in adsorbing the NH$_3$ produced in the NO$_x$ trap in the reduction stage, a dual-layer configuration (two layers with different composition) outperforms the dual-brick (consecutive washcoat formulations along the axial length), largely because the NH$_3$ generated in the NO$_x$ trap layer is better utilized in an adjacent SCR layer.[130] The optimal configuration in terms of relative thickness of the layers depends on the operating temperatures and SCR site density.

Plasma-assisted catalytic NO$_x$ reduction

Non-thermal, atmospheric pressure plasma has been widely studied for the removal of VOC from waste gas streams for almost 20 years.[132]

Non-thermal plasma (NTP) discharges in exhaust gas have been investigated as a potential technology to reduce NO$_x$ and PM emissions in Diesel exhaust as well as NO$_x$ and cold start hydrocarbons in lean gasoline exhaust.

NTP is generated, by means of an electrical excitation, to induce thermodynamically unstable NO$_x$ species; the latter should decompose into N$_2$ and O$_2$.

In automotive applications, the reduction of NO$_x$ is preferable. On the other hand, oxidative NO$_x$ reactions dominate the subsequent kinetics, converting most flue gas NO to NO$_2$ and HNO$_3$[133–135] through reactions with plasma-produced oxygen and hydroxyl (equations (33) and (34)).

$$O + NO \leftrightarrow NO_2 \qquad (33)$$

$$OH + NO_2 \leftrightarrow HNO_3 \qquad (34)$$

Several other byproducts can be produced, including N$_2$O, N$_2$O$_3$, N$_2$O$_4$, HONO, and HONO$_2$, and the selectivity of the plasma aftertreatment therefore remains an important drawback, as does the large power requirement. The electric power supply has similar characteristics for each experimental setup: the ΔV range is from 5 to 35 kV and the frequency is about 1 kHz. Park et al. investigated NO$_x$ abatement for Diesel

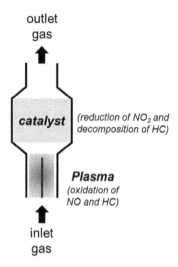

outlet
gas

catalyst *(reduction of NO₂ and decomposition of HC)*

Plasma
(oxidation of NO and HC)

inlet
gas

Figure 12 Scheme of the plasma-assisted catalytic reduction system developed at Lawrence Livermore National laboratory
Source: U.S. Patent N° 5711147. Issued January 27, 1998. Adapted from Ref.[136]

exhaust, for which pulsed voltages (namely 24 kV and 600 Hz) were shown to be more effective than a direct current (about 80% NO_x conversion).[134] HNO_3 (and other acidic byproducts) can be removed easily at a stationary combustion facility in the form of solid nitrates using a downstream scrubber. However, neither chemical scrubbing nor the conversion of NO to HNO_3 constitutes a practical approach for mobile engine NO_x control, due to the requirements of onboard scrubbing chemicals, the periodic disposal of accumulated solid waste, and the requirement of an acid scrubber. Yamamoto et al.[135] have proposed a plasma reactor followed by a Na_2SO_3 solution scrubber to reduce volumes, even though this technology can only be used for Diesel engine trucks.

Combinations of plasma with solid catalysts, referred to as "plasma-assisted catalysts" or simply "plasma catalysts", have been suggested for NO_x reduction.[135]

As a whole, plasma reactors are pulsed corona-type reactors and the catalyst is generally placed after the plasma device (Fig. 12).[139] Although plasma aftertreatment favors oxidation over reduction, a useful synergism can be observed when it is combined with a catalyst.[134] The reducing capability of some catalysts is enhanced considerably when NO_x is presented to catalytic surfaces as NO_2 (peroxyl radicals favor the conversion of NO to NO_2) rather than NO.[137-140] This means that the NTP technology may significantly improve catalyst selectivity and removal efficiency. Thus, lower temperatures can be used in NO_x reduction with plasma-catalyst technology than with NTP. An increase in temperature improves catalytic activity but the action of plasma is only retained from room temperature to 300 °C. Moreover, the presence of water vapor and oxygen promotes the NO removal rate, as a consequence of synergistic phenomena on the catalyst surface.[137]

Several catalysts, including γ-Al_2O_3, TiO_2, MnO_2/TiO_2, Ag/TiO_2, V_2O_5-WO_3/TiO_2, HZSM-5, NaX, Cu-ZSM and Na-Y and Ba-Y, have been proposed to be active for this technology.[135, 138, 139] The best performance for NO_x removal at low temperatures

(150–270 °C) has been achieved using Na- and Ba-doped zeolite Y. However, promising results (90% NO_x removal) have also been obtained with In/γ-Al_2O_3 catalysts in the presence of sulfur oxides.[140]

Over the years, several strategies have been proposed to improve NO_x abatement via plasma-assisted catalytic technologies. Among these, the multiple-step treatment strategy, whereby two or more plasma-catalyst reactors are utilized in series, has been shown to increase the maximum NO_x conversion.[141] Moreover, this technology should reduce the energy and/or hydrocarbon supplies for fixed conversion efficiency. Similarly, the HC-SCR process occurs via oxidation of NO, followed by the reduction of NO_2 with hydrocarbons (*vide supra*); hence, NTP can oxidize NO without depleting the number of hydrocarbons available for the reduction of NO_2 to N_2.[142] This means that the function of SCR catalysts could be greatly simplified by focusing on the reduction of NO_2 by hydrocarbons.

A different approach has been studied by Bittenson et al.,[143] who proposed generating atomic nitrogen in an electric discharge external to the exhaust stream, followed by rapid injection and mixing into exhaust gas to achieve an NO_x chemical reduction.

$$N + NO \leftrightarrow N_2 + O \qquad (35)$$

This approach leads to the elimination of the generation of very reactive oxygen and hydroxyl species by the discharge, thus preventing contact between soot and wet particles and the electrode of the plasma reactor; in this case, the electrical power supply has a larger frequency: 40–60 kHz.

Simultaneous NO_x and soot removal systems

The stringent regulation limits on both NO_x and PM are forcing the automotive industry, one hand, to maximize engine control to reduce pollutant emissions, and on the other hand to pile up a number of costly catalytic converters; this results in rather high pressure drops, complex control features, inefficiency linked to considerable weight and space consumption and elevated costs. In this context, Diesel particulate traps appear to be necessary. Our research group has been involved in various European R&D programs (e.g. CATATRAP, DEXA-cluster, TOP-Expert, ATLANTIS) for several years, with the aim of developing novel multifunctional catalytic traps.[144] The simultaneous removal of soot and NO_x in a single catalyzed filter (single brick solution) represents the most ambitious strategy in this field, in view of the considerable advantages in terms of both investment costs and pressure drop reduction.

The possibility of obtaining a contemporary reduction in NO_x and soot from Diesel engine exhausts has clearly been pointed out in several kinetic studies (Fig. 13).[14, 15, 145-149] Fino et al.[145] studied the kinetics of a soot-NO-O_2 reacting system over perovskite-type catalysts and formulated two reaction mechanisms (Fig. 14). In *mechanism 1*, soot combustion leads to the formation of two oxygen vacancies on the catalytic surface, which become active centers for the chemisorption of two NO molecules. Hence, the adsorbed NO dissociates into N (ads) and O (ads) with subsequent formation of either N_2 or N_2O. However, this mechanism does not account for the positive effect of molecular oxygen, which should conversely

Mechanism 1:	Mechanism 2:
$C + \text{-[O][O]-} \rightarrow \text{-[ ][ ]-} + CO_2$	$NO(g) \rightarrow NO_{ad}$
$NO + \text{-[ ][ ]-} \rightarrow \text{-[O]-[ ]-}$	$C_f + NO_{ad} \rightarrow C[N,O]$
$NO + \text{-[O]-[ ]-} \rightarrow \text{-[O]-[O]-}$	$C[N,O] + NO_{ad} \rightarrow CO_2(g) + N_2(g) (+ C_f)$
$\text{-[O]-[O]-} \rightarrow N_2O + \text{-[O]-[ ]-}$	$C[N,O] + NO(g) \rightarrow CO_2(g) + N_2(g)$
$\text{-[O]-[O]-} \rightarrow N_2 + \text{-[O]-[O]-}$	$C[N,O] + NO_{ad} (+ C_f) \rightarrow CO_2(g) + C[N,N]$
	$C[N,N] \rightarrow C + N_2(g)$
	$NO(g) + \tfrac{1}{2}O_2 \rightarrow NO_2$
	$NO_2(g) \leftrightarrow NO_{ad} + O_{ad}$
	$C[N,N] + O_{ad} \rightarrow N_2O(g) + C$

Figure 13 Reaction mechanisms proposed for the soot/NO$_x$/ O$_2$ reacting system over perovskite-type catalysts
Source: Adapted from Ref.[142]

lower the reaction rate by filling up the oxygen vacancies over the oxidation catalysts.[145–147] On the other hand, according to *mechanism 2*, carbon plays a role in the reduction of NO through the formation of C(N,O) adducts. The latter are formed through the combination of reactive free carbon (C$_f$) and the NO molecule adsorbed on the catalyst surface (NO$_{ad}$). Finally, N$_2$ can be formed through the reaction of C(N,O) with further NO$_{ad}$ (LH-type kinetics) or directly with a gaseous NO molecule (ER-type kinetics). However, it is also possible for N$_2$ to be formed through the decomposition of C(N,N) adducts. According to this mechanism, the promoting effect of oxygen can be ascribed to the formation of active centers for NO chemisorption on C$_f$ as a consequence of carbon combustion. The other beneficial effect is the easy formation of NO$_2$, which is much more active than NO for the reduction of soot. Thus, the presence of soot has a beneficial effect in a combined SCR + CSF at low temperature, since it may promote the reduction of NO$_2$ (via the "fast SCR reaction"). On the other hand, the NO$_2$/NO ratio has to be kept close to one for effective SCR + SCF systems.

DeSoot-DeNO$_x$ Catalysts

La-K-Cu-V-based perovskites[145] have shown their potential application as active catalysts for the simultaneous removal of NO$_x$ and soot. The role of vanadium species in the perovskite lattice has been found to be prevalent in obtaining outstanding NO$_x$ abatement efficiencies. Nevertheless, several other complex reaction pathways involving reaction intermediates, either present on the catalyst surface or on that of the carbon particulate itself, could take place.[145] On the other hand,

nanostructured spinel-type oxide catalysts, that is, AB$_2$O$_4$ (where A = Co and Mn, B = Cr and Fe) have proved to be particularly active in the simultaneous removal of soot and NO$_x$ in the 350–450 °C temperature range.[146, 147] The best compromise between soot and NO abatement below 400 °C has been shown by the CoCr$_2$O$_4$ catalyst. The relevant activity of chromite catalysts could be explained by their higher amount of suprafacial, weakly chemisorbed oxygen (α-species), which contributes actively to soot combustion through spillover in the 300–500 °C temperature range. Similarly, nanostructured perovskite-type lanthanum ferrites, that is, La$_{1-x}$A$_x$Fe$_{1-y}$B$_y$O$_3$ (where A = Na, K, Rb and B = Cu) have displayed high catalytic activity toward carbon combustion and NO conversion in the same temperature range.[144] Other multicomponent catalysts, characterized by the highest possible α-oxygen type concentration, have been prepared by our group over the last few years, with careful attention being paid to their compatibility with the substrate material or to the poisoning components present in the Diesel exhaust gas (e.g. sulfur oxides). Using a standard protocol on an engine bench, promising results have been obtained with the La$_{0.9}$K$_{0.1}$Cr$_{0.9}$O$_{3-\delta}$ + 1 wt% Pt catalyst over a wall-flow SiC trap.[150] The presence of Pt in fact aids the oxidization of NO and therefore allows more NO-NO$_2$-NO cycles before it leaves the catalyst (the efficiency of NO$_x$ utilization can be as high as 140% on average). Li et al.[151, 152] and Kotarba et al.[153] have reported that K-promoted CoMgAlO hydrotalcite can favor soot combustion, but can also lead to a 30% conversion of NO$_x$ under "real" conditions. Lin et al.[154] have shown high catalytic activity of BaAl$_2$O$_4$ systems for the simultaneous removal of soot and NO$_x$ under "loose" contact conditions.

It is worth noting that ceria-based catalysts have received a great deal of attention because CeO$_2$ alone, or in combination with other metals/metal oxides, may exhibit promising oxidation activity under either O$_2$ or in a NO$_x$/O$_2$ atmosphere.[155] Thus, zirconium and many rare earth elements (e.g. La, Pr, Sm, Y, Gd, Tb, Lu, Hf and Nd) have been introduced into the ceria framework to improve the oxidation activity (OSC and redox properties) of CeO$_2$-based materials and their structural stability. The redox behavior and the availability of chemisorbed oxygen (α-species) are important features for/of these materials.[156–158] Atribak et al.[149] have reported interesting results with Ce–Zr mixed oxides for soot combustion under NO$_x$/O$_2$. Among the different mixed oxides, Ce$_{0.76}$Zr$_{0.24}$O$_2$ provided the best performance.[159] The same authors observed that the most active mixed oxides are those which combine high surface areas

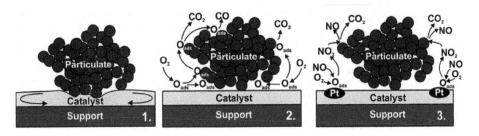

Figure 14 Simultaneous abatement of NO$_x$ and soot over different solid catalysts: 1 = mobile catalyst; 2 = catalyst promoting oxygen spillover; 3 = catalyst coupling a NO → NO$_2$ functionality[148]

with a homogeneous distribution of cerium and zirconium throughout the particles (Ce/Zr surface ratio about 3.2).[149]

The number of soot-catalyst contact points also plays a significant role in complex solid–solid–gas systems, just as the overall activity for NO_x and soot removal depend on the interaction between the two solids and the gas mixture.[160, 161] Nanostructured CeO_2-based materials, or other mixed oxides, are particularly interesting because of their small-featured size, which endows them with size- and shape-dependent properties due to the high surface-to-volume ratio (= higher number of coordinated unsaturated sites), and their unique electronic features (quantum size effects).[162–165] Moreover, the effects of cooperation between active sites (or between different phases) are favored, thus leading to a simultaneous NO_x and soot abatement.

Some issues pertaining to Diesel NO_x aftertreatment catalysts

One of the main current issues for NO_x aftertreatment catalysts is that no material seems to ensure a high NO_x abatement throughout the whole Diesel exhaust temperature range (i.e. 200–500 °C).[2–4, 8, 14, 30] As reported above, many multicomponent catalysts have been designed over the years to increase their NO_x abatement activity and to cover a broader temperature range. This is often possible since synergistic (beneficial) effects may arise through phase-cooperation (namely, the structural relations among atoms/ions/electrons) and spillover phenomena.[166, 167] Moreover, solid catalysts must be active and stable during operating conditions. As far as the former aspect is concerned, catalytic materials are usually tested under laboratory conditions, which can be sometimes be different from real operating conditions. For instance, researchers may perform the activity tests with relatively low GHSV or O_2 concentrations; however, high values of these parameters are known to adversely affect NO_x abatement. Similarly, two features of Diesel exhausts can be critical for catalyst stability, that is, the sulfur content and the presence of water vapor associated with high temperatures. Sensitivity to sulfur poisoning can also be a major problem for metal-exchanged zeolites.[95] Feeley et al.[102, 168] have found that long-term exposure of Cu-ZSM-5 to SO_2 leads to its permanent deactivation. Conversely, Pt-based catalysts may exhibit good resistance to poisoning by SO_2. Nevertheless, these catalysts favor the oxidation of SO_2 and thus cause an increase in the total mass of the particulate emitted. Moreover, phosphorous from lubricating oil can cause a decline in catalyst performance.

In the case of low-temperature DeNO$_x$ catalysts, a further problem still has to be solved, that is, the formation of N_2O instead of N_2 with a subsequent negative environmental impact, due to the fact that N_2O is a greenhouse gas.[169]

Sensitivity to water vapor is a very critical limitation for the automotive application of solid catalysts. In fact, since combustion exhaust streams are rich in water vapor (10–16%), catalysts should be able to operate in the presence of high levels of water vapor. Given the fact that NO_x levels are often <2% (typically at ca. 50 ppm level), it is very difficult to find a catalyst with active sites, which are necessary for NO sorption, that will not be flooded by water. Because of the high exhaust velocities (GHSV > 30,000 h⁻¹), it is impractical to try

to separate such large quantities of water from such a huge volume of gas. Therefore, the thermal and hydrothermal stability of catalysts is pivotal for any successful NO_x decomposition technology. The effect of water on the stability of Cu-ZSM-5 has been widely investigated (vide supra).[47–50] Zeolite activity decreases to various extents in the presence of water, according to the amount of water and the reaction temperature. This effect is probably due to the fact that water, like NO_x, is a good Lewis base and competes for the same active sites on which NO_x reduction must occur.[95] Conversely, Corma et al.[9] observed that the deactivation of Cu-ZSM-5 occurs through the partial dealumination of the zeolite, the reversible migration of copper species, and the irreversible formation of catalytically inactive and stable Cu–Al clusters, which have some resemblance to $CuAl_2O_4$, but without the symmetry of Cu in the spinel structure. In order to improve the hydrothermal resistance of zeolites, several methods have been proposed over the years, but none of them seems so far to have completely solved the problem. On the other hand, as previously mentioned, Pt-ZSM-5 (Pt nanoparticles dispersed on the zeolite surface) is more resistant to water vapor.[170] Nevertheless, the low selectivity, and hence the relatively high amounts of N_2O formed in the presence of this catalyst are important issues that still need to be solved. Nanostructured mixed metal oxides, on the other hand, can be considered suitable candidates for NO_x abatement, due to their unique electronic features and (usually) high hydrothermal stability.

This analysis is consistent also in the case of the use of biofuels: as far as emissions are concerned, biodiesel produces substantially smaller amounts of CO, hydrocarbons (HC), and soot, which has been reported to dramatically decrease by 50, 70, and 50%, respectively, for pure biodiesel and to moderately decrease by 12, 20, and 12% for a diesel with a 20% biodiesel content.[171] These are average values, but a more than 80% particulate matter emission reduction was recorded in tests with pure biodiesel[172] with respect to fossil one. These emission reductions are reached at the price of a modest 10% emission increase in nitrogen oxides (NO_x) for pure biodiesel, which is close to zero for a 20% biodiesel content,[171] therefore not entailing radically different catalytic formulations.

Moreover, as in the case of innovative biofuels like farnasane, which often require a re-calibration of the engine map to achieve combustions efficiencies comparable to the ones of conventional diesel, beneficial results in terms of emissions can be achieved at the same time, thus eliminating the problem of increased pollutant emissions.[173]

Conclusions

The progressive requirements for fuel-efficient diesel cars highlight the problem of the necessity of removing NO_x under lean-burn conditions in order to satisfy current legislation. Since direct NO decomposition cannot be applied successfully to control Diesel engine emissions, researchers have begun to investigate alternative approaches, such as the selective catalytic reduction of NO_x by means of ammonia/urea or hydrocarbons. The SCR of NO_x by ammonia/urea has become a key technology for the aftertreatment of exhaust emitted from Diesel and other lean-burn engines. In fact, a high efficiency of NO_x removal can be obtained with NH_3-SCR (namely 70–98%).

The uniqueness of this technology is that the reaction can occur in the presence of excess O_2. Thus, the NH_3-SCR approach has received a great deal of interest for Diesel-engine vehicles. On the other hand, the HC-SCR of NO_x has been investigated extensively since the early 1990s, and many catalysts have been tested. Although promising results have been obtained with this technology (more than 80% NO_x conversion), most of the research was conducted in the absence, or low presence, of sulfur. Moreover, it has been revealed that the narrow activity temperature range and the poor activity below 300 °C are relevant drawbacks for the application of HC-SCR catalysts.

Recently, growing interest has been shown in the adsorption of NO_x (NO_x traps) from lean exhaust, followed by release and catalytic reduction under rich conditions. This strategy has been shown to achieve ca. 70–90% NO_x reduction, and seems the most promising approach for NO_x abatement in diesel engines, since it does not require a reducing agent. NTP has been proposed as a promising technology to reduce NO_x and PM emissions in Diesel exhaust. At present, however, this approach appears more complicated than other advanced technologies for NO_x abatement.

The stringent regulation limits for both NO_x and PM emissions are forcing the automotive industry to pile up a number of costly catalytic converters, which results in rather high pressure drops, sophisticated control systems, inefficiency linked to considerable weight and space consumption and hence elevated costs. Therefore, the simultaneous removal of soot and NO_x in a single filter catalyzed represents the most ambitious strategy, in view of the great advantages that can be obtained in terms of both investment costs and pressure drop reduction. In this scenario, synergistic effects play a key role in catalytic converters. This means that it is necessary to obtain more detailed knowledge of the surface phenomena (reaction mechanisms, cooperative effects, etc.) in order to be able to develop effective catalytic systems.

References

1. J. M. Pardiwala, F. Patel and S. Patel: Proc. International Conference on Current Trends in Techology, Nuicone, 2011.
2. R. M. Heck, R. J. Farrauto and S. T. Gulati: 'Catalytic air pollution control: commercial technology', 3rd edn; 2009, New Jersey, NJ, (Wiley-VCH), Hooken.
3. G. Ertl, H. Knözinger, F. Schüth and J. Weitkamp: 'Handbook of heterogeneous catalysis', 2nd edn; ; 2008, Weinheim: (Wiley-VCH).
4. P. Eastwood: 'Critical topics in exhaust gas aftertreatment'; 2000, Baldock, Research Studies Press Ltd.
5. I. Fechete, Y. Wang and J. C. Védrine: Catal Today, 2012, 189, 2–27.
6. O. Deutschmann: 'Modeling and Simulation of Heterogeneous Catalytic Reactions', 330–331; 2012, Weinheim, Wiley-VCH.
7. A. G. Konstandopoulos, M. Kostoglou and N. Vlachos: Int. J. Vehicle Des., 2006, 41, 256–284.
8. D. Duprez and F. Cavani: 'Handbook of advanced methods and processes in oxidation catalysis', 25–50; 2014, London, Imperial College Press.
9. V. Vestreng, L. Ntziachristos, S. Semb, I. S. A. Isaksen and L. Tarrason. Atmos. Chem. Phys., 2009, 9, 1503–1520.
10. P. Barbaro and C. Bianchini: 'Catalysis for sustainable energy production'; 2009, 393–438, Weinheim, Wiley-VCH.
11. I. Nova and E. Tronconi: 'Urea-SCR technology for deNO$_x$ after treatment of diesel exhausts'; 2014, Verlag, Springer.
12. S. Yashnik and Z. Ismagilov: Appl. Catal. B Environ., 2015, 170–171, 241–254.
13. M. Rutkowska, U. Díaz, A. E. Palomares and L. Chmielarz: Appl. Catal. B: Environ., 2015, 168–169, 531–539.
14. P. Granger and V. L. Parvulescu: Chem. Rev., 2011, 111, 3155–3207.
15. Z. Liu and S. I. Woo: Catal. Rev., 2006, 48, 43–89.
16. J. Hagen: 'Industrial catalysis: a practical approach', 2nd edn, 317–328; 2006, Weinheim, Wiley-VCH .
17. R. A. van Santen and M. Neurock: 'Molecular heterogeneous catalysis'; 2006, Weinheim, Wiley-VCH.
18. H. Falsig, T. Bligaard, C. H. Christensen and J. K. Nørskov: Pure Appl. Chem., 2007, 79, 1895–1903.
19. C. T. Goralski and W. F. Schneider: Appl. Catal. B, 2002, 37, 263–277.
20. V. Johánek, S. Schauermann, M. Laurin, C. S. Chinnakonda, S. Gopinath, J. Libuda and H.-J. Freund: J. Phys. Chem. B, 2004, 108, 14244–14254.
21. Q. Ge and M. Neurock: J. Am. Chem. Soc., 2004, 126, 1551–1559.
22. J. K. Nørskov, F. Studt, F. Abild-Pedersen and T. Bligaard: 'Fundamental concepts in heterogeneous catalysis'; 2014, Veinheim, Wiley.
23. H. Hamada, Y. Kintaichi, M. Sasaki and T. Ito. Chem. Lett., 1990, 7, 1069–1070.
24. A. Ogata, K. Obuchi, K. Mizuno, A. Ohi, H. Aoyama and H. Ohuki. Appl. Catal., 1990, 65, L11–L15.
25. H. Huang, Y. Xu, Q. Feng and D. Y. C. Leung: Catal. Sci. Technol., 2015, 5, 2649–2669.
26. R. J. Wu, T. Y. Chou and C. T. Yen: Appl. Catal. B Environ., 1995, 6, 105–116.
27. B. Frank, G. Emig and A. Renken: Appl. Catal. B Environ., 1998, 19, 45–57.
28. B. Frank, R. Lubke, G. Emig and A. Renken: Chem. Eng. Technol., 1998, 21, 494–502.
29. M. Iwamoto and H. Hamada: Catal. Today, 1991, 10, 57–71.
30. S. Roy, M. S. Hegde and G. Madras: Appl. Energy, 2009, 86, 2283–2297.
31. M. Iwamoto, H. Yahiro, Y. Torikai, T. Yoshioka and N. Mizuno: Catal. Lett., 1990, 11, 1967–1970.
32. Z. Say, E. I. Vovk, V. I. Bukhtiyarov and E. Ozensoy: Appl. Catal. B Environ., 2013, 142–143, 89–100.
33. Sharma V, P. A. Crozier, R. Sharmaa and J. B. Adams: Catal. Today, 2012, 180, 2–8.
34. X. Zhang, A. B. Walters and M. A. Vannice: Appl. Catal. B, 1996, 7, 321–336.
35. P. Esteves, Y. Wu, C. Dujardin, M. K. Dongare and P. Granger: Catal. Today, 2011, 176, 453–457.
36. D. Marrocchelli, N. H. Perry and S. R. Bishop: Phys. Chem. Chem. Phys., 2015, 17, 10028–10039.
37. L. F. Liotta, M. Ousmane, G. Di Carlo, G. Pantaleo, G. Deganello, G. Marcì, L. Retailleaud and A. Giroir-Fendler: Appl. Catal. A, 2008, 347, 81–88.
38. C. Tofan, D. Klvana and J. Kirchnerova: Appl. Catal. A, 2002, 223, 275–286.
39. Y. Shimada, S. Miyama and H. Kuroda: Chem. Lett., 1998, 10, 1797–1800.
40. H. Iwakuni, Y. Shinmyou, H. Yano, H. Matsumoto and T. Ishihara: Appl. Catal. B, 2007, 74, 299–306.
41. Y. Teraoka, H. Ogawa, H. Furukawa and S. Kagawa: Catal. Lett., 1992, 12, 361–366.
42. C. H. Kim, G. Qi, K. Dahlberg and W. Li: Science, 2010, 327, (5973), 1624–1627.
43. V. Blasin-Aubé, J. Belkouch and L. Monceaux: Appl. Catal. B Environ., 2003, 43, (2), 175–186.
44. S. Banerjee and V. R. Choudhary: J. Chem. Sci., 2000, 112, 535–542.
45. A. Corma: J. Catal., 2003, 216, 298–312.
46. J. Čejka, A. Corma and S. Zones: 'Zeolites and catalysis'; 2010, Weinheim, Wiley-VCH.
47. H. Sjövall, L. Olsson, E. Fridell and R. J. Blint: Appl. Catal. B Environ., 2006, 64, 180–188.
48. J. Park, H. J. Park, J. H. Baik, I. Nam, C. Shin, J. Lee, B. K. Cho and S. H. Oh: J. Catal., 2006, 240, 47–57.
49. P. N. Panahi, D. Salari, A. Niaei and S. M. Mousavi: J. Ind. Eng. Chem. 2013, 19, 1793–1799.
50. C. Seo, B. Choi, H. Kim, C. H. Lee and C. B. Lee: Chem. Eng. J., 2012, 191, 331–340.
51. L. Pang, C. Fan, L. Shao, K..Song, J. Yi, X. Cai, J..Wang, M. Kang.and T. Li: Chem. Eng. J., 2014, 253, 394–401.
52. M. Yu. Kustova, S. B. Rasmussen, A. L. Kustov and C. H. Christensen: Appl. Catal. B, 2006, 67, 60–67.
53. W. Weisweiler and R. Wunsch: Chem. Eng. Process, 1998, 37, 229–232.
54. P. N. R. Vennestrøm, T. V. W. Janssens, A. Kustov, M. Grill, A. Puig-Molina, L. F. Lundegaard, R. R. Tiruvalam, P. Concepción and A. Corma: J. Catal., 2014, 309, 477–490.
55. Q. Ye, L. Wang and R. T. Yang: Appl. Catal. A Gen., 2012, 427–428, 24–34.
56. G. Busca: 'Heterogeneous catalytic materials', 478; 2014, Amsterdam, Elsevier.
57. R. D. Reitz: Combust. Flame, 2013, 160, 1–8.

58. M. V. Twigg: *Appl. Catal. B Environ.*, 2007, **70**, 2–15.
59. B. Guan, R. Zhan, H. Lin and Z. Huang: *J. Environ. Manag.*, 2015, **154**, 225–258.
60. J. Yan: 'Handbook of clean energy systems', Vol. 2, 1083–1109; 2015, New York, NY, Wiley.
61. I. Lezcano-Gonzalez, U. Deka, B. Arstad and A. Van Yperen-De: *Phys. Chem. Chem. Phys.*, 2014, **16**, 1639–1650.
62. M. Koebel, M. Elsener and G. Madia: *SAE Tech. Pap. 2001-01-3625*, 2001.
63. I. Gekas, L. Nyengaard.and T. Lund: *SAE Tech. Pap. 2002-01-0289*, 2002.
64. W. Mathes, F. Witzel and S. Schnapp: International patent application WO 99/05402, 1999.
65. J. T. Tarabulski and J. D. Peter-Hoblyn, US Patent 5809775, 1998.
66. www.dieselnet.com
67. B. K. Yun and M. Y. Kim: *Appl. Therm. Eng.*, 2013, **50**, (1), 152–158.
68. Y. Shu, H. Sun, X. Quan and S. Chen: *J. Phys. Chem. C*, 2012, **116**, (48), 25319–25327.
69. J. A. Dumesic, N. Y. Topsøe, H. Topsøe, Y. Chen and T. Slabiak: *J. Catal.*, 1996, **163**, 409–417.
70. 'Advanced Clean-Energy Vehicles (ACEVs), Project Summary'; 2004, Tokyo, Japan, Japan Automobile Research Institute JARI.
71. 'Chemicals known to the state to cause cancer or reproductive toxicity'; 2005, Oakland, CA, California Environmental Protection Agency, OEHHA.
72. O. Kröcher, M. Devadas, M. Elsener, A. Wokaun, N. Söger, M. Pfeifer, Y. Demel and L. Mussmann: *Appl. Catal. B*, 2006, **66**, 208–216.
73. J. Gieshoff, M. Pfeifer, A. Schafer-Sindlinger, P. Spurk, G. Garr and T. Leprince: *SAE Tech. Pap. 2001-01-0514*, 2001.
74. S. J. Schmieg, S. H. Oh, C. H. Kim, D. B. Brown, J. H. Lee, C. H. F. Peden and D. H. Kim: *Catal. Today*, 2012, **184**, 252–261.
75. V. Bacher, C. Perbandt, M. Schwefer, R. Siefert, S. Pinnow and T. Turek: *Appl. Catal. B*, 2015, **162**, 158–166.
76. M. A. Zamudio, N. Russo and D. Fino: *Ind. Eng. Chem. Res.*, 2011, **50**, 6668–6672.
77. P. Zelenka, W. Cartellieri and P. Herzog: *Appl. Catal. B*, 1996, **10**, 3–28.
78. A. Frobert, S. Raux, S. Rousseau and G. Blanchard: *Top. Catal.*, 2013, **56**, (1–8), 125–129.
79. J. M. Herreros, P. George, M. Umar and A. Tsolakis: *Chem. Eng. J.*, 2014, **252**, 47–54.
80. V. Houel, P. Millington, R. Rajaram and A. Tsolakis: *Appl. Catal. B Environ.*, 2007, **73**, (1–2), 203–207.
81. A. Frobert, S. Raux, A. Lahougue, C. Hamon, K. Pajot and G. Blanchard: *SAE Int. J. Fuels Lubr.*, 2012, **5**, (1), 389–398.
82. H. Gu, K. M. Chun and S. Song: *Int. J. Hydrogen Energy*, 2015, **40**, 9602–9610.
83. B. Guan, R. Zhan, H. Lin and Z. Huang: *Appl. Therm. Eng.*, 2014, **66**, 395–414.
84. R. Mrad, A. Aissat, R. Cousin, D. Courcot and S. Siffert: *Appl. Catal. A*, 2014, available at http://dx.doi.org/10.1016/j.apcata.2014.10.02.
85. A. Obuchi, I. Kaneko, J. Oi, A. Ohi, A. Ogata, G. R. Bawenda and S. Kushiyama: *Appl. Catal. B*, 1998, **15**, 37–47.
86. A. Obuchi, I. Kaneko, J. Oi, A. Ohi, A. Ogata, G. R. Bawenda and S. Kushiyama: *Appl. Catal. B Environ.*, 1998, **15**, 37–47.
87. J. P. Breen and R. Burch: *Top. Catal.*, 2006, **39**, 53–58.
88. P. S. Kim, M. K. Kim, B. K. Cho, I. S. Nam and S. H. Oh: *J. Catal.*, 2013, **301**, 65–76.
89. G. G. De Soete: *Comb. Sci. Technol.*, 1996, **121**, 103–121.
90. L. J. Lobree, A. W. Aylor, J. A. Reimer and A. T. Bell: *J. Catal.*, 1997, **169**, 188–193.
91. A. W. Aylor, L. J. Lobree, J. A. Reimer and A. T. Bell: *J. Catal.*, 1997, **170**, 390–401.
92. L. J. Lobree, I. Hwang, J. A. Reimer and A. T. Bell: *Catal. Lett.*, 1999, **63**, 233–240.
93. L. J. Lobree, A. W. Aylor, J. A. Reimer and A. T. Bell: *J. Catal.* 1998, **181**, 189–204.
94. M. Iwamoto and H. Yahiro: *Catal. Today*, 1994, **22**, 5–18.
95. T. Maunula, J. Ahola and H. Hamada: *Appl. Catal. B Environ.*, 2006, **64**, 13–24.
96. F. Bin, C. Song, G. Lv, J. Song, X. Cao, H. Pang and K. Wang: *J. Phys. Chem. C*, 2012, **116**, (50), 26262–26274.
97. S. Erkfeldt, A. Palmqvist and M. Petersson: *Appl. Catal. B*, 2011, **102**, 457–554.
98. A. Sultana, M. Sasaki, K. Suzuki and H. Hamada: *Catal. Commun.*, 2013, **41**, 21–25.
99. G. Landi, L. Lisi, R. Pirone, G. Russo and M. Tortorelli: *Catal. Today*, 2012, **191**, 138–141.
100. H. K. Shin, H. Hirabayashi, H. Yahiro, M. Watanabe and M. Iwamoto: *Catal Today*, 1995, **26**, 13–21.
101. J. S. Feeley, M. Deeba, R. J. Farrauto, D. Dang: US Patent 5776423 A, 1998.
102. M. Deeba, J. Feeley, R. Farrauto, N. Steinbock and A. Punke: *SAE Tech. Pap. 952491*, 1995.
103. T. N. Burdeinaya, V. A. Matyshak, V. F. Tretyakov, L. S. Glebov, A. G. Zakirova and M. A. Carvajal: *Appl. Catal. B*, 2007, **70**, 128–137.
104. M. K. Neylon, M. J. Castagnola, N. B. Castagnola and C. L. Marshall: *Catal. Today*, 2004, **96**, 53–60.
105. T. Miyadera: *Appl. Catal. B*, 1997, **13**, 157–165.
106. K. Arve, K. Svennerberg, F. Klingstedt, K. Eränen, L. R. Wallenberg, J.-O. Bovin, L. Čapek and D Yu Murzin: *J. Phys. Chem. B*, 2006, **110**, 420–427.
107. K. Masuda, K. Shinoda, T. Kato and K. Tsujimura: *Appl. Catal. B*, 1998, **15**, 63–73.
108. H. Kannisto, H. H. Ingelten and M. Skoglundh: *Top. Catal.*, 2009, **52**, 1817–1820.
109. G. R. Bamwenda, A. Ogata, A. Obuchi, J. Oi, K. Mizuno and J. Skrzypek: *Appl. Catal. B*, 1995, **6**, 311–323.
110. Y. Miyahara, M. Takahashi, T. Masuda, S. Imamura, H. Kanai, S. Iwamoto, T. Watanabe and M. Inoue: *Appl. Catal. B*, 2008, **25**, 289–296.
111. Y. Zhang, R. W. Cattrall, I. D. McKelvie and S. D. Kolev: *Gold Bull.*, 2011, **41**, 145–153.
112. A. Ueda, T. Oshima and M. Haruta:. *Appl Catal. B*, 1997, **12**, 81–93.
113. A. Ueda and M. Haruta: *Appl. Catal. B*, 1998, **18**, 115–121.
114. V. G. Milt, C. A. Querini, E. E. Mirò and M. A. Ulla: *J. Catal.*, 2003, **220**, 424–432.
115. N. Rankovic, C. Chizallet, A. Nicolle, D. Berthout and P. Da Costa: *Oil Gas Sci. Technol.* 2013, **68**, (6), 951–1113.
116. W. S. Epling, L. E. Campbell, A. Yezerets, N. W. Currier and J. E. Parks II: *Catal. Rev.*, 2004, **46**, 163–245.
117. A. Lindholm, N. W. Currier, E. Fridell, A. Yezerets and L. Olsson: *Appl. Catal. B*, 2007, **75**, 78–87.
118. L. J. Gill, P. G. Blakeman, M. V. Twigg and A. P. Walker: *Top. Catal.*, 2004, **28**, 157–164.
119. V. G. Milt, A. Querini, E. E. Mirò and M. A. Ulla: *J. Catal.*, 2003, **220**, 424–432.
120. L. Lietti, P. Forzatti, I. Nova and E. Tronconi: *J. Catal.*, 2001, **204**, 175–191.
121. I. Nova, L. Lietti, L. Castoldi, E. Tronconi and P. Forzatti: *J. Catal.*, 2006, **239**, 244–254.
122. K. Krutzsch, D. Webster, E. Chaize, S. Hodjati, C. Petit, V. Pitchon, A. Kiennemann, R. Loenders, O. Monticelli, P. A. Jacobs, J. A. Martens, B. Kasemo, M. Weibel and G. Wenninger: *SAE Tech. Pap. 982593*, 1998.
123. J. S. Hepburn and W. Watkins, European Patent Application, EP 0 857 510 A1, 1998.
124. D. Dou and J. Balland: *SAE Tech. Pap. 2002-01-0734*, 2002.
125. W. A. Cutler and J. P. Day: *SAE Tech. Pap. 1999-01-3500*, 1999.
126. I. Hachisuka, T. Yoshida, H. Ueno, N. Takahashi, A. Suda and M. Sugiura: *SAE Tech. Pap. 2002-01-0732*, 2002.
127. K. S. Kabin, R. L. Muncrief, M. P. Harold and Y. Li: *Chem. Eng. Sci.*, 2004, **59**, 5319–5327.
128. N. Takahashi, K. Yamazaki, H. Sobukawa and H. Shinjoh: *Appl. Catal. B*, 2007, **70**, 198–204.
129. Y. Bisaiji, K. Yoshida, M. Inoue, K. Umemoto and T. Fukuma, JSAE 20119272, SAE *Tech. Pap. 2011-01-2089*.
130. B. M. Shakya, M. P. Harold and V. Balakotaiah: *Chem. Eng. J.*, 2014, **237**, 109–122.
131. F. Can, X. Courtois, S. Royer, G. Blanchard, S. Rousseau and D. Duprez: *Catal. Today*, 2012, **197**, 144–154.
132. D. Duprez and F. Cavani: 'Handbook of advanced methods and processes in oxidation catalysis', 155–172; 2014, London, Imperial College Press.
133. T. Hammer and S. Broer: *SAE Tech. Pap. 982428*, 1998.
134. M. C. Park, D. R. Chang, M. H. Woo, G. J. Nam and S. P. Lee: *SAE Tech. Pap. 982514*, 1998.
135. T. Yamamoto and C. L. Yang: *SAE Tech. Pap. 982432*, 1998.
136. B. M. Penetrante and R. C. Brusasco: SERDP Project Cp-1077, 2001.
137. G. Yu, Q. Yu, K. Zeng and X. Zhai: *J. Environ. Sci.*, 2005, **17**, 846–848.
138. J. Van Durme, J. Dewulf, C. Leys and H. Van Langenhove: *Appl. Catal. B*, 2008, **78**, 324–333.
139. J. H. Kwak, J. Szanyi and C. H. F. Peden: *J. Catal.*, 2003, **10**, 291–298.
140. D. N. Tran, C. L. Aardahl, K. G. Rappe, P. W. Park and C. L. Boyer: *Appl. Catal. B*, 2004, **48**, 155–164.
141. R. G. Tonkyn, S. E. Barlow and J. W. Hoard: *Appl. Catal. B*, 2003, **40**, 207–217.

142. P. Talebizadeh, M. Babaie, R. Brown, H. Rahimzadeh, Z. Ristovski and M. Arai: *Renewable Sustainable Energy Rev.*, 2014, **40**, 886–901.

143. S. N. Bittenson and F. E. Becker: *SAE Tech. Pap. 982515*, 1998.

144. D. Mescia, J. C. Caroca, N. Russo, N. Labhsetwar, D. Fino, G. Saracco and V. Specchia: *Catal. Today*, 2008, **137**, 300–305.

145. D. Fino, P. Fino, G. Saracco and V. Specchia: *Appl. Catal. B*: 2003, **43**, 243–259.

146. D. Fino, N. Russo, G. Saracco and V. Specchia: *J. Catal.*, 2006, **242**, 38–47.

147. D. Fino, N. Russo, G. Saracco and V. Specchia: *Power Technol.*, 2008, **180**, 74–78.

148. D. Fino and V. Specchia: *Power Technol.*, 2008, **180**, 64–73.

149. I. Atribak, A. Bueno-Lopez and A. Garcia-Garcia: *Top. Catal.*, 2009, **52**, 2088–2091.

150. E. Cauda, D. Fino, G. Saracco and V. Specchia: *Top. Catal.* 2004, **30**, (31), 299–303.

151. Q. Li, M. Meng, N. Tsubaki, X. Li, Z. Li, Y. Xie, T. Hu and J. Zhang: *Appl. Catal. B*, 2009, **91**, 406–415.

152. Q. Li, M. Meng, Z. Q. Zou, X. G. Li and Y. Q. Zha: *J. Hazard. Mater.*, 2009, **161**, 366–372.

153. B. Ura, J. Trawczynski, A. Kotarba, W. Bieniasz, M. J. Illá-Gómez, A. Bueno-López and F. E. López-Suárez: *Appl. Catal. B*, 2011, **101**, 169–175.

154. H. Lin, Y. Li, W. Shangguan and Z. Huang: *Combust. Flame*, 2009, **156**, 2063–2070.

155. A. Trovarelli and P. Fornasiero: 'Catalysis by ceria and related materials', 2nd edn, 565–621; 2013, London, Imperial College Press

156. E. Aneggi, M. Boaro, C. de Leitenburg, G. Dolcetti and A. Trovarelli: *J. Alloy Compd.* 2006, **408–412**, 1096–1102.

157. S. Bernal, G. Blanco, J. M. Pintado, J. M. Rodrìguez-Izquierdo and M. P. Yeste: *Catal. Commun.*, 2005, **6**, 582–585.

158. S. Carrettin, P. Concepción, A. Corma, J. M. López-Nieto and V. F. Puntes: *Angew Chem. Int. Ed.*, 2004, **43**, 2538–2540.

159. I. Atribak, A. Bueno-López and A. García-García: *J. Catal.*, 2008, **259**, 123–132.

160. S. Bensaid, N. Russo and D. Fino: *Catal. Today*, 2013, **216**, 57–63.

161. M. Piumetti and S. Bensaid, N. Russo and D. Fino: *Appl. Catal. B*, 2015, **165**, 742–751.

162. S. J. Tans, A. R. M. Verschueren and C. Dekker: *Nature*, 1998, **393**, 49–52.

163. A. K. Geim and K. S. Novoselov: *Nat. Mater.*, 2007, **3**, 183–191.

164. D. L. Feldheim and C. A. Foss: 'Metal nanoparticles', 352; 2002, New York, NY, Marcel Dakker.

165. J. M. Thomas and R. Raja: *Top. Catal.*, 2010, **53**, 848–858.

166. J. C. Védrine: *Appl. Catal. A*, 2014, **474**, 40–50.

167. L. T. Weng and B. Delmon: *Appl. Catal. A*, 1992, **81**, 141–213.

168. J. S. Feeley, M. Deeba and R. J. Farrauto: *SAE Tech. Pap. 950747*, 1995.

169. C. Lambert, D. Dobson, C. Gierczak, G. Guo, J. Ura and J. Warner: *Int. J. Powertrains*, 2014, **3**, (1), 4–25.

170. C. Chen, F. Chen, L. Zhang, S. Pan, C. Bian, X. Zheng, X. Meng and F. S. Xiao: *Chem. Commun.*, 2015, **51**, 5936–5938.

171. 'A comprehensive analysis of biodiesel impacts on exhaust emissions'. U.S. Environmental Protection Agency, EPA420-P-02-001, October 2002.

172. J. C. Caroca, F. Millo, D. Vezza, T. Vlachos, A. De Filippo, S. Bensaid, N. Russo and D. Fino: *Ind. Eng. Chem. Res.*, 2011, **50**, 2650–2658.

173. F. Millo, S. Bensaid, D. Fino, S. J. Castillo Marcano, T. Vlachos and B. K. Debnath: *Fuel*, 2014, **138**, 134–142.

17

On the role of CoO in CoO$_x$/TiO$_2$ for the photocatalytic hydrogen production from water in the presence of glycerol

M.A. Khan, M. Al-Oufi, A. Tossef, Y. Al-Salik, and H. Idriss*

SABIC-Corporate Research and Development (CRD), KAUST, Thuwal, Saudi Arabia

Abstract The photocatalytic water splitting activity of nanocomposite photocatalysts of TiO$_2$ with CoO$_x$ was studied under UV and visible light, and the catalysts were characterized by XRD, XPS, and UV–vis techniques. The presence of CoO$_x$ enhances the hydrogen production activity of TiO$_2$ by five times at an optimal loading of . 2 wt. %. To investigate the role of CoO$_x$, the photocatalytic activity was also studied under visible light and with different amounts of sacrificial agent. Our results indicate that the increasing activity was not due to increasing absorption of the visible light but most likely due to the role of CoO$_x$ nanoparticles as hole scavengers at the interface with TiO$_2$. XPS Co2p analyses of CoO/TiO$_2$ showed a considerable decrease in their signal after prolonged reaction time (44 h) when compared to that of the fresh catalyst. Because part of Co^{2+} cations is dissolved in solution, in neutral or acidic pH, the possible increase in the reaction rate upon their addition to TiO$_2$ under UV excitation was investigated. No change in the reaction rate was observed upon, on purpose, addition Co^{2+} cations to TiO$_2$ under UV excitation. Thus, one may rule out the reduction of Co^{2+} to Co^0 with excited electrons within TiO$_2$. In order to further increase the reaction rate, we have synthesized and tested a hybrid system composed of CoO and Pd nanoparticles (Pd wt. % = 0.1, 0.3, 0.5, and 1 wt. %) where 0.3 wt. % Pd – 2 wt. % CoO/TiO$_2$ showed the highest rate.

Keywords Photocatalysis, hydrogen production, heterojunction, water splitting, titanium oxide, oxygen evolution

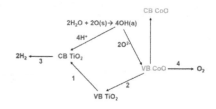

Introduction

Hydrogen is poised to play an important role in a sustainable energy system because it is storable, transportable, and can be converted into electricity efficiently using fuel cells when necessary. Moreover, it is an essential reactant in chemical industry and making it from renewables will contribute in recycling of carbon dioxide via chemical processes, such as the Fischer–Tropsch, methanol synthesis, and ammonia synthesis reactions. Photocatalytic water splitting using sunlight is considered a promising route to clean and renewable hydrogen production. The current efficiency of the process is still below what is needed for commercialization. The water splitting reaction is an uphill reaction in which the Gibbs free energy increases by 237 kJ mol^{-1}.[1,2] Water splitting reaction is a multistep process involving (i) light absorption (ii) charge separation and transfer, and (iii) redox reactions on the surface.[1–3] The water splitting process can be envisaged as two-half reactions: water oxidation and proton reduction to hydrogen fuel. Water oxidation is more challenging than hydrogen ion reductions because the

generation of one molecule of gaseous oxygen requires four holes and occurs on a timescale approximately five orders of magnitude slower than that of H$_2$ evolution.[3–7]

Various semiconductors such as TiO$_2$, CdS, ZnO, C$_3$N$_4$, and WO$_3$ have been explored for water splitting.[1–3,6,7] TiO$_2$ remains the leading semiconducting material for water splitting with its good conversion efficiency (of UV light: ca. 4–5% of the solar spectrum) and stability. Improving the light absorption and charge carrier separation in TiO$_2$ remains the biggest challenge in the efficiency of the water splitting process. Loading of cocatalysts such as metal nanoparticles or secondary semiconductors, acting as either electron or hole acceptors for improved charge separation, is needed. Various noble metals such as Pt, Au, Ag, and Pd have been studied in some details on TiO$_2$ for their water splitting activity.[8–12] These metal particles act primarily as reduction cocatalysts/electron sinks, therefore preventing electron-hole recombination and improving the H$_2$ production rates. In order to improve the oxidation half reaction, semiconductors such as PdS, RuO$_x$, IrO$_x$, and CoO$_x$ have been investigated.[6,7] Among these, CoO$_x$ has been reported to be a very efficient oxidation cocatalyst with variety of oxide and

*Corresponding author. Email: IdrissH@SABIC.com

oxy-nitride semiconductors. Domen and coworkers reported the use of CoO as an oxygen evolution promoter when used with GaZnInON for photocatalytic water splitting.[13] The as-prepared material was unstable under visible light with continuous production of N_2 yet upon loading CoO, the self-decomposition of the oxynitride catalyst was suppressed and O_2 evolution increased 7-fold. Barroso and coworkers studied the photo-electrochemical activity of α-Fe_2O_3/CoO_x nanocomposite electrodes for water oxidation.[14] Using transient absorption spectroscopy, the authors observed a 3-order of magnitudes increase of lifetime of photo-generated holes upon addition of CoO_x. A possible explanation for this increase in lifetimes was the formation of Schotky-type heterojunction leading to reduced recombination of electrons and holes. Li and coworkers also investigated CoO_x and CoPi as cocatalysts for water oxidation together with $BiVO_4$ and reported improvements in oxygen evolution.[15] Upon using a dual cocatalyst system of Pt/CoPi with yttrium-doped $BiVO_4$, the authors were able to achieve overall water splitting with production of both H_2 and O_2. Domen and coworkers also reported a CoO_x-modified $LaTiO_2N$ photocatalyst for water oxidation. Under visible light illumination, the O_2 evolution dramatically increased from 25 to 736 μmolh^{-1}.[16] Recently, Li and coworkers also reported record quantum efficiency of ~11.3% (when excited with light having a wave length between 400 and 500 nm) for water oxidation using CoO_x/Ta_3N_5 photocatalyst; with $AgNO_3$ as an electron scavenger.[17]

Cobalt oxide has also been used in conjunction with TiO_2 for photocatalytic water splitting.[18] Doping of Co leads to improvement in photocatalytic activity; however, the mechanism of charge carriers separation and the reason for activity enhancement in these composites is not well understood. It was proposed that possible charge transfer between Co^{2+} and TiO_2 lead to improvement in H_2 production.[18] Moreover, the activity under UV light decreased with time due to leaching of Co metal ions into the solution. Du and coworkers also reported the study of CoO_x-loaded titanium dioxide/cadmium sulfide (TiO_2/CdS) semiconductor composites.[19] The purpose of CoO_x is to prevent the photocorrosion of CdS. Using sodium sulfide (Na_2S)/sodium sulfite (Na_2SO_3) as hole scavengers under visible light irradiation, the maximum rate of hydrogen evolution achieved was 660 μmol g^{-1} h^{-1}, which was about seven times higher than TiO_2/CdS and CdS photocatalysts under the same conditions. The mechanism proposed was that electrons excited in CdS are transferred to the conduction band of TiO_2 and subsequently to the CoO_x cocatalyst where redox reactions take place to produce hydrogen.[19] However, this mechanism is unlikely as the conduction band edge of CoO_x is higher than that of TiO_2.[20]

In this study, we evaluate nanocomposite photocatalysts of TiO_2 and CoO_x for water splitting and study the role of CoO_x. Catalysts were characterized using UV–vis absorption, BET, and XPS. We evaluate the photocatalytic activity for H_2/O_2 evolution under UV and visible light. We have also addressed the possible Co^{2+} reduction to Co metal upon photoexcitation of TiO_2. Lastly, we demonstrate an active system consisting of dual cocatalysts of Pd and CoO_x nanoparticles on TiO_2 that can lead to efficient charge carrier separation and result in very high H_2 evolution rates of ca. 11,000 μmolg^{-1} h^{-1}.

Experimental

CoO_x–TiO_2 were prepared by wet impregnation. Anatase TiO_2 from Hombikat was used as the support catalyst. Different loadings of Co (0.5, 1, 2 and 4 wt. %) on TiO_2 support were prepared by adding known amount of $Co(NO_3)_2 \cdot 6H_2O$ salt solution to 500 mg of TiO_2 support. Excess water was evaporated under constant stirring with slow heating at 80 °C. The dried photocatalysts was calcined at 400 °C for 5 h. Photocatalysts with dual cocatalysts of Pd and CoO_x were prepared by sequential impregnation of Pd on CoO/TiO_2, starting from ($PdCl_2$).

UV–vis absorbance spectra of the powdered catalysts were collected over the wavelength range of 250–900 nm on a Thermo Fisher Scientific spectrophotometer equipped with praying mantis diffuse reflectance accessory. Absorbance (A) and reflectance (% R) of the samples were measured. The reflectance (% R) data were used to calculate the band gap of the samples using the Tauc plot (Kubelka–Munk function). XRD spectra were recorded using a Bruker D8 Advance X-ray diffractometer. A 2θ interval between 20 and 90° was used with a step size of 0.010° and a step time of 0.2 s/step. Based on the (1 0 1) diffraction line, the crystallite size is of the order of 8 nm. This was also further confirmed by TEM measurements. XPS was conducted using a Thermo scientific ESCALAB 250 Xi; the base pressure of the chamber was typically in the low 10^{-9} to high 10^{-10} mbar range. Charge neutralization was used for all samples (1 eV). Spectra were calibrated with respect to C1s at 285.0 eV. Quantitative analyses were conducted using the following sensitivity factors: Co2p (3.8), Ti2p (1.8), and O1s (0.66). Ar ion bombardment was performed with an EX06 ion gun at 1 kV beam energy and 10 mA emission current. The sputtered area of 900 × 900 μm^2 was larger than the analyzed area: 600 × 600 μm^2. Data acquisition and treatment were done using the Avantage software. BET surface areas of catalysts were measured using Quantachrome Autosorb analyzer by N_2 adsorption with surface areas of 133 m^2 g^{-1} for pure anatase TiO_2 and 131 m^2 g^{-1} for 2 wt. % CoO–TiO_2, both calcined at 400 °C.

Photocatalytic reactions were evaluated in a 135-mL-volume Pyrex glass reactor using 4 mg of catalyst. 30 mL of 5 vol. % glycerol aqueous solution or 30 mL of a 0.05 M $AgNO_3$ aqueous solution were used to evaluate the H_2 and O_2 evolution, respectively. The final slurry was purged with N_2 gas to remove any O_2 and subjected to constant stirring. The reactor was then exposed to the UV light, a 100 Watt ultraviolet lamp (H-144GC-100, Sylvania par 38) with a flux of ~5 mW cm^{-2} at a distance of 5 cm. Similarly, to evaluate the UV + visible light activity a Xenon lamp (Asahi spectra MAX-303) with a total flux of 26 mW cm^{-2} (UV ~ 3.3 mW cm^{-2} and visible (up to 600 nm) ~ 22.7 mW cm^{-2}) was used. Product analyses were performed by gas chromatography (GC) equipped with thermal conductivity detector (TCD) connected to Porapak Q packed column (2 m) at 45 °C and N_2 was used as a carrier gas. O_2 analysis was performed by GC equipped with TCD connected to packed molecular sieve (5A) column and He was used as a carrier gas.

Results and discussion

The band gaps and absorption properties of the photocatalysts were studied using diffuse reflectance UV–vis spectroscopy.

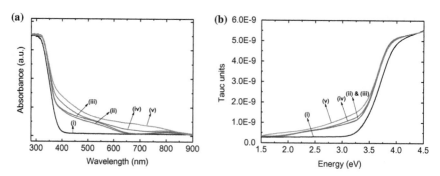

Figure 1   (a) UV–vis absorption spectra of TiO$_2$ photocatalyst with different loadings of Co (in wt. %) (i) 0 (ii) 0.5 wt. % (iii) 1 wt. % (iv) 2 wt. % and (v) 4 wt. % and (b) plots of Tauc units vs. (eV) for the same series

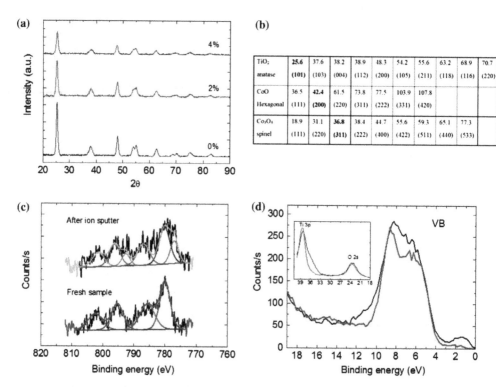

Figure 2   (a) XRD spectra of TiO$_2$ with different loadings of Co (in wt. %); (b) XRD diffraction lines in degrees (2Θ) for TiO$_2$ (anatase), hexagonal CoO, and spinel Co$_3$O$_4$. Diffraction lines are in (-) and bolded numbers represents the dominant line in each oxide; (c) XPS spectra of Co 2p peak in CoO$_x$-loaded TiO$_2$ containing 1.0 wt. % cobalt before and after Ar ions sputtering and (d) valence band of the same samples of Figure 2(b), before sputtering (red line) and after ion sputtering (black line). The inset represents the corresponding Ti3p and O2s lines

The UV–vis spectra of CoO$_x$–TiO$_2$ are recorded in the range of 250–900 nm as shown in Fig. 1(a). Spectra show typical absorption from anatase TiO$_2$ with a band edge around 370–380 nm (E_g ~ 3.2 eV) due to the charge transfer from the valence band formed by O2p orbitals to the conduction band formed by 3d t2 g orbitals of the Ti^{4+} cations.[18,19] Spectra of CoO$_x$–TiO$_2$ nanocomposite photocatalysts showed absorption in the visible region. One can see, in particular, for the 0.5 and 1 wt. % of Co an absorption peak in the region of 500 nm (~2.5 eV) which can be attributed to Co^{2+} → Ti^{4+} charge-transfer interaction, as indicated in earlier reports.[18,19] Another absorption peak near

800 nm (~1.5 eV) is caused by the transition of electrons from the occupied Co 3d states below the Fermi level to the unoccupied Co 3d states which form the conduction band of CoO$_x$.[21,22]

The Kubelka–Munk theory is generally used for the analysis of diffuse reflectance spectra obtained from weakly absorbing samples. It provides a correlation between reflectance and concentration. The concentration of an absorbing species can be determined using the Kubelka–Munk equation: $F(R) = (1 − R)^2/2R = k/s = Ac/s$, where R is reflectance, k is absorption coefficient, s is scattering coefficient, c is concentration of the absorbing species, and A is the absorbance. The optical

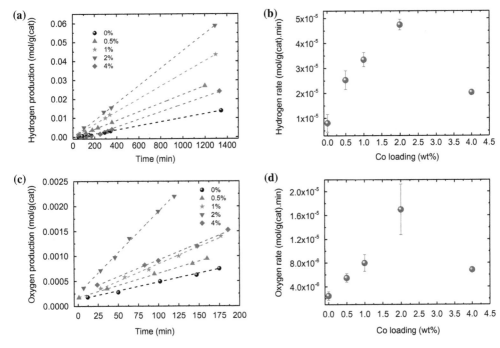

Figure 3   (a) H$_2$ production as a function of time over TiO$_2$ with different loadings of Co (in wt. %); (b) H$_2$ production rates (extracted from Figure 3(a)) as function of Co loading. Reaction conditions: 4 mg catalyst, 30 mL H$_2$O and 5 vol. % glycerol under UV lamp (375 nm) at a flux = 4–5 mW cm^{-2}; (c) O$_2$ evolution from TiO$_2$ with different loadings of Co (in wt. %) using 0.05 M AgNO$_3$ solutions and (d) O$_2$ evolution rates as function of Co loading. Note the 2 to 1 ratio in the hydrogen to oxygen rates (in separate experiments): Figure 3(b) and (d)

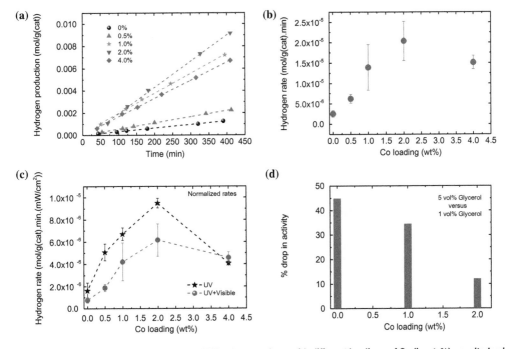

Figure 4   (a) H$_2$ production as a function of time of TiO$_2$ photocatalysts with different loadings of Co (in wt. %) – excited using a Xe lamp giving both UV and visible light; (b) H$_2$ production rates as a function of Co loading. Reaction conditions: 4 mg catalyst, 30 mL H$_2$O and 5 vol. % glycerol under Xenon lamp (250–650 nm) with a total flux of 26 mW cm^{-2} (UV ~ 3.3 ± 0.2 mW cm^{-2}, visible ~ 22.7 mW cm^{-2}); (c) H$_2$ production rates (normalized to UV flux) for UV light vs. (UV + visible) light as a function of Co loading and (d) % drop in activity upon using 1% glycerol as function of Co loading

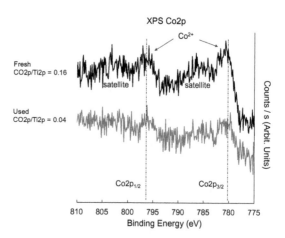

Figure 5   XPS Co2p of the 4 wt. % CoO/TiO$_2$ before and after photoreaction for 44 h. [Catal.] = 42 mg/80 mL (liquid composed of water and 5 vol. % of glycerol) = 0.53 g L^{-1}; UV flux = ca. 5 mW cm^{-2}; total reactor volume = 200 mL; initial solution pH 7.2 and final solution pH < 5. Used catalyst was filtered and dried before XPS measurements

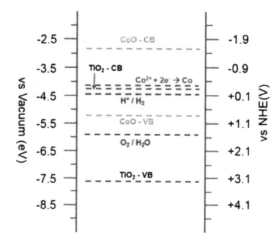

Figure 6   Electronic diagram of CoO and TiO$_2$ oxide systems; the redox potential of H$_2$O to H$_2$ and O$_2$ is also included

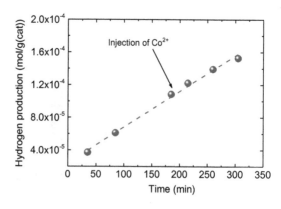

Figure 7   Photocatalytic hydrogen over TiO$_2$ (anatase) in the presence of 5 vol. % glycerol under UV lamp (375 nm) at a flux = 4 mW cm^{-2}. Reactor volume: 135 mL, liquid volume: 30 mL, catalyst concentration = 3 mg/30 mL = 0.1 g L^{-1}; [Co^{2+}] = 0.45 mg/30 mL

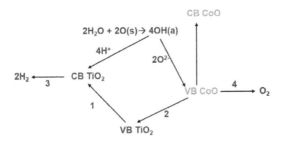

Figure 8 Events diagram of water splitting on the CoO/TiO$_2$ photocatalyst; CB = conduction band, VB = valence band. The 4H$^+$ and 2O^{2-} is a schematic description of 2 H$_2$O + 2 O(s) → 4OH(a); where (a) stands for adsorbed and (s) for surface

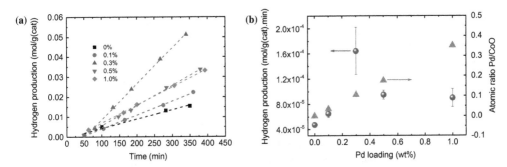

Figure 9 (a) H$_2$ production as a function of time of 2 wt. % Co–TiO$_2$ photocatalyst with different loadings of Pd (in wt. %) and (b) H$_2$ production rates of 2 wt. % Co–TiO$_2$ with different loadings of Pd (in wt. %)

band gap of semiconductors can be determined by plotting $(F(R) \times E)^{1/r}$ against the radiation energy in (eV), using $r = 2$ for indirect allowed transitions of charge carriers (indirect band gap material) or $r = \frac{1}{2}$ for direct allowed transition (direct band gap material). The resulting plot has a distinct linear regime, which denotes the onset of absorption. Thus, extrapolating this linear region to the abscissa yields the energy of the optical band gap of the material. Tauc plots of our catalysts are shown in Fig. 1(b) ($r = 2$) which shows a slight decrease in optical bad gap of TiO$_2$ with increasing Co loading.

The effect of Co loading on crystal structure of TiO$_2$ support was studied using XRD. Fig. 2(a) shows the X-ray diffraction patterns of TiO$_2$ with different loadings of Co (in wt. %). The XRD patterns with the characteristic planes of anatase phase at $2\theta = 25.5°$ (1 0 1), 37.7° (0 0 4), and 48.2° (2 0 0) are seen. The XRD pattern does not show any cobalt phase (up to 4 wt. % loading), indicating that cobalt ions are uniformly dispersed on the TiO$_2$ support. This was also indicated by others where at low loadings the CoO$_x$ diffraction peaks could not be detected.[18,19,23,24] The XRD peaks positions of anatase TiO$_2$ also do not show any change upon Co loading, confirming there is no change in structure/crystal phase of TiO$_2$ or doping of Co ions into TiO$_2$. A broadening of the TiO$_2$ diffraction peaks is, however, observed with the addition of Co, larger FWHM. This broadening could be due to either smaller TiO$_2$ crystallites and/or lattice strain on TiO$_2$ due to the presence of CoO nanoparticles. Because BET surface areas measurements did not show change upon CoO (within experimental errors) loading this change might be due to lattice strain. A table of the diffraction lines of TiO$_2$ (anatase), CoO, and Co$_3$O$_4$ is added, Fig. 2(b).

In order to further analyze the chemical composition of CoO$_x$ and electronic state of the composites, the CoO-loaded TiO$_2$ sample was analyzed by XPS and X-ray valence regions. Fig. 2(c) shows the Co 2p spectra from 1 wt. % CoO–TiO$_2$ samples calcined at 400 °C for 5 h. The XPS Co 2p of Co, CoO, and Co$_3$O$_4$ has been studied in some details by many workers.[25] Co2p of Co^{2+} has its characteristic satellites, reduction of Co^{2+} leads to Co^0 which results in a shift in the binding energy by about 2 eV. The binding energy of Co^{3+} is very close to that of Co^{2+} but Co^{3+} satellites are much more attenuated, and therefore, the presence of strong satellites can gauge the extent of Co^{2+} contribution.[26] In Fig. 2(c), XPS Co2p before and after Ar ion sputtering is presented. The binding energies for Co 2p$_{3/2}$ and Co 2p$_{1/2}$ are at 781.4 eV and 797.1 eV. These positions and the spin orbit splitting of 15.5 eV and satellites presence at about 7 eV above the main peaks (ca. 788 and 804 eV) are consistent with those reported for Co^{+2} of CoO.[18,23,26] Argon ions sputtering results in the preferential removal of oxygen anions and consequently the reduction of metal cations to lower oxidation states.[27] We use to ascertain that our as-prepared sample is mainly composed of Co^{2+} and not Co^0. This can be seen in Fig. 2(c) and (d). In Fig. 2(c), a shoulder at the lower binding energy side is seen at about 778 eV that is attributed to Co^0. The appearance of Co^0 is associated with the decrease of the signal of Co^{2+}. In Fig. 2(d), the valence band region is presented for the fresh and Ar ions sputtered surfaces. The appearance of the lines at about 1 eV below the Fermi level is indicative of 3d electrons due to both Ti cations in reduced states and metallic Co.[28–30] The inset in the Fig. presents the Ti3p and O2s for the same samples. The broad structure at the low binding energy side of the Ti3p is due to the presence of Ti cations in lower oxidation state than +4 due to preferential removal of oxygen anions upon sputtering. Quantitative analyses of the Co2p, Ti2p, O1s indicated that Co is present in about 0.1 at. % on the TiO_2 surface.

The H_2 and O_2 production activities of CoO_x–TiO_2 photocatalysts under UV excitation are presented in Fig. 3. The photocatalytic activity was evaluated over 24 h and was stable and reproducible. Pure anatase TiO_2 calcined at 400 °C showed H_2 production rates of ~10 μmolg^{-1} min^{-1}. The loading of CoO_x resulted in improvement in the H_2 evolution. The highest H_2 production rates of ~47 μmolg^{-1} min^{-1} was obtained when the Co metal concentration was 2 wt. % relative to TiO_2. The H_2 production rates as a function of Co loading is plotted in Fig. 3(b). One can notice that increasing the Co loading above 2 wt. % decreased the photocatalytic activity. The highest photocatalytic activity of 2 wt. % cobalt loaded samples could be due to the optimum dispersion of CoO_x particles over TiO_2 photocatalyst. This trend is similar to other systems such as CuO/TiO_2, Cu_2O/TiO_2, and NiO/TiO_2 as reported earlier.[31–33] Actually, this trend is also seen for noble metals where a maximum efficiency occurs for a narrow range of concentrations depending on the nature of the metal and of the semiconductor.[9,34–37] In order to further probe into this trend, we have monitored the O_2 evolution activity of the same catalysts under UV excitation and in the same reactor but with 0.05 M $AgNO_3$ solutions to scavenge electrons. While this method has its drawback (such as the possible deposition of the metallic Ag that is reduced on the surface of the catalyst),[38,39] it does give a good indication on the potential of the semiconductor for the hole transfer and therefore for comparative reasons the method has its merit.[40–42] As shown in Fig. 3(c), the O_2 evolution monitored in a separate experiment with $AgNO_3$ as sacrificial agent was linear as a function of time and was observed to be in a stoichiometric ratio of 1:2 to the H_2 production seen earlier with glycerol as the sacrificial agent. This may tell that the potential of the catalyst is indeed non negligible and the rates of electron and hole transfer are comparable. The O_2 evolution also showed a similar trend to the H_2 production as shown in Fig. 3(d) with the 2 wt. % Co loading had the highest activity (~21 μmolg^{-1} min^{-1}).

To further investigate the contribution of CoO_x in the enhancement of photocatalytic activity, the reaction was also tested under UV + visible light irradiation under a total flux of ~26 mW cm^{-2} (UV ~ 3.3 mW cm^{-2}, visible ~ 22.7 mW cm^{-2}). This is indeed important, as numerous reports have indicated that the addition of CoO may enhance the reaction due to further absorption of light.[18,19,43] As shown in Fig. 4(a), similar to the trend under UV lamp, the loading of cobalt on TiO_2 resulted in improvement in the H_2 evolution. The highest H_2 production rates were also achieved when the Co concentration was 2 wt. % as seen in Fig. 4(b). Fig. 4(c) presents the H_2 production rates normalized to UV flux where similar trends of activity, under UV light only and UV + visible light, are seen. *This indicates that there is no contribution of visible light charge carriers in the photocatalytic water splitting process when both UV and visible light are present; in other words, visible light excitation effect is negligible in a catalytic system.* Earlier reports have mentioned the possibility of charge carriers generated in CoO_x being

pumped into CB of TiO_2 leading to enhancement of activity.[19] Our results indicate that any charge carriers being generated in CoO_x from visible light do not participate in the photocatalytic water splitting process. Other reports have indicated that CoO in nanoparticles has activity for hydrogen production yet the catalyst deactivates very fast within 1 h. It is likely that in that reported study, a non-catalytic surface reaction occurred; in other words, this was a stoichiometric and not a catalytic reaction.[22]

CoO is a p-type semiconductor with a relatively low work function (~4.4 eV) making it attractive as an oxidation cocatalyst.[44] The experimentally measured valence band edge formed of O 2p orbitals is at 0.7 V vs. NHE while the conduction band edge is at −1.7 V vs. NHE.[44] It is possible that the enhanced photocatalytic activity of our composite catalysts is due to the formation of a Schottky-type heterojunction, leading to efficient charge carrier separation. The high valence band edge in CoO_x is ideal for trapping photogenerated holes in TiO_2. To confirm this hypothesis, we tested the photocatalysts by changing the concentration of the "hole scavenging" sacrificial agent. The photocatalytic activity under the same conditions was conducted with lowering the glycerol concentration from 5 to 1 vol. %. As shown in Fig. 4(d), in pure anatase TiO_2, the H_2 evolution rate drops by ~45% when glycerol concentration is reduced to 1 vol. %. In contrast, samples with 2 wt. % of CoO_x show better activity, a drop of ~10%. This result may indicate that CoO_x nanoparticles function similar to the sacrificial agent i.e., as an oxidation cocatalyst/hole trapping agent.

In order to test the possibility of Co^{2+} dissolution into the liquid phase, we have conducted long-term photoreaction experiment (44 h) over 4 wt. % CoO/TiO_2 under UV excitation and analyzed the Co content by XPS. We have used a high loading of 4 wt. % CoO in order to decrease the errors associated with computing the total peak area of the Co2p signal. Fig. 5 presents the XPS Co2p region of the fresh and used catalysts. The signal is typical of the Co2p of Co^{2+} cations as indicated in Fig. 2. Qualitatively, the XPS Co2p after reaction is identical to that before. However, the Co2p/Ti2p ratio decreased from 0.16 to 0.04 during the reaction time indicating considerable loss of Co cations.

While the reduction to Co metal (similar to that of Cu^{2+} to Cu^0)[33] by excited electrons is possible, one may refute it based on the redox potential. The redox potential of Cu is within the band gap of TiO_2 (+0.34 V), and therefore, its reduction by conduction band electrons of TiO_2 occurs. The redox potential of Co is −0.28 V, therefore above TiO_2 conduction band[45]. Still, because the difference between the redox potential of Co^{2+}/Co and that of TiO_2 conduction band is very small (within the limits of measurements (Fig. 6), alterations due to changes in dipole moments[46] may occur. We have conducted an experiment in which the photoreaction on TiO_2 alone was monitored and where Co^{2+} (from $Co(NO_3)_2$) cations were introduced during the run. As shown in Fig. 7, no deviation in hydrogen production is seen upon the addition of Co^{2+} cations, indicating that their introduction did not alter the activity of TiO_2. This result may exclude the possibility of Co^{2+} reduction by the CB band electrons of TiO_2. It is also not possible under the reaction conditions to reduce CoO to Co metal by hydrogen produced since thermodynamics indicate that at room temperature the partial pressure of O_2 needs to be equal to 10^{-30} torr.[47]

Fig. 8 presents the events diagram, to complement the electronic diagram of Fig. 6. Upon contact between CoO and TiO_2 and photoexcitation with UV light, electrons are transferred from the VB to the CB of TiO_2 (leaving holes in the VB) – step 1. Electrons can then be transferred from the VB of CoO to the empty states of the VB of TiO_2 – step 2. This results in increasing the likelihood of electrons in the CB of TiO_2 to reduce H^+ to ½ H_2 – step 3. At the same time, O anions of OH (a) (surface hydroxyls) donate electrons to the VB of CoO – step 4. Based on the work's results (in particular, the absence of enhancement of the rate when visible light is added), the rate of reaction can be explained without invoking visible light excitation of CoO, in the presence of UV light equivalent to that provided from the sun. It is worth mentioning that because no change in the reaction rate is seen between excitation under UV and excitation under UV + vis. light, in addition to the absence of hydrogen production under vis. light alone, electron excitation from the VB of CoO to its CB does not contribute into the reaction. Within this context, electron transfer from CoO CB to TiO_2 CB, as often invoked in many work, may have no physical meaning. This is simply because under UV excitation, the TiO_2 CB is populated by excited electrons and therefore less poised to receive excited electrons from the CoO CB.

The role of dual cocatalysts i.e., oxidation and reduction cocatalysts has been investigated by many workers including those reported in Refs.[6,48,49]. To investigate the role of another cocatalyst, we loaded Pd metal as the reduction cocatalyst on top of 2 wt. % Co–TiO_2 which showed the best activity. In order to assess the activity, we also changed the concentration of Pd on top of the catalyst. The H_2 production activity of 2 wt. % CoO_x–TiO_2 photocatalysts impregnated with Pd metal is shown in Fig. 9(a). Further improvement of the H_2 evolution reaction was observed. The highest H_2 production rates were achieved when the Pd concentration was 0.3 wt. % as seen in Fig. 9(b), with H_2 production rates of ~180 μmolg^{-1} min^{-1}. Thus it seems that a system where a dual semiconductor-based cocatalyst i.e., CoO as an oxidation cocatalyst and Pd as reduction cocatalyst can function and is stable. The best performing Pd/Co ratio is = ca. 0.1.

It is worth extracting a few points from this study. The optimum amount of CoO on top of TiO_2 is found to be 2 wt. %. Because we do not see a change in the slope of hydrogen production and because the reduction potential is above the CB of TiO_2, the chemical state of the cobalt oxide(s) is plausibly maintained as CoO. The BET surface area of TiO_2 is = 133 m^2/g$_{Catal.}$. We have calculated the expected surface area of a 2 wt. % of CoO, as an example. It is equal to 3×10^{-4} mole/g$_{Catal.}$ = 16 m^2 assuming monolayer dispersion (as an upper limit). This would cover a small fraction of the surface; agglomeration would decrease it further. Therefore, the finding of an optimal reaction rate so narrowly dependent on the fraction of CoO may not be linked to a geometric effect (blocking of available sites) but most likely to an electronic effect where electrons transfer are optimized at the interface TiO_2 and CoO.

Conclusions

Nanocomposite photocatalysts by impregnating anatase TiO_2 with different amounts of Co salt solutions were prepared, characterized, and tested. The presence of CoO enhances the activity of TiO_2 with optimal loading determined to be ca. 2 wt. %, and the rate of hydrogen evolution was about five times higher than that of TiO_2 alone. The increase in activity was not due to Co^{2+} reduction by TiO_2 CB electrons. The increase in activity was also not due to increasing absorption of the visible light. It is most likely due to the role of CoO nanoparticles as hole scavengers at the interface with TiO_2. The addition of Pd (as hydrogen ion reduction sites) further improved the reaction rate ca. 4 times compared to that of the composite system, to 180 μmolg^{-1} min^{-1}. No catalytic deactivation was seen for prolonged reaction time (up to ca. 24 h). A schematic description of the events is given in which electron transfer occur from the VB of CoO to that of TiO_2 (upon photoexcitation), this results in increasing water oxidation upon electron transfer from hydroxyl oxygen anions to the VB of CoO.

References

1. A. Kudo and Y. Miseki: Chem. Soc. Rev., 2009, **38**, 253–278.
2. S. J. A Moniz, S. A. Shevlin, D. J. Martin, Z.-X. Guo and J. Tang: Energ. Environ. Sci., 2015, **8**, 731–759.
3. T. Hisatomi, J. Kubota and K. Domen: Chem. Soc. Rev., 2014, **43**, 7520–7535.
4. A. J. Cowan, C. J. Barnett, S. R. Pendlebury, M. Barroso, K. Sivula, M. Grätzel, J. R. Durrant and D. R. Klug: J. Am. Chem. Soc., 2011, **133**, 10134–10140.
5. J. Tang, J. R. Durrant and D. R. Klug: J. Am. Chem. Soc., 2008, **130**, 13885–13891.
6. J. Yang, D. Wang, H. Han and C. Li: Acc. Chem. Res., 2013, **46**, 1900–1909.
7. L. Yang, H. Zhou, T. Fan and D. Zhang: Phys. Chem. Chem. Phys., 2014, **16**, 6810–6826.
8. S. Bashir, A. K. Wahab and H. Idriss: Catal. Today, 2015, **240**, (Part B), 242–247.
9. V. Jovic, W.-T. Chen, D. Sun-Waterhouse, M. G. Blackford and H. Idriss, G. I. N. Waterhouse: J. Catal., 2013, **305**, 307–317.
10. M. Murdoch, G. I. N. Waterhouse, M. A. Nadeem, J. B. Metson, M. A. Keane, R. F. Howe, J. Llorca and H. Idriss: Nat. Chem., 2011, **3**, 489–492.
11. R. Su, R. Tiruvalam, A. J. Logsdail, Q. He, C. A. Downing, M. T. Jensen, N. Dimitratos, L. Kesavan, P. P. Wells, R. Bechstein, H. H. Jensen, S. Wendt, C. R. A. Catlow, C. J. Kiely, G. J. Hutchings and F. Besenbacher: ACS Nano, 2014, **8**, 3490–3497.
12. Z. Zhang, A. Li, S.-W. Cao, M. Bosman, S. Li and C. Xue: Nanoscale, 2014, **6**, 5217–5222.
13. K. Kamata, K. Maeda, D. Lu, Y. Kako and K. Domen: Chem. Phys. Lett., 2009, **470**, 90–94.
14. M. Barroso, A. J. Cowan, S. R. Pendlebury, M. Grätzel, D. R. Klug and J. R. Durrant: J. Am. Chem. Soc., 2011, **133**, 14868–14871.
15. D. Wang, R. Li, J. Zhu, J. Shi, J. Han, X. Zong and C. Li: J. Phys. Chem. C, 2012, **116**, 5082–5089.
16. F. Zhang, A. Yamakata, K. Maeda, Y. Moriya, T. Takata, J. Kubota, K. Teshima and S. Oishi, K. Domen: J. Am. Chem. Soc., 2012, **134**, 8348–8351.
17. S. Chen, S. Shen, G. Liu, Y. Qi, F. Zhang and C. Li: Angew. Chem., 2015, **127**, 3090–3094.
18. G. Sadanandam, K. Lalitha, V. D. Kumari, M. V. Shankar and M. Subrahmanyam: Int. J. Hydrogen Energ., 2013, **38**, 9655–9664.
19. Z. Yan, H. Wu, A. Han and X. Yu, P. Du: Int. J. Hydrogen Energ., 2014, **39**, 13353–13360.
20. X. He, X. Song, W. Qiao, Z. Li, X. Zhang, S. Yan, W. Zhong and Y. Du: J. Phys. Chem. C, 2015, **119**, 9550–9559.
21. K. Deori and S. Deka: CrystEngComm, 2013, **15**, 8465–8474.
22. L. Liao, Q. Zhang, Z. Su, Z. Zhao, Y. Wang, Y. Li, X. Lu, D. Wei, G. Feng, Q. Yu, X. Cai, J. Zhao, Z. Ren, H. Fang, F. Robles-Hernandez, S. Baldelli and J. Bao: Nat. Nanotechnol., 2014, **9**, 69–73.
23. P. Jiang, W. Xiang, J. Kuang, W. Liu and W. Cao: Solid State Sci., 2015, **46**, 27–32.

24. L. Samet: *Mater. Charact.*, 2013, **85**, 1–12.

25. S. C. Petitto, E. M. Marsh, G. A. Carson and M. A. Langell: *J. Mol. Catal. A: Chem.*, 2008, **281**, 49–58.

26. H. Idriss, C. Diagne, J. P. Hindermann, A. Kiennemann and M. A. Barteau: *J. Catal.*, 1995, **155**, 219–237.

27. H. Idriss and M. A. Barteau: *Catal. Lett.*, 1994, **26**, 123–139.

28. H. Idriss, V. S. Lusvardi and M. A. Barteau: *Surf. Sci.*, 1996, **348**, 39–48.

29. M. C. Biesinger, L. W. M. Lau, A. R. Gerson and R. S. C. Smart: *Appl. Surf. Sci.*, 2010, **257**, 887–898.

30. R. Riva, H. Miessner, R. Vitali and G. Del Piero: *Appl. Catal. A: Gen.*, 2000, **196**, 111–123.

31. W.-T. Chen, A. Chan, D. Sun-Waterhouse, T. Moriga, H. Idriss and G. I. N. Waterhouse: *J. Catal.*, 2015, **326**, 43–53.

32. W.-T. Chen, V. Jovic, D. Sun-Waterhouse, H. Idriss and G. I. N. Waterhouse: *Int. J. Hydrogen Energ.*, 2013, **38**, 15036–15048.

33. L. Sinatra, A. P. LaGrow, W. Peng, A. R. Kirmani, A. Amassian, H. Idriss and O. M. Bakr: *J. Catal.*, 2015, **322**, 109–117.

34. G. R. Bamwenda, S. Tsubota, T. Nakamura and M. Haruta: *J. Photochem. Photobiol, A: Chem.*, 1995, **89**, 177–189.

35. M. Bowker, L. Millard, J. Greaves, D. James and J. Soares: *Gold Bull.*, 2004, **37**, 170–173.

36. Z. H. N. Al-Azri, W.-T. Chen, A. Chan, V. Jovic, T. Ina, H. Idriss and G. I. N. Waterhouse: *J. Catal.*, 2015, **329**, 355–367.

37. V. Jovic, Z. N. Al-Azri, W.-T. Chen, D. Sun-Waterhouse, H. Idriss and G. N. Waterhouse: *Top. Catal.*, 2013, **56**, 1139–1151.

38. A. Bhardwaj, N. V. Burbure, A. Gamalski and G. S. Rohrer: *Chem. Mater.*, 2010, **22**, 3527–3534.

39. J. Giocondi, P. Salvador and G. Rohrer: *Top. Catal.*, 2007, **44**, 529–533.

40. R. M. Navarro Yerga, M. C. Álvarez Galván, F. del Valle,, J. A. Villoria de la Mano and J. L. G. Fierro: *ChemSusChem*, 2009, **2**, 471–485.

41. E. M. Sabio, R. L. Chamousis, N. D. Browning and F. E. Osterloh: *J. Phys. Chem. C*, 2012, **116**, 3161–3170.

42. R. Li, Y. Weng, X. Zhou, X. Wang, Y. Mi, R. Chong, H. Han and C. Li: *Energ. Environ. Sci.*, 2015, **8**, 2377–2382.

43. Y.-F. Wang, M.-C. Hsieh, J.-F. Lee and C.-M. Yang: *Appl. Catal. B: Environ.*, 2013, **142–143**, 626–632.

44. M. T. Greiner, M. G. Helander, W.-M. Tang, Z.-B. Wang, J. Qiu and Z.-H. Lu: *Nat. Mater.*, 2012, **11**, 76–81.

45. http://www.chemeddl.org/services/moodle/media/QBank/GenChem. Tables/EStandardTable.htm.

46. D. Zhang, M. Yang and S. Dong: *J. Phys. Chem. C*, 2015, **119**, 1451–1456.

47. M. Ivill, S. J. Pearton, S. Rawal, L. Leu, P. Sadik, R. Das, A. F. Hebard, M. Chisholm, J. D. Budai and D. P. Norton: *New J. Phys.*, 2008, **10**, 065002.

48. K. Maeda, D. Lu, K. Teramura and K. Domen: *Energ. Environ. Sci.*, 2010, **3**, 470–477.

49. K. Maeda and K. Domen: *J. Phys. Chem. Lett.*, 2010, **1**, 2655–2661.

18

Pd deposition on TiO_2(110) and nanoparticle encapsulation

Michael Bowker*[1,2] and Ryan Sharpe[1]

[1]Cardiff Catalysis Institute, School of Chemistry, Cardiff University, Cardiff CF10 3AT, UK
[2]UK Catalysis Hub, Research Campus at Harwell (RCaH), Rutherford Appleton Laboratory, Harwell, Oxon, OX11 0FA, UK

Abstract The effect of sputtering, annealing and oxidation on the surface properties of TiO_2(110), and on the same surfaces with nanoparticles present, has been investigated. Sputtering the crystal clean gives a much reduced surface with Ti^{2+} as the dominant species. This surface is mainly $Ti^{3+,4+}$ after annealing in vacuum. Oxidation reduces the surface Ti^{3+} considerably. When Pd nanoparticles are annealed on any of the investigated titania surfaces, the particles become encapsulated by a film of titanium oxide. This is

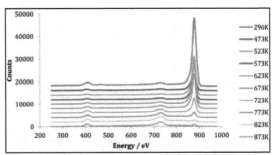

particularly noticeable in ion scattering spectroscopy (ISS) where the Pd:Ti ratio drops by a factor of 300 after annealing to 750 K, indicating complete coverage of the Pd nanoparticles by the oxide film. This happens most easily for the nanoparticles deposited on the reduced surfaces (beginning at ~673 K) but also occurs for the very oxidized surface at ~773 K. Thus, reduced Ti from the subsurface region can migrate onto the Pd surface to form the sub-oxide, the sub-oxide being a thin TiO-like layer.

Keywords Nanoscience, Catalysis, Model catalysts, Nanoparticles, Encapsulation

Introduction

The nature of the surface of titania has become of great interest in the last 25 years[1] because of its use in a variety of applications, especially in surface treatments, coatings with special properties and, particularly with respect to the current paper, in catalysis. As regards the latter it is of considerable importance in photocatalysis, titania still being the most photo-hydro-stable material for light absorption and utilization. Many other possible photocatalysts with narrower band gaps, which would allow them to work in the visible (CdS for instance) oxidize and corrode in the presence of light. When metal nanoparticles are deposited on them, such photocatalysts have been shown to be active for sacrificial water splitting (in which holes are trapped by a sacrificial organic such as methanol[2,3]) and hydrogen production.[2–11]

There is also an important effect in catalysis called SMSI (the strong metal–support interaction), which is usually detrimental to the catalyst performance, and generally occurs for transition metal nanoparticles anchored onto reducible supports after high-temperature (~600K +)

reduction. In relation to the current paper, this is because of the encapsulation of the metal particles by a film of titanium oxide.[12–15] This, in turn, has been shown to be because of the inherent reducibility of titania, especially in the ultra-high vacuum (UHV) treatment environment.[1,16–18]

Recently, O'Shea et al.[19] obtained transmission electron microscopy (TEM) images of Co nanoparticles covered by a few atomic layers of TiO_x ($x < 2$) after a reduction treatment (TPR). X-ray photoelectron spectroscopy (XPS) showed that this might be Ti^{3+}. Majzik et al.[20] used Auger electron spectroscopy (AES) and scanning tunneling microscopy (STM) to show SMSI on Rh nanoparticles on TiO_2(110). Scanning tunneling microscopy showed ordered decoration of TiO_x on Rh, exhibiting a 'wheel structure', which is very similar to that reported for other supported metal systems, such as Pt[14] and Pd islands,[21] and for thin films of oxidized titanium deposited on metal single crystals.[22] This decoration could be removed with a short sputtering procedure[20] (100 s, 0.5 keV). These pinwheel structures have also been observed by Castell on Pd clusters supported on reduced $SrTiO_3$[23] and in TiO_x on Au(111)[24] using STM and Auger. Linsmeier[25] used ISS, XPS and thermal desorption experiments to investigate SMSI after high-temperature reduction treatment. Thermal desorption spectra showed significant CO adsorption on a clean TiO_2 sample with Rh nanoparticles (~1 mL), and very little adsorption after SMSI had been induced. X-ray photoelectron spectroscopy showed Ti^{3+} present in the SMSI state. This is in contrast to Bennett et al.[21,26] who showed the layer on Pd to be Ti^{2+}-like by using take-off angle variations in XPS, though the effect on CO adsorption[27] was similar to that observed by Linsmeier and Taglauer.[25] Reactivity measurements by Bonanni[28] suggest that high-temperature annealing of Pt on reduced TiO_2 results in the encapsulation of Pt by a reduced titania layer, agreeing with a mechanism proposed by Fu et al.[29,30]

Materials and methods

All experiments were performed in a UHV system built by Omicron Vacuum Physik capable of performing STM, XPS, low energy ion scattering/ISS and low energy electron diffraction (LEED). The system comprises three separate chambers pumped by a combination of four turbomolecular pumps, three titanium sublimation pumps and two ion pumps, resulting in a base pressure of $<1 \times 10^{-9}$ mbar.

The TiO_2 sample was cleaned by Ar^+ bombardment at 1 keV, followed by annealing in UHV or oxygen, typically at 773 K, although temperatures in the range 673–873 K were used for certain experiments. Surface cleanliness was monitored by XPS and ISS, and gas purity was analyzed using a quadrupole mass spectrometer. X-ray photoelectron spectra were recorded using an Al $K\alpha$ photon source and an analyzer pass energy of 50 eV unless stated otherwise and were recorded at room temperature. All XPS data were analyzed with CasaXPS[31] and binding energies were calibrated to the O(1 s) peak at 530.4.[32]

The titanium dioxide sample used was a TiO_2(110) single crystal (Pi-Kem Ltd. (Salisbury, UK)). The sample was mounted on a standard Omicron molybdenum plate via spot-welded Ta strips. A thermocouple was attached to the sample plate holder for temperature measurement.

Palladium films were grown by metal vapor deposition. The source of Pd was a W filament (0.25 mm, Advent, 99.95%) tightly wrapped with Pd wire (0.125 mm, Goodfellow, 99.95%), which was resistively heated by passing a current of 3 A through the coil in UHV.

Results

Sputtering and annealing TiO_2(110)

The TiO_2(110) surface was prepared in several different ways before the deposition of Pd, Fig. 1, as follows

a. sputtered at ambient temperature
b. sputtered then annealed to 873 K for 10 min, and
c. sputtered then annealed to 873 K in the presence of 2×10^{-6} mbar of O_2 for 60 min.

Considering Fig 1c first, here the surface is highly oxidized with main peaks at 459 eV and 465 eV binding energy (b.e.) for the $2p_{3/2,1/2}$ states characteristic of the 4 + oxidation state of Ti. After sputtering at ambient temperature (Fig. 1a), the peaks are very poorly defined and broadened on the low b.e. side, characteristic of reduction of the Ti states. This is because of the well-known effect of sputtering on titania surfaces, namely that the lighter atom, the oxygen, is

depleted in the surface relative to Ti.[33] This leaves a much reduced surface, which has mainly $Ti^{3+/2+}$ states present.[1,16,34] After annealing such a surface in vacuum (Fig. 1b), the peaks have sharpened up, but nevertheless, when compared with Fig. 1c, the surface is clearly still in a considerably reduced state. The Ti^{2+} state is significantly lowered in intensity and the reduced states are now dominated by Ti^{3+}, with peaks at 457.5 and ~462.3 eV. Figure 2 shows the difference spectrum obtained by subtracting the spectrum of the sputter-annealed surface, from that obtained by sputtering (that is, between states c and a above), as detailed in the Supplementary Material. There is loss of intensity of the peaks at ~455.1 and 461.0 eV because of the annealing, consistent with a major component of Ti^{2+} after sputtering, but with some Ti^{3+} too. Figure 3 shows the difference between spectra b and c, indicating the dominance of the 3 + state as the reduced species after annealing the sputtered surface.

Figure 4 is a plot of the effect of annealing on the signals for two of these surfaces, indicating the changes in the oxidized and sputtered states with anneal temperature. Clearly, although annealing in UHV has some oxidative effect, annealing in oxygen, even at such low pressures, is much more effective, as also indicated by the data of Table 1 showing the ISS Ti:O ratios for the three surfaces.

Dosing of Pd

Figure 5 shows the evolution of the $Pd_{3d}:Ti_{2p}$ signal ratio, for dosing Pd stepwise onto the three titania surfaces at 300 K (the individual spectra are shown in the Supplementary Material Fig. 1a–f). As can be seen, the ratio generally increases in a similar fashion as the coverage increases for the three types of surface. It is important to note that we do not have an independent way of measuring the flux of Pd at the surface and, since these experiments were carried out at different times, with different life history of the filament used, we cannot say with confidence that there is a significant change in morphology of the growing film. That is, the higher ratio of Pd:Ti for the sputtered/annealed layer seen in Fig. 5 could simply be because of a higher flux of Pd at the surface, hence the inset shows a version normalized to the dose time for the sputtered surface. It is clear that the shape is identical for all three curves. The curves indicate an initial period of near linearity (up to ~150 s dosing in the inset), followed by an increased slope, which could indicate initial growth of the Pd as a few monolayers, as proposed by Kaden et al.,[35] with multilayer island growth following that initial phase.

Thermal evolution of the Pd layer

The authors then annealed these variously prepared Pd–TiO_2 surfaces for 10 min at a variety of temperatures to examine the effect on the Pd and Ti XPS signals, and these data are summarized in Figs. 6 and 7. All the curves show a decrease in the Pd/Ti ratio with increasing temperature, beginning at about 523 K, most likely because of some de-wetting/sintering in the deposited film,[36] followed by a bigger decrease in ratio at higher temperatures. There is little apparent change in the shape or position of the Pd 3d XPS peaks (Fig. 7b), but there is a considerable change in the Ti line shape (Fig. 7a) with an

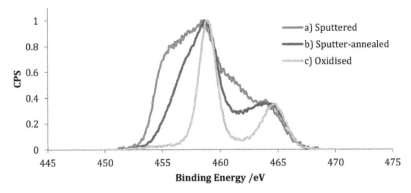

Figure 1   Normalised Ti 2p spectra from the TiO$_2$(110) surface after *a* sputtering, *b* sputtering and annealing to 873 K for 10 min and *c* sputtering and then annealing at 873 K in oxygen (2 × 10^{-6} mbar) for 60 min

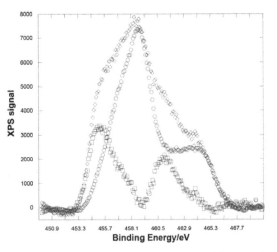

Figure 2   Showing the difference spectrum (blue curve) between spectrum c (green curve, sputtered surface) and spectrum b (red curve, sputtered surface annealed) in Fig. 1 of the main text. The dominant peaks in the difference spectrum are at 455.3 and 461.0 eV assigned to the Ti^{2+} state

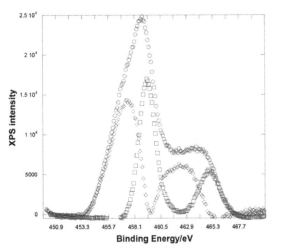

Figure 3   Showing the difference spectrum (green curve) between spectra b and c in Fig. 1. The dominant peaks are at 457.5 and ~462.5 eV, characteristic of mainly Ti^{3+}

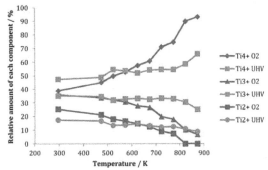

Figure 4   Variation of the various Ti states as a function of anneal temperature for a sputtered surface being annealed in O$_2$ and in ultra-high vacuum (UHV), derived from the Ti$_{2p}$ X-ray photoelectron spectroscopy (XPS)

Table 1   Ti/O ion scattering spectroscopy (ISS) integral ratios for the three prepared surfaces

Surface	Ti/O ISS ratio
Sputtered	2.82
Sputter-annealed	2.72
Oxidized	1.97

increase in the low binding energy, reduced Ti states. However, what we particularly wish to focus on here is the effect seen in Fig. 8, which shows a remarkable decrease in the Pd ISS signal with increasing anneal temperature and an increase in the Ti and O signals. The Pd:Ti ISS ratio decreases by a factor of ~300, largely as a result of the catastrophic loss of Pd signal. Together with this, there is a significant change in the Ti$_{2p}$ XPS peaks, that is, an increase in intensity at ~456 eV, beginning to occur above 673 K and consistent with the appearance of Ti^{2+} on the surface of the sample. This is the opposite of the behavior observed for the surface without Pd present. The full XPS and ISS data are shown in the Supplementary Material Figs. 2 and 3.

Discussion

TiO$_2$(110)

It is clear from the above data that the treatment method during surface preparation has a substantial effect on the

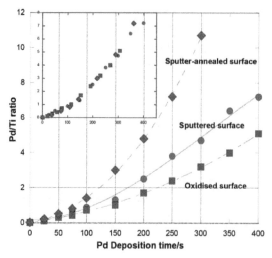

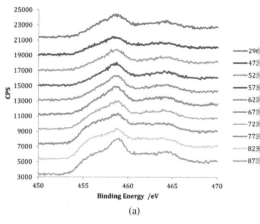

Figure 5 Pd 3d/Ti 2p X-ray photoelectron spectroscopy (XPS) area ratios for each surface during deposition. Red symbols are for deposition onto the sputtered surface, green for the sputter–annealed surface, and blue for the surface sputtered and annealed in oxygen and (inset) normalized to the Pd/Ti ratio for the sputtered surface at 300 s

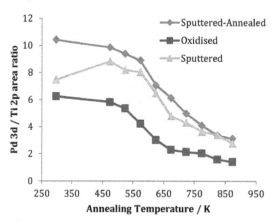

Figure 6 The change in Pd3d/Ti$_{2p}$ area ratio with annealing temperature for the three prepared surfaces

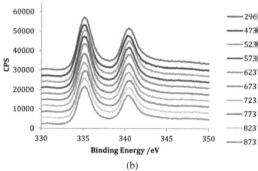

Figure 7 a The variation in the Ti$_{2p}$ X-ray photoelectron spectroscopy (XPS) signal with increasing temperature for the sputtered surface with Pd post-dosed. b The variation in the Pd 3d XPS with increasing anneal temperature for the sputtered surface post-dosed with Pd

nature of the surface. Sputtering has the effect of removing contaminants from the surface layer, but also removes Ti and O. However, as might be expected, more O is removed during this process and so the Ti:O ratio increases, as seen in Table 1. Concomitant with this, XPS indicates a decrease in the cation charge at the surface, that is, reduction because of the loss of oxygen from the surface. It appears that in the present case these states are mainly Ti^{2+} immediately after sputtering, though after annealing that state diminishes, with an increase in Ti^{3+} and Ti^{4+}. Nolan et al.[37] showed that addition of metallic Ti to the TiO$_2$(110) surface results in conversion of that adatom to Ti^{2+} with distribution of the additional electrons around adjacent sites in the adlayer, rather than the formation of only reduced cations immediately beside the anion vacancy, or of polarons. The electrons do not remain located in the anion vacancy for titania, but neither are they itinerant, as might be inferred from the fact

that they are d-band (conduction band) electrons. Rather they are located in defect states within the band gap, as shown by Nolan et al.[37] However, here we have clearly shown that for a heavily reduced surface low oxidation states are formed, though here the dominant species appears to be Ti^{2+}. Annealing that layer in vacuum results in partial reoxidation of the surface, mainly because of loss of the 2 + state and diffusion of Ti^{3+} into subsurface interstitial sites.[16,21,34]

Table 1 indicates that annealing the sputtered surface has relatively little effect on the Ti:O ratio under these circumstances, but oxidation has a much larger effect. This is owing to the fact that new, stoichiometric titania layers grow over the surface during oxidation, as we showed previously by STM imaging,[16,21,34] occurring by migration of Ti^{3+} from subsurface sites back to the surface.

Pd/TiO$_2$(110)

The authors have previously described the evolution of the structure of Pd films on annealed single crystal TiO$_2$(110) with metal dosing and have shown that the film grows in a Volmer–Weber fashion at ambient temperature,[12,13] that is by nucleation of small nanoparticles, which eventually grow and merge into a holey film.[21,26] After annealing to high temperature such films can be flat[26] and show reasonably good LEED patterns.[21,26] However, annealing to very high-temperature results

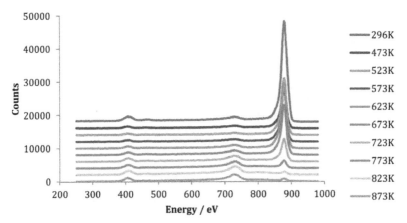

Figure 8 The effect of heating on the ion scattering spectroscopy (ISS) for the sputtered surface post-dosed with Pd

in the encapsulation of the metal with a TiO$_x$ film.[21,26] The authors previously proposed that this film on the Pd was TiO-like, since varying the photoelectron take-off angle in XPS indicated that the Ti^{2+} state was dominant at the surface.[21,26,38]

In the present case, we have looked at the effect of pretreating the surface on the growth of the Pd layer and have shown, surprisingly, that the evolution of signal increases in a similar form for the three different surfaces, indicating relatively little change in the growth mode with changes in TiO$_2$ surface composition.

However, of more interest is the effect of annealing on such samples. What is clearly shown is that an encapsulation layer forms over the Pd particles, beginning at ~700K (see Figs. 8 and 9, and Figs. 2 and 3 of Supplementary Material). Here the Pd signal in ISS declines to near zero by ~800 K, and the Pd:Ti ratio decreases by a factor of ~300, while in the XPS, the Pd:Ti signal ratio has only declined by ~60%. Thus, the Pd is still present at the surface of the titania crystal, but has been covered by a layer of TiO$_x$. The XPS confirms an increase in the state at ~455 eV b.e., associated with Ti^{2+}, and so the encapsulating layer on the Pd appears to be TiO-like, as suggested by the take-off angle results reported previously,[21,26,38] though obviously this layer is also bonded to Pd, and so is unlike bulk TiO.

The effect of encapsulation is seen for all three forms of treatment of the titania surface, Fig. 9. However, what is clear is that it occurs at lower temperature (by ~80 K) for the sputtered and the sputter–annealed surface than for the oxidized surface. This implies that there are still reduced states in the material even for the oxidized surface, but they reach the surface, and so to the Pd nanoparticles, with more difficulty than for the reduced crystal, which has reduced forms of Ti already at the surface.

Dulub et al.[14] carried out similar experiments with Pt on TiO$_2$ using ISS and STM and also observed SMSI. They dosed ~25 monolayers of Pd, completely eliminating the Ti signal in ISS and then annealed to >773 K. This caused the Pt signal to disappear from the ISS and the Ti signal to reappear. Fu et al.[30] found a strong dependence of the encapsulation process on the electron density in the conduction band of TiO$_2$. Encapsulation of Pd clusters was observed on TiO$_2$ crystals, which were heavily sputtered or reduced, but not on unreduced or slightly sputtered crystals.

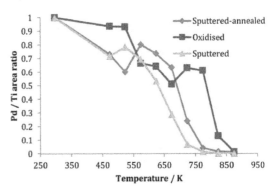

Figure 9 The change in Pd/Ti ISS ratio against annealing temperature of the three prepared surfaces. Note that in the case of the sputtered surface much more Pd was dosed onto the surface (450 s of dose), compared with 300 s for the annealed and 400 s for the oxidized surface

In some cases, they found a mix of Ti^{2+} and Ti^{3+} in the XPS after annealing Pd/TiO$_2$ to 823 K.

It is likely that the encapsulation is strongly related to the state of the reduction of the crystal used in all these cases. Usually, the way the crystal is treated in UHV changes its colour to (at least) pale, transparent blue,[1,34] reflecting the presence of Ti^{3+} in the bulk. In turn this can be useful since the conductivity of the sample increases enormously in such cases,[34] enabling good STM images to be obtained for titania samples. Presumably, there are enough such reduced states in the bulk for our apparently oxidized samples (which are likely to be oxidized in the near-surface region only) to still enable the SMSI effect to occur, even if at higher temperatures, while perhaps in the case of Fu et al.[30] this was not the case. Indeed, the authors showed earlier a fourth-order dependence of the oxidation of reduced titania crystals upon the amount of Ti^{3+} in the bulk,[21] which would lead to a dramatic dependence of these effects on the average crystal stoichiometry.

Conclusions

In conclusion, the authors have shown that sputtering of clean TiO$_2$(110) can result in reduction of the surface such that Ti^{2+} can dominate the cation states. Thermal annealing results in some reoxidation of the surface layers, especially $Ti^{2+} \rightarrow Ti^{3+}$, but oxidative treatment is more effective. Surprisingly, the growth mode of Pd on the surface appears to be little affected by the reduction state of the oxide. However, upon annealing of such surfaces, encapsulation of the Pd occurs (the so-called SMSI state), and this is strongly affected by the nature of the oxidation state of the underlying titania surface. This layer is TiO-like.

Conflicts of interest

The authors declare that there are no conflicts of interest.

Acknowledgments

The authors are grateful to Dr Rob Davies of Imperial College London and Dr Dave Morgan of Cardiff University for their technical assistance. The authors acknowledge the EPSRC for partial financial support for a studentship to RS.

References

1. U. Diebold: Surf. Sci. Rep., 2003, 48, 53–229.
2. M. Bowker: Green Chem., 2011, 13, 2235–2246.
3. M. Bowker: Catal. Lett., 2012, 142, 923–929.
4. A. Kudo and Y. Miseki: Chem. Soc. Rev., 2009, 38, 253–278.
5. V. Jovic, P. H. Hsieh, W. T. Chen, D. X. Sun-Waterhouse, T. Soehnel and G. I. N. Waterhouse: Int. J. Nanotechnol., 2014, 11, 686–694.
6. Z. H. N. Al-Azri, V. Jovic, W. T. Chen, D. X. Sun-Waterhouse, J. B. Metson and G. I. N. Waterhouse: Int. J. Nanotechnol., 2014, 11, 695–703.
7. M. C. Wu, I. C. Chang, W. K. Huang, Y. C. Tu, C. P. Hsu and W. F. Su: Thin Solid Films, 2014, 570, 371–375.
8. M. V. Dozzi, G. L. Chiarello and E. Selli: J. Adv. Oxid. Technol., 2010, 13, 305–312.
9. R. Su, R. Tiruvalam, A. J. Logsdail, Q. He, C. A. Downing, M. T. Jensen, N. Dimitratos, L. Kesavan, P. P. Wells, R. Bechstein, H. H. Jensen, S. Wendt, C. R. A. Catlow, C. J. Kiely, G. J. Hutchings and F. Besenbacher: Acs Nano, 2014, 8, 3490–3497.
10. M. A. Henderson: Surf. Sci. Rep., 2011, 66, 185–297.
11. A. A. Ismail and D. W. Bahnemann: Sol. Energ. Mat. Sol. Cells, 2014, 128, 85–101.
12. R. A. Bennett, P. Stone and M. Bowker: Catal. Lett., 1999, 59, 99–105.
13. R. A. Bennett, P. Stone and M. Bowker: Faraday Discuss., 1999, 114, 267–277.
14. O. Dulub, W. Hebenstreit and U. Diebold: Phys. Rev. Lett., 2000, 84, 3646–3649.
15. F. Netzer and S. Surnev: 'Scanning tunneling microscopy in surface science'; 2010, Weinheim, Wiley-VCH.
16. P. Stone, R. Bennett and M. Bowker: New J. Phys., 1999, 1, 1.1–1.12.
17. M. A. Henderson: Surf. Sci., 1995, 343, L1156–L1160.
18. M. A. Henderson: Surf. Sci., 1999, 419, 174–187.
19. V. A. D. O'Shea, M. C. A. Galvan, A. E. P. Prats, J. M. Campos-Martin and J. L. G. Fierro: Chem. Commun., 2011, 47, 7131–7133.
20. Z. Majzik, N. Balazs and A. Berko: J. Phys. Chem. C, 2011, 115, 9535–9544.
21. R. A. Bennett, C. L. Pang, N. Perkins, R. D. Smith, P. Morrall, R. I. Kvon and M. Bowker: J. Phys. Chem. B, 2002, 106, 4688–4696.
22. G. Barcaro, E. Cavaliere, L. Artiglia, L. Sementa, L. Gavioli, G. Granozzi and A. Fortunelli: J. Phys. Chem. C, 2012, 116, 13302–13306.
23. F. Silly and M. R. Castell: J. Phys. Chem. B, 2005, 109, 12316–12319.
24. C. Wu, M. S. J. Marshall and M. R. Castell: J. Phys. Chem. C., 2011, 115, 8643–8652.
25. C. Linsmeier and E. Taglauer: Appl. Catal., 2011, 391, 175–186.
26. M. Bowker, P. Stone, P. Morrall, R. Smith, R. Bennett, N. Perkins, R. Kvon, C. Pang, E. Fourre and M. Hall: J. Catal., 2005, 234, 172–181.
27. M. Bowker, P. Stone, R. Bennett and N. Perkins: Surf. Sci., 2002, 497, 155–165.
28. S. Bonanni, K. Ait-Mansour, H. Brune and W. Harbich: ACS Catal., 2011, 1, 385–389.
29. Q. Fu and T. Wagner: Surf Sci Rep., 2007, 62, 431–498.
30. Q. Fu, T. Wagner, S. Olliges and H. D. Carstanjen: J. Phys. Chem. B, 2005, 109, 944–951.
31. N. Fairley: 'CasaXPS, Version 2.3.15, Casa Software Ltd, 1999-2011', CasaXPS, Version 2.3.15'; 1999–2011, Casa Software Ltd.
32. U. Diebold and T. Madey: Surf. Sci. Spectra., 1997, 4, 227–231.
33. R. Kelly and N. Lam: Radiat. Eff., 1973, 19, 39–48.
34. M. Bowker and R. A. Bennett: J. Phys. Condens. Matter, 2009, 21, 9.
35. W. E. Kaden, T. P. Wu, W. A. Kunkel and S. L. Anderson: Science, 2009, 326, 826–829.
36. R. A. Bennett, D. M. Tarr and P. A. Mulheran: J. Phys. Condens. Matter, 2003, 15, S3139–S3152.
37. M. Nolan, S. D. Elliott, J. S. Mulley, R. A. Bennett, M. Basham and P. Mulheran: Phys. Rev. B, 2008, 77, 235424.
38. M. Bowker and E. Fourre: Appl. Surf. Sci., 2008, 254, 4225–4229.

19

Selectivity determinants for dual function catalysts: applied to methanol selective oxidation on iron molybdate

Michael Bowker[*1,2], **Matthew House**[1], **Abdulmohsen Alshehri**[1], **Catherine Brookes**[1,2], **Emma K. Gibson**[2] and **Peter P. Wells**[2]

[1]Cardiff Catalysis Institute, School of Chemistry, Cardiff CF10 3AT, UK
[2]Rutherford Appleton Laboratory, UK Catalysis Hub, Research Complex at Harwell (RCaH), Harwell, Oxon OX11 0FA, UK

Abstract Evolution of the IRAS spectrum with temperature after adsorbing methanol at room temperature. The bands at 2930 and 2820 cm^{-1} are due to the methoxy species C–H stretches, while that at 2870 is due to the formate.

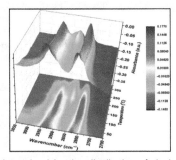

Here, we report a simple, quantitative model to describe the behaviour of bi-cationic oxide catalysts, in terms of selectivity variation as a function of increased loading of one cation into a sample of the other. We consider its application to a particular catalytic system, namely the selective oxidation of methanol, which proceeds with three main C1 products, namely CO_2, CO, and H_2CO. The product selectivity varies in this order as Mo is added in increasing amounts to an iron oxide catalyst, and the product selectivity is determined by the distribution of dual sites and single sites of each species.

Keywords Methanol oxidation, Oxide catalysis, Iron oxide, Molybdenum oxide, Reaction modelling, Selectivity, Formaldehyde

Introduction

A number of catalyst systems are based on oxidic materials with two or more cations present in the active phase, and these are often applied in selective oxidation reactions. Commercial examples include the oxidation and ammoxidation of propene to acrolein and acrylonitrile,[1] where mixed oxides of Bi and Mo or Fe and Sb were originally used, butane conversion to maleic anhydride[2] and methanol oxidation to formaldehyde.[3]

The latter reaction is carried out on either Ag or iron molybdate catalysts, and for the latter case it has been found that there is a very strong dependence of the product selectivity on the cation ratio in the bulk.[3,4] In turn, these changes are then because of changes in that ratio in the top layer, which is the active layer for reaction with the gas phase. This has lead to a few studies where researchers have made surface doped materials in the hope of learning what is the active configuration and what it is that makes that special for high selectivity to formaldehyde.[5–10] Coverages of $<6\,nm^{-2}$ (ca. 70% coverage) of MoO_3 appear to be supported as a monolayer, whereas at higher amounts

aggregation of oxide particles was observed.[5–7] These studies all show that with molybdenum present, improved selectivity towards formaldehyde is achieved for methanol oxidation. An increase in methanol oxidation turn-over frequency was observed with increasing molybdenum oxide coverage on the surface (up to $\sim6\,Mo\,nm^{-2}$), and this was proposed to be because of the necessity for adjacent molybdenum sites for the selective reaction (one for methanol adsorption, and one for hydrogen abstraction from the methoxy).[5] The structure of the monolayer was seen to be similar to that of bulk $Fe_2(MoO_4)_3$, with good methanol conversion and formaldehyde selectivity, though we only took conversion up to 60%.

Similar catalysts have also been created by simply heating a mixture of MoO_3 with Fe_2O_3[6] In that case, the dispersion capacity was calculated to be about $5\,nm^{-2}$ after calcination at 420°C. Calcination at higher temperatures lead to the reaction of MoO_3 with the bulk oxide to produce ferric molybdate. This $Fe_2(MoO_4)_3$ remained at the surface, effectively encapsulating the Fe_2O_3 inside. Similarly, Huang et al. made a material consisting of iron oxide and molybdenum ground together, and calcined to various temperatures;[7] they reported high selectivity performance, though again conversion was only taken to $\sim70\%$.

*Corresponding author, email bowkerm@cardiff.ac.uk.

In our laboratories, we have put some considerable effort into the understanding of selective oxidation reactions and especially methanol oxidation. Our main conclusion is that it is Mo, which is the most important component in ferric molybdate catalysts for high selectivity; it has very high selectivity on its own to formaldehyde, and is proposed to be present in commercial catalysts as an active Mo monolayer on top of ferric molybdate.[8–12]

Here, we examine how the selectivity of such catalysts is affected by the relative surface compositions of different cations that have intrinsically different properties from one another.

Experimental

The experimental methods used have been described in detail previously.[9,13,14] To summarise, the reactor was a tubular reactor, 1/4″ OD, operating in TPD (temperature programmed desorption) mode with a continuous flow of helium at a flow-rate of 30 mL min^{-1}, with a usual loading of 0.5 g of catalyst. Methanol was dosed onto the surface before heating by manual injection. Part of this flow was taken into an on-line mass spectrometer (Hiden Hiden Analytical quadrupole Hal 201) and the gas products were analysed as the catalyst was heated by ramping the temperature of the fan oven in which it was housed, typically at 12°C min^{-1}. The thermocouple measuring the temperature was placed directly in the catalyst bed. In reaction mode, the methanol was injected periodically (typically every 2 min) into a continuous flow of 10% oxygen in helium while ramping the temperature.

Site Distribution, general considerations

It is imagined that a catalyst consists of two components A and B, and further that these two components have a novel property when intimately mixed. That is, when like atoms are beside each other, then the reactivity is different from when they are isolated. So pure component A has the property of yielding product X, pure component B gives Y, while the mixed material yields product Z from isolated sites. It is assumed that these properties are then purely a function of the nature of *dual sites* in the system, then the following applies for random site occupation, for a catalytic surface which is changing from pure A to pure B:

- Fraction dual sites A–A, $AA = N_A^2/N$
- Fraction dual sites B–B, $BB = N_B^2/N$
- Fraction isolated sites A and B, $AB = 1 - AA - BB$

where $N_{A,B}$ are the number of A and B sites out of the total sites N, and adsorption of B results in loss of an A site (that is, $A + B = 1$)

By way of example, Fig. 1 shows three extremes of such behaviour for a dilute layer of B in A, a dilute layer of A mixed in mostly B, while approximately equal amount of both is shown in the middle panel. In a mixed layer then, the fraction of isolated sites is maximised at half coverage of A and B on the surface.

More explicitly, the result of this is shown in Fig. 2, which shows the variation in the distribution of homogeneous double sites, and of isolated sites as a function of increasing fraction of B in the mixed surface of A and B sites. There are parabolic declines in A, and rises in B double sites with increasing B

coverage, and during this process the occurrence of sites which do not have a like neighbour shows a broad maximum at 1/2 monolayer coverage, when there are 50% of such sites, and 50% of sites with like neighbours. Thus, if there is a specific product Z from the mixed surface, then it is expected that product will be maximised in the intermediate coverage range.

Site distribution, a specific example

We have examined the oxidation of methanol on iron molybdate catalysts and a major conclusion from this is that the surface of such catalysts is completely dominated by Mo.[4,8–10] This is the main reason that they are so selective, since we[4,13,15] and others[16] have shown that MoO$_3$ itself is highly selective for this reaction. However, what is of somewhat more interest to us is the effect of going to lower Mo:Fe ratios than that which is used commercially (typically 1.6:1–2.2:1) and lower than the stoichiometric ratio of 1.5:1 for ferric molybdate [Fe$_2$(MoO$_4$)$_3$] itself.

Figure 3 shows a result to indicate the kinds of reaction profile which we see as a function of temperature for methanol oxidation on ferric molybdate (in this case Mo:Fe 2.2:1). Here, is seen that conversion begins at ~160°C, and there is some evolution of dimethyl ether at very low conversion. When conversion becomes significant, then formaldehyde is seen as the major product, being formed at very high peak yield (~95%), while at higher temperatures and conversion CO dominates. At very high conversion and temperature CO$_2$ is seen, and can become dominant, but is likely to be because of secondary CO oxidation. This behaviour is similar to what is observed for 1.5:1 Mo:Fe catalysts, and for MoO$_3$ itself, which are highly selective to formaldehyde over a wide temperature range,[4,8,13] though the conversion for MoO$_3$ itself is much lower than for iron molybdate and only reaches 100% at much higher temperatures.[4,13] Data for iron oxide are in contrast to those in Fig. 3, since it is a complete combustor of methanol to CO$_2$ and H$_2$O[4,13]. For a catalyst with a sub-stoichiometric loading of Mo, as in Fig. 4, for example, the major product is CO, very different from the results of the individual oxides and the iron molybdate catalyst shown in Fig. 3.

It is useful to examine the results of TPD experiments on such materials, since TPD often gives very clear results, relating to the presence and nature of surface intermediates. They also relate closely to the reactor results in flowing oxygen and methanol. Figure 5 shows the TPD for these materials, again contrasting the behaviour of the different types. Iron oxide gives only CO$_2$ as the carbon product, while MoO$_3$ gives only formaldehyde. The full catalyst behaves very similarly to MoO$_3$ alone, yielding only formaldehyde, but with a slightly lower peak temperature, perhaps indicative of a slightly lower activation energy for methoxy dehydrogenation. However, a low Mo:Fe ratio catalyst gives CO as a major product, as shown in Fig. 6.

The important point here is that as the Mo concentration in molybdate catalysts increases, so the selectivity changes from combustion to selective oxidation, as follows

Low Mo loadings
$$CH_3OH + 3/2\ O_2 \rightarrow CO_2 + 2H_2O$$

Intermediate Mo loadings
$$CH_3OH + O_2 \rightarrow CO + 2H_2O$$

Figure 1   Schematic diagrams of site distribution as type B (black squares) is added to a lattice of sites A (white squares). Left low B coverage, middle ~ half a monolayer of type B, right near saturation of B.

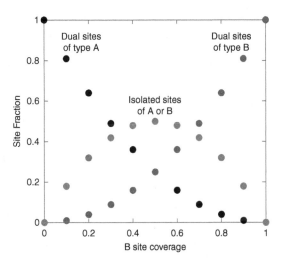

Figure 2   The variation of dual sites and single sites as a function of coverage of type B on the surface, assuming a randomised distribution model

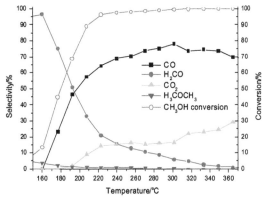

Figure 4   Product selectivity and conversion for a Mo:Fe ratio of 0.5, showing the dominance of CO production

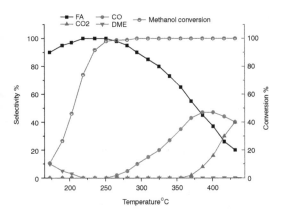

Figure 3   The temperature dependence of selectivity and conversion for Fe$_2$(MoO$_4$)$_3$. FA – formaldehyde, DME – dimethyl ether

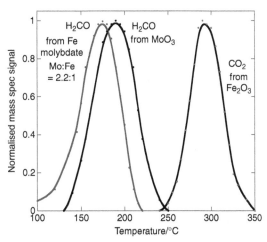

Figure 5   Normalised TPD spectra from MoO$_3$, ferric molybdate with Mo:Fe = 2.2:1, and Fe$_2$O$_3$. The latter shows predominantly carbon dioxide as the methanol oxidation product, whereas the former two materials give only formaldehyde

High Mo loadings

$$CH_3OH + 1/2 \ O_2 \rightarrow H_2CO + H_2O$$

In connection with the considerations in section 3 above, the results for product yield from iron oxide as the level of Mo doping varies from 0 to the stoichiometric material ferric molybdate, and this variation is shown in Fig. 7. It is apparent that, as shown above, CO_2 is dominant at high Fe levels, while H_2CO is dominant at low Fe. However, over a wide range of the intermediate concentrations, CO is the dominant product, and this behaviour is somewhat similar to that in Fig. 2 above. The question is, what exactly is it that dictates the shape of the curve in Fig. 7? It is our contention that double sites here are important for the selective reaction, as has already been proposed for Mo sites and formaldehyde production by others,[5,17,18] and as discussed further below.

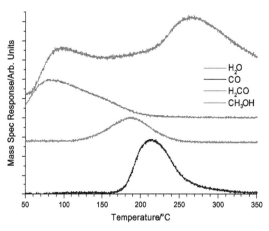

Figure 6   TPD from iron molybdate catalyst with a Mo:Fe ratio of 0.5. For such a sample, CO is a major product

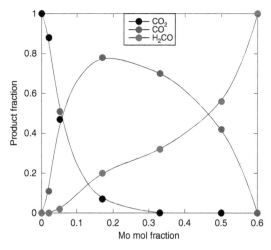

Figure 7   The yields of the three main products of methanol oxidation seen in TPD, as a function of the amount of Mo in the catalyst, note that a mole fraction of 0.6 corresponds with the stoichiometric materials $Fe_2(MoO_4)_3$

Although the data of Fig. 7 are broadly similar to those in Fig. 2, the results are not presented in the same way, so cannot be accurately compared. Figure 2 presents products as a function of *surface loading*, whereas Fig. 7 is a function of *bulk loading* and we do not at this point know if the latter is the same as the surface concentration of Mo. Indeed, we[4,8,10] and others[5–7,11,12] have shown that there is a strong tendency of Mo to segregate to the surface of these materials, that is, the surface concentration is higher than the bulk concentration. This may especially be important for the very low mole fraction of Mo in Fig. 7, where a very dramatic effect of low bulk loadings of Mo on knocking out CO_2 production in TPD. By only 0.28 of stoichiometry in the bulk, the CO_2 production has been reduced by a factor of 20. Further 100% selectivity to formaldehyde has been achieved at a mole fraction of 0.6, the stoichiometry of compound ferric molybdate.

We have some evidence related to this from XPS. For very low bulk loadings of Mo (Mo:Fe, 0.02:0.05) in Fig. 7 above, the surface ratio by XPS is higher (at 0.09 and 0.13, respectively), whereas at high Mo ratios, the XPS value is similar to that for the bulk loading. Thus, this would shift the low Mo level data points a little higher along the abscissa. However, this would not really have a dramatic effect on the shape of the curve; it would still show a severe decline in CO_2 at low Mo loadings. We must also remember that XPS measures the surface *region* (~ 5 layers), and does not give the top surface layer concentration. Studies with doping Mo onto the surface[9,10] shows that it tends to remain there and does not diffuse into the bulk. Indeed, if the reverse is attempted (Fe doping onto the surface of MoO_3), then Fe diffuses away from the surface into the bulk.[9]

There is another likely reason for the difference between Figs. 2 and 7, and that relates to the assumption that two sites are needed for methanol combustion and selective oxidation to formaldehyde, whereas one is needed for CO production. That two Mo sites are needed for formaldehyde production has been proposed earlier by a number of authors.[5,17,18] But when it comes to the combustion of methanol, albeit thermodynamically favoured, it is clearly a rather more complex reaction than selective oxidation, since it proceeds via a number of steps, such as

$$CH_3OH + O_s \rightarrow CH_3O + OH_s$$

$$CH_3O + 3O_s \rightarrow HCOO + 2OH_s + O_v$$

$$HCOO + 3OH_s \rightarrow CO_2 + 2H_2O + O_s + 3O_v$$

$$Overall: \quad CH_3OH + 3O_s \rightarrow CO_2 + 2H_2O + 3O_v$$

where O_s is a surface lattice oxygen and O_v is an anion vacancy at the surface.

Thus, it is not difficult to envisage that at least one of these steps may require a bigger ensemble than just two sites. The most stable intermediate in this sequence is the formate, and the presence of both methoxy and formate is shown by *in situ* IRAS (Fig. 8). Here, the methoxy (bands at 2820, 2930 cm^{-1}) dominates in the temperature range between 100 and 230°C, but it converts to formate (2870 cm^{-1}) above that temperature, which in turn is reduced as surface decomposition occurs. As shown above, the decomposition of this intermediate may require the involvement of three

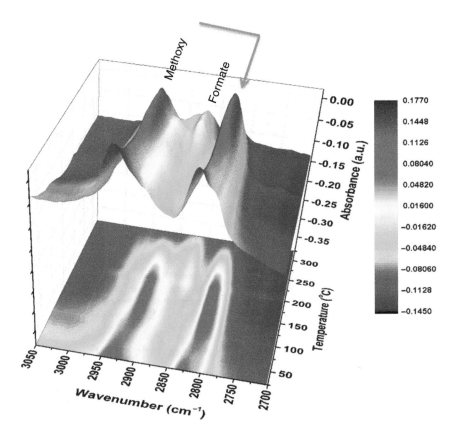

Figure 8   Evolution of the IRAS spectrum with temperature after adsorbing methanol at room temperature. The bands at 2930 and 2820 cm^{-1} are because of the methoxy species C–H stretches, while that at 2870 is because of the formate

oxygen functions, and the adsorbed system may require a large number of Fe sites for the hydroxyls and bidentate formate. If such large ensembles are required, then this could explain why the CO_2 yield goes down so quickly with increasing bulk loading of Mo. For instance, a model with the requirement for a larger ensemble adjacent Fe sites to be required, rather than dual sites, has just this effect, as shown in Fig. 9, and shows much closer agreement with the data of Fig. 7. It is interesting to note that in another selective oxidation reaction, namely the oxidation of ethylene to ethylene oxide, it has long been considered that CO_2 production is more 'demanding' than selective oxygen insertion to ethene, and that the role of Cl in hugely improving selectivity of Ag catalysts for this reaction is to restrict the size of sites available.[19] It is considered that the active site for combustion requires many more sites in the ensemble than does the selective oxidation. Thus, the plot in Fig. 8 would appear to be a reasonable approach to the methanol oxidation problem and we believe that this model gives a general description of the behaviour of such materials.

What may be a little more contentious is the single site hypothesis for CO production, and even if that hypothesis is true, there are two different single sites to be considered, namely Mo and Fe single sites. It would be difficult to imagine that such single sites have the exactly same chemistry or stability for the intermediate (probably methoxy), which produces it. Indeed, if we examine the desorption of CO in particular, as the Mo loading is varied (Fig. 10), this peak shifts

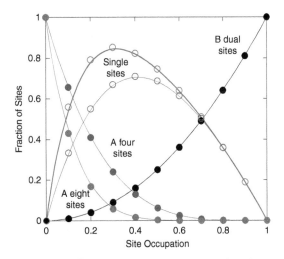

Figure 9   The distribution of surface sites with higher ensemble knock-out of species A sites by the adsorption of species B. In the specific case here, A are iron sites and B are Mo sites. If eight sites, for instance, were required for combustion, then they are affected as shown by the red curve, full red data points

significantly with increasing Mo coverage, from ~260°C at 0.05 to 215°C at 0.5 Mo:Fe ratio. This is in complete contrast to formaldehyde production and the formate decomposition, which stay at fixed decomposition temperatures.

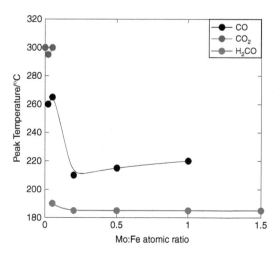

Figure 10   The variation in peak temperatures for CO_2, CO, and H_2CO desorption from catalysts with varying ratios of Mo:Fe. There is little variation for CO_2 and H_2CO, but a significant shift for CO at low Mo loadings

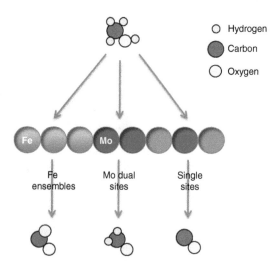

Figure 11   Model of the active sites on the Fe–Mo oxide materials used in this work, showing the variation in products from different site ensembles

Of course, these considerations then relate in an important way to the industrial catalyst typically used, which consists of iron molybdate with super-stoichiometric amounts of Mo present. The stoichiometry of ferric molybdate [$Fe_2(MoO_4)_3$] is Mo:Fe = 1.5, whereas industrial catalysts usually have the ratio in the range 1.7–2.2. In terms of our discussion above, it is clearly unwise to have any Fe in the surface layer of such catalysts. There would be a danger of that being the case if pure ferric molybdate were used because of two reasons: (i) possibilities from batch to batch of being slightly sub-stoichiometric and (ii) loss of Mo during time on stream, which certainly occurs over a long period under reaction conditions. Thus, high ratios of Mo:Fe compensate for any possible sub-stoichiometry and maintain the surface layers of the catalyst as purely Mo oxide.

Thus, we envisage this model pictorially, as shown in Fig. 11, which shows sites for CO, formaldehyde and carbon dioxide production.

In conclusion, we have shown that methanol reaction with iron oxide loaded with increasing amounts of surface Mo changes from combustion, to selective oxidation to CO, to selective oxidation to formaldehyde in sequence. Over a wide range of Mo loading CO is the dominant product and we propose that this is because of a requirement for a different site ensemble, requiring only one cation to be involved in the rate determining step. For formaldehyde, two Mo sites are required and it is likely that an even bigger ensemble is required for combustion on iron oxide via the formate intermediate. A simple ensemble model appears to show the general trend of behaviour observed and may well be applicable to a range of catalytic reactions and to other bi-cationic materials.

Acknowledgements

The authors would like to acknowledge the EPSRC for funding (EPI019693/1 and EP/K014714/1). Thanks also go to Diamond Light Source for provision of beamtime (SP8071-3) and for the contribution to the studentship of C. Brookes. The RCaH are also acknowledged for use of facilities and support of their staff.

References

1. R. K. Grasselli: Top. Catal., 2002, 21, 79.
2. G. J. Hutchings: Appl. Catal., 1991, 72, 1.
3. A. P. Soares, M. F. Portela and A. Kiennemann: Catal. Rev. Sci. Eng., 2005, 47, 125.
4. M. P. House, A. F. Carley, R. Echeverria-Valda and M. Bowker: J. Phys. Chem. C, 2008, 112, 4333–4341.
5. H. Yamada, M. Niwa and Y. Murakami: Appl. Catal. A, 1993, 96, 113.
6. F. Xu, Y. Hu, L. Dong and Y. Chen: Chin. Sci. Bull., 2000, 45, 214.
7. Y. Huang, L. Cong, J. Yua, P. Eloy and P. Ruiz: J. Mol. Catal. A, 2009, 302, 48.
8. M. P. House, M. D. Shannon and M. Bowker: Catal. Lett., 2008, 122, 210–213.
9. M. Bowker, C. Brookes, A. F. Carley, M. P. House, M. Kosif, G. Sankar, I. Wawata, P. P. Wells, P. Yaseneva and Yaseneva: Phys. Chem. Chem. Phys., 2013, 15, (29), 12056–12067.
10. C. Brookes, P. P. Wells, G. Cibin, N. Dimitratos, W. Jones and M. Bowker: ACS Catal., 2014, 4, (1), 243–250.
11. L. E. Briand, A. M. Hirt and I. E. Wachs: J. Catal., 2001, 202, 268.
12. L. J. Burcham, L. E. Briand and I. E. Wachs: Langmuir, 2001, 17, 6175.
13. M. Bowker, R. Holroyd, A. Elliott, P. Morrall, A. Alouche, C. Entwistle and A. Toerncrona: Catal. Lett., 2002, 83, 165–176.
14. M. P. House, A. F. Carley and M. Bowker: J. Catal., 2007, 252, 88–96.
15. M. Bowker, A. F. Carley and M. P. House: Catal. Letts, 2008, 120, 34–39.
16. T. Mizushima, Y. Moriya, N. H. H. Phuc, H. Ohkita and N. Kakuta: Catal. Commun., 2011, 13, 10–13.
17. J. N. Allison and W. A. Goddard III: J. Catal., 1985, 92, 127.
18. T. Waters, R. A. O'Hair and A. G. Wedd: J. Am. Chem. Soc., 2003, 125, 3384–3396.
19. C. T. Campbell and B. E. Koel: J. Catal., 1985, 92, 272–283.

20

Origin of catalytic activity in sponge Ni catalysts for hydrogenation of carbonyl compounds

Glenn Jones*

Johnson Matthey Technology Centre, Blounts Court, Sonning Common, Reading RG4 9LJ, UK

Abstract Computational results are presented from density function theory (DFT) that describe the reactivity of Al and early transition metal doped, sponge Ni catalysts. To develop an understanding of the catalytic activity of these materials, the direct reduction of acetaldehyde has been studied as a test system. Use of the scaling paradigm proposed by Norskov and co-worker shows the influence of dopant atoms upon the atomic adsorption energy of C and O and consequently adsorption energies of reaction intermediates. Construction of a simple kinetic model (parameterized from DFT) demonstrates that the presence of Al improves catalytic performance of carbonyl hydrogenation by increasing the reactivity towards O containing molecules, whilst at the same time decreasing the affinity towards C. Comparison is made to acetylene hydrogenation, where the activity is dependent on the C affinity of the catalysts. It is thus suggested that should

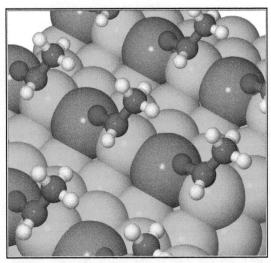

one desire to selectively hydrogenate a carbonyl group in the presence of an alkene then the use of early transition metal dopants may facilitate this selectivity. Alternatively, one could use a dopant that is able to reduce the affinity for C but maintains a high O affinity. The origin of the activity change due to doping is shown to be the intrinsic electronic structure of the dopant rather than a perturbation of the lattice constant due to the dopant atom.

Keywords DFT, Hydrogenation, Transition metal, Heterogenous, Screening, Electronic structure, Computational, Selectivity

Introduction

Sponge Ni is an archetypal example of an important class of industrial hydrogenation catalysts whose origin can be traced back to the early Twentieth Century and their discovery by Murray Raney.[1−3] Sponge Ni finds use in a wide variety of organic reductions and plays a major role in the industrial hydrogenation of vegetable oils. Despite this long history and a significant amount of research, relatively little work has focused on the molecular level mechanisms governing the catalytic activity of this material. This paper aims

*Corresponding author, Email: gjones@matthey.com

to address this gap in knowledge by presenting results that answer the question of how Al and other dopants, influence the activity of sponge Ni catalysts.

Herein computational results probing the reactivity of doped sponge-Ni type containing catalysts are presented. The objective of this work is to develop an understanding of the observed reactivity and use the insight gained as an aid in the prediction of improved, doped sponge Ni catalysts. The work can (broadly speaking) be broken down into four sections. We begin by investigating the stability of Ni:Al surface alloys supported on fcc-Ni. We then proceed to molecular adsorption studies of reactive intermediates involved in the hydrogenation of acetaldehyde on Ni:Al surface alloys. The adsorption studies form the basis of kinetic modeling studies and finally we draw the work together and predict dopant materials that can potentially optimize the intrinsic catalytic activity.

The stability of Ni:Al has been studied from a thermodynamic perspective with the primary objective being to identify reasonable models for the reactivity studies. Kinetic Monte Carlo studies conducted within the IMPRESS project by Barnard et al. have suggested very strongly that the activated sponge Ni is not a bulk like Ni_xAl_y alloy, but instead a Ni and Al containing surface alloy supported on fcc-Ni.[4] Supporting evidence of this fact is found from X-ray photoelectron spectroscopy and neutron diffraction studies, which show that the bulk structure of the activated particle is predominantly fcc-Ni (with some Al impurities) and that Al is present at the surface in (Al:Ni) ratios varying from approximately 30–50%.[4] The thermodynamic studies presented herein corroborate with this picture, suggesting that Al is present as a metastable, low concentration surface alloy.

Direct hydrogenation of acetaldehyde, whilst interesting in its own right, is studied as a prototype aldehyde from which the results can be extrapolated to the industrially relevant butyraldehyde hydrogenation. Using the Ni:Al near surface alloy that our calculations find to be most stable as a model of the activated catalyst, adsorption studies have been conducted of the various reactive intermediates involved in the reduction of acetaldehyde. The adsorption studies provide the raw data for the subsequent reaction studies, which start with the construction of free energy diagrams for the acetaldehyde reaction. This allows us to gain valuable insight that we can use when developing a micro-kinetic model. Solution of the resultant micro-kinetic model provides a theoretical measure of the activity to which we can measure the quality of novel catalyst candidates.

The predicted activity of materials makes use of the scaling paradigm proposed by Norskov and co-workers.[5] In this work we consider the adsorption of alkoxy and carbonyl species at a metal surface. It is found that the alkoxy binds via an O atom, and consequently the adsorption energy is a function of the atomic O adsorption energy. Whereas, the carbonyl binds via both a C and an O atom, leading to the adsorption energy being a linear function of both the atomic C and the atomic O adsorption energies. By combining these descriptors with the micro-kinetic model an activity map can be produced which shows where the theoretical optimal catalyst lies. This knowledge is then used to screen dopant materials.

The paper concludes by discussing the physical origin of differences in reactivity as due to two primary influences. The introduction of a dopant material has both a geometric and an (intrinsic) electronic influence upon catalytic activity. It is shown that the electronic properties of the dopant are dominant in the case where they are present in low concentrations. The question of whether a heterogeneous metal catalyst can be used to selectively hydrogenate a carbonyl group in the presence of an alkene is also addressed. There is some hope that this may be achieved by the introduction of a dopant to alter C and O

affinity or possibly by engineering the surface facets to exploit surface sensitivity of the carbonyl reduction. We close with a discussion on the limitations of the modeling, in particular the complexity of the sponge-Ni morphology illustrating that whilst we can make a lot of progress in our understanding, the ubiquitous materials and conditions gap between theory and experiment has yet to be fully closed.

Computational approach

The starting point for the work presented herein is the generation of raw data from density functional theory calculations. Adsorption studies were conducted on stepped surfaces using 2×1-fcc{211} slabs consisting of 10 atomic layers and a vacuum region of 10 Å, whereas calculations on flat surfaces were carried out in similar fashion but utilizing a 2×2-fcc{111} slabs with 3 atomic layers*. Total energies were calculated using density function theory (DFT), as implemented in the computer code DACAPO, using the RPBE functional to describe electron exchange and correlation.[6] The valence electrons were modeled using a plane-wave cut off of 340 eV and a density grid of 680 eV, with the Brillouin zone being sampled by a $4 \times 4 \times 1$ Monkhorst-Pack k-point mesh. Pseudopotentials of the Vanderbilt type were used to model the core electron regions.

For clarity we have defined adsorption energies (E_{ads}^{DFT}) of intermediates ($E_{Metal/C_xH_{y-n}O_z}^{DFT}$) referenced to the final product of the overall reaction in question ($E_{C_xH_yO_z}^{DFT}$) and hydrogen ($E_{H_2}^{DFT}$) in the gas phase as

$$E_{ads}^{DFT} = E_{Metal/C_xH_{y-n}O_z}^{DFT} - \left(E_{C_xH_yO_z}^{DFT} + \frac{n}{2}E_{H_2}^{DFT} + E_{Metal}^{DFT} \right)$$

Whereas binding energies of atomic species (X) have been defined by referencing to the isolated unbound atoms

$$E_{bind}^{DFT} = E_{Metal/X}^{DFT} - \left(E_{Metal}^{DFT} + E_X^{DFT} \right)$$

Transition states are located using a constrained minimization algorithm. For (de)hydrogenation reactions the reaction coordinate is well defined and the computational cost associated with more elaborate methods is unnecessary, the derived linear energy relationships for transition states are shown in the supplementary material.

The influence of reaction conditions upon the reactivity has been modeled using thermodynamic corrections obtained from the statistical thermodynamics of a classical ideal gas. Once the free energies are obtained it is then straightforward to obtain equilibrium constants and rate constants for use in a micro-kinetic model. The methodology has been outlined in previous work.[7,8]

Experimentally the synthesis of a porous sponge-Ni is facilitated by NaOH ($\sim$20% weight) leaching at moderate temperatures (50–80°C).[4] Determining the relative stability of different Ni and Al surface alloy configurations requires

---

* Convergence testing has illustrated this is sufficient to capture trends between different transition metal surfaces, whereas convergence of absolute values does require a deeper slab.

a model that captures the influence of the solvent on the ions formed during this leaching process. This could potentially be carried out *ab-initio* on non-metallic substrates; however it lays a heavy demand on computational resources.[9] To achieve this on a metallic surface is a significant computational challenge in it own right, therefore an empirical approach that starts by considering a general chemical equation describing the leaching process is used

$$Ni_{y-x}^{\{ijk\}}Al_x + xNi_{(bulk)} + xOH_{(aq.)}^- + xNa_{(aq.)}^+ + 3xH_2O_{(l.)} \rightleftharpoons$$
$$Ni_y^{\{ijk\}} + xAl(OH)_{4(aq.)}^- + xNa_{(aq.)}^+ + \frac{3x}{2}H_{2(g.)}$$

where $\{ijk\}$ are the Miller indices of the surface facet in question, y is the number of Ni atoms in a computational supercell of pure Ni and x is the number of Al atoms in the surface alloy. By considering the thermodynamic processes linking the products to reactants one can calculate the overall Gibbs free energy of reaction $\Delta G_{reac.}^{overall}$. To achieve this, a route must be identified that allows a description of the energy changes in terms of either experimental literature values or those that can be calculated rapidly within DFT. This route is not necessarily a reflection of the true chemistry that occurs, but rather an application of Hess' Law.[10] In this case it is the overall energy change we interested in, that can be obtained by considering the following set of reactions

$$Ni_{y-x}^{\{ijk\}}Al_x + xNi_{(bulk)} + xOH_{(aq.)}^- + xNa_{(aq.)}^+ + 3xH_2O_{(l.)}$$
$$\Updownarrow (A)$$
$$Ni_{y-x}^{\{ijk\}}Al_x + xNi_{(bulk)} + 3[\tfrac{x}{2}O_{2(g.)} + xH_{2(g.)}] + xOH_{(aq.)}^- + xNa_{(aq.)}^+$$
$$\Updownarrow (B)$$
$$Ni_y^{\{ijk\}} + xAl_{(bulk)} + 3[\tfrac{x}{2}O_{2(g.)} + xH_{2(g.)}] + xOH_{(aq.)}^- + xNa_{(aq.)}^+$$
$$\Updownarrow (C)$$
$$Ni_y^{\{ijk\}} + xAl(OH)_{3(s.)} + xOH_{(aq.)}^- + xNa_{(aq.)}^+ + \tfrac{3x}{2}H_{2(g.)}$$
$$\Updownarrow (D)$$
$$Ni_y^{\{ijk\}} + xAl(OH)_4^- + xNa_{(aq.)}^+ + \tfrac{3x}{2}H_{2(g.)}$$

which in turn are associated with the following Gibbs free energy changes

$$\Delta G_{reac.}^A = -3x\Delta G_{form.}(H_2O_{(l)})$$
$$\Delta G_{reac.}^B = [G(Ni_y^{\{ijk\}}) + xG(Al_{(bulk)})] - [G(Ni_{y-x}^{\{ijk\}}Al_x) + xG(Ni_{(bulk)})]$$
$$\Delta G_{reac.}^C = x\Delta G_{form.}(Al(OH)_{3(s.)})$$
$$\Delta G_{reac.}^D = xG(Al(OH)_{4(aq.)}^-) - (xAl(OH)_{3(s.)} + xOH_{(aq.)}^-)$$

It can be seen that step A requires the Gibbs free energy of formation for liquid water, this is readily available from tabulated values.[11] Step B can be obtained directly from DFT calculations (either bulk or surface values), assuming the phonon contribution to the entropy is cancelled out between the initial and final state. The free energy change can thus be rewritten as $\Delta G_{reac.}^B = [E(Ni_y^{\{ijk\}}) + xE(Al_{(bulk)})] - [E(Ni_{y-x}^{\{ijk\}}Al_x) + xE(Ni_{(bulk)})]$. The values for step C and D are also readily available from literature.[12,13]

There is currently no published experimental, atomic resolved information to be used as a precedent for the structure of the Ni and Al atoms at the surface of a sponge-Ni catalyst. X-ray photoelectron spectroscopy studies,[4] have shown that the average Ni:Al ratio at the surface varies between 30 and 50% for different samples, in keeping with recent Monte Carlo studies from the same work. The stability of Ni and Al surface alloys supported on an fcc-Ni lattice was calculated for the 2 × 1-{211} and 2 × 2-{111} surface facets (Figs. 1 and 2). Figure 3 shows the free energy of reaction per Al plotted against the number of Al present in the surface alloy. The corresponding pure Ni surface is used as the reference level. Figure 3 shows that the alloys containing more Al atoms are less thermodynamically stable towards leaching than those containing fewer Al atoms. It also shows that thermodynamic arguments favor the formation of a pure Ni surface, although this is most likely not achieved in practice due to kinetic constraints.[4]

The structures identified in this study are necessarily ordered alloys, due to the periodic boundary conditions; however the experimental observation of broad diffraction peaks show the structure is inhomogeneous[14] which is supported by kinetic Monte Carlo studies.[4] The contribution to stability from configurational entropy (S_{conf}), has been estimated using a simple approximation,[15] $S_{conf} = k_b[-c \ln c - (1 - c) \ln (1 - c)]N$, where c is concentration (in mole fractions) of Al atoms in Ni and N is the number of sites. ΔTS_{conf} per cell for different coverages ($T = 313$ K) is found to be in the order of hundredths of an eV and therefore does not influence our stability trends presented herein.

These surface stability calculations have served two purposes. Firstly to see if there is a simple reason for the experimentally observed Ni:Al ratios in the surface region and secondly to justify the choice of surface model to be

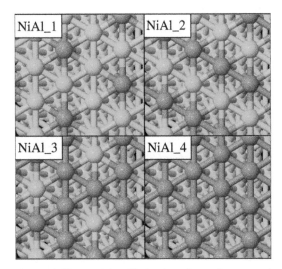

Figure 1 Al containing Ni {111} surface alloys, tested within this study. Top layer Ni:Al substitution ratios considered herein; (NiAl_1) 3:1, (NiAl_2) 2:2, (NiAl_3) 1:3, (NiAl_4) 0:4. Green: Ni atoms, grey: Al atoms, nomenclature corresponds to labeling used in Fig. 3

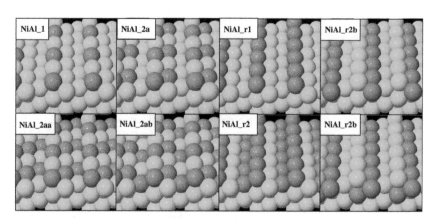

Figure 2 Al containing Ni {211} surface alloys, tested within this study. Structures correspond to single Al atom substitution per atomic layer of Ni, or to single row Al substitution per atomic Ni layer, corresponding to Ni:Al ratios between 3:1, 2:2, 1:3. Green: Ni atoms, grey: Al atoms, nomenclature corresponds to labeling used in Fig. 3

used in the reactivity studies. This has been partially successful on the first count where they have shown low Al concentrations are more favorable than high Al concentrations. Thermodynamics also predicts Al to lie at the surface of the catalyst; however the results are limited by our choice of periodicity. Therefore it is not possible say conclusively, what the thermodynamically most stable Ni:Al surface is. The most stable alloy at the periodicities described above (2×1-{211} and 2×2-{111}) correspond to one Al per repeat unit, at an Al concentration of 0.5 and 0.25 in the top surface layer, for the {211} and {111} surface respectively. Thus, in the absence of atomically resolved experimental results to guide us the above discussion serves as justification of the choice of surface model for the reactivity studies.

Adsorption of reactive intermediates

The starting point is to consider an idealized reaction profile of the acetaldehyde reduction. This then leads to a reaction profile containing two significant surface intermediates, the adsorped actaldehyde and the partially hydrogenated ethoxy species. The next step is to establish the relationship between atomic and molecular adsorption energies. This is achieved by plotting the adsorption energies of the molecular species on a given metal surface against those of an atomic species. In doing so, this allows one to think in terms of reactivity trends that can be used in the catalytic screening process. A number of adsorption energy trends have been published which provide the starting point for this study.[5,7,16,17] However, for some of the species of interest in this study no such precedent exists.

Fig. 4 shows a plot of the adsorption energy of the ethoxy species plotted against the atomic oxygen adsorption energy on the {211}-surface for a series of metals. It shows, that for ethoxy intermediate, the scaling predicted by Norskov et al. (which is based upon valence arguments) still holds true when considering pure metals.[16] Furthermore, for doped Ni alloys (where Al has been replaced by a transition metal, which will be

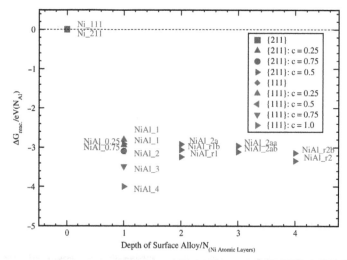

Figure 3 Free energy of reaction for dissolution of Al into NaOH solution. Negative value indicates that dissolution is more favorable. Results are shown for fractional Ni substitions in terms of Al coverage (c = 0.0, 0.25, 0.5, 0.75 ML)

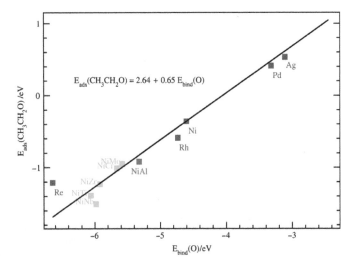

Figure 4 Plot of relationship between O binding energy and adsorption energy of ethoxy on pure metals (blue), NiAl (red) and doped Ni surfaces (green)

discussed further), there still remains a good fit. If we now turn our attention to acetaldehyde adsorption the situation is slightly more complex. The carbonyl group of the molecule possesses a pi-bond, in order to determine the valance parameter for the adsorbed species we consider whether the surface is reactive enough to cleave this pi-bond (thereby facilitating adsorption) or not (in which case the molecule does not bind). Recent work has shown that adsorption energies of alkenes and alkynes can be predicted on the basis of a simple valence argument. Using this argument it is possible to construct a descriptor by analogy that takes into account both the adsorption energy of C and O. Figure 5 shows a plot of the acetaldehyde adsorption energy vs reactivity descriptor.

It can be seen for pure metals this works quite well, however for our alloys we are not performing so well, this is most likely due to the d-band model, upon which the valence model is built, being less successful in the regime of strong binding. However despite only achieving semi-quantitative accuracy we still capture the reactivity trends, which for a screening study is of primary importance.

We are now in the position to consider the free energy diagrams for the reduction of acetaldehyde to ethanol on both the {111} and {211} surface (Figs. 6 and 7). A brief analysis shows that the {111} terrace is unreactive to acetaldehyde adsorption and will therefore be a poor catalyst, even for the most reactive metals. If we now consider the {211} surface, we see that adsorption becomes exothermic when we reach Ni

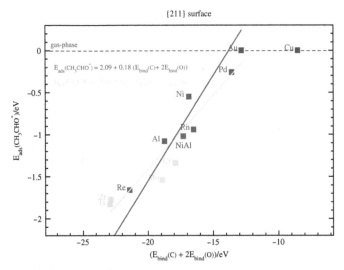

Figure 5 Plot of relationship between $[E_{bind}(C) + 2E_{bind}(O)]$ and adsorption energy of acetaldehyde. It can be seen that Cu and Au (violet) do not bind species and that a good fit can be achieved to pure metal data (blue). If an ideal slope of 0.25 (as predicted by a simple valence model)[16] is used, better agreement is achieved for doped Ni alloys (green) and NiAl (red)

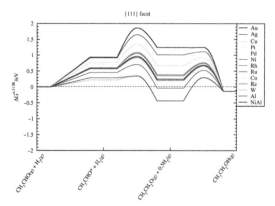

Figure 6 Free energy diagram for acetaldehyde reduction over {111} surface

and Rh and even then formation of ethoxy is still endothermic, implying that any metal less reactive is likely to be a poor catalyst for this reaction. For clarity this study will now focus on the relative reactivity of the stepped {211} surface.

Kinetic model and screening of dopant candidate

A good understanding can be established on the basis of thermodynamic arguments alone. However, further information can be extracted through consideration of a simple kinetic model. On the basis of the free energy diagrams (Figs. 6 and 7) it is shown that for acetaldehyde reduction either of two hydrogenation steps could be the rate-determining step, therefore the following model is been used to accommodate this

(1) $C_2H_4O(g) + * \xrightleftharpoons{\text{Equilibrium}} C_2H_4O^*$

(2) $\frac{1}{2}H_2(g) + * \xrightleftharpoons{\text{Equilibrium}} H^*$

(3) $C_2H_4O^* + \frac{1}{2}H_2(g) \xrightarrow{\text{RDS.}} C_2H_5O^*$

(4) $C_2H_5O^* + \frac{1}{2}H_2(g) \xrightarrow{\text{RDS.}} C_2H_5OH(g) + *$

The model can be solved within the steady state approximation to give an analytical solution. By using the

scaling relationships between molecular adsorption energies and atomic adsorption energies, in combination with linear relationships for transition states (see Supplementary Material),[18,19] we are now able to obtain rates of reaction for the above scheme, for any combination of C and O adsorption energy. As expected from reaction profile in Fig. 6 the {111} surface exhibits an extremely low turn-over frequency and can be considered to be inert, whereas the {211} surface gives a good turn over frequency. The volcano plot obtained from this analysis for the {211} surface is shown in Fig. 8. The peak of the volcano can be seen to lie at a low carbon adsorption and relatively high O adsorption energies. This can be understood in terms of the 'Sabatier principle', increasing the reactivity of the catalyst improves the adsorption of the reactant, however, if the reactivity increases too much, then the surface will be poisoned by the ethoxy intermediate.[20,21] It can also be seen within the model that the dependence of C and O affinity of our alloy affects the acetaldehyde adsorption whereas only O affinity effects the ethoxy intermediate, thereby giving us hope that we can tune the adsorbates adsorption properties independently.

Overall the plots show that Ni would be a reasonable catalyst for this reaction, however by adding Al we increase the O affinity whilst decreasing the C affinity. In this case, the increase in the O affinity increases the adsorption energy of acetaldehyde by over compensating for the lost C affinity. This leads to an increase in turn-over due to the increased coverage of reactant material and the increased affinity for the ethoxy species. This is in stark contrast to the role of other p-block elements, which when used as dopants show a reduction in activity; suggesting that Al is unique amongst the p-block in its properties.

A qualitative explanation for our predicted activity lies in consideration of the local density of states (Fig. 9). Here we can see that the d-band of Ni contracts and rises in energy when a Ni atom is adjacent to Al. This is presumably because Al has no d-band for Ni to couple to. Thereby creating a situation in which the Ni d-band is effectively under-coordinated. However, when we move to the lower 3D elements,

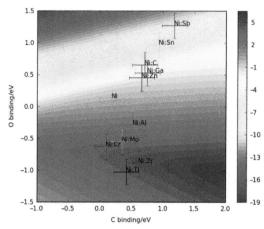

Figure 8 Reactivity plot of {211} surface, uncertainties of ±0.2 eV are shown. Model conditions are; total pressure: 42 bar (40:2 H_2:C_2H_4O), temperature: 313 K, color bar shows TOF per active site/s^{-1}

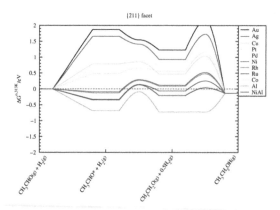

Figure 7 Free energy diagram for acetaldehyde reduction over {211} surface

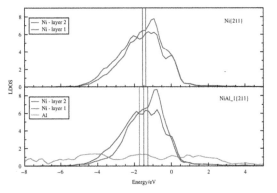

Figure 9 Sum of spin up and spin down density of states projected onto atomic orbitals for surface atoms of Ni{211} and Ni:Al {211} surface alloy considered within this work (structure NiAl_1, Fig. 1). d-band center of top 2 layer Ni atoms are shown, along with p-band of Al atom. Vertical lines indicate d and p-band centers

there is likely to be greater coupling between Ni p-states and these elements, thereby leading to a lowering of the valance d-band due to the more delocalized nature of the core electronic states.

Ni doped with an early transition metal shows, in general an increase in the activity of the catalyst. These early transition metals have a very strong affinity for oxygen and the origin of the increased activity lies in the increase in oxygen affinity by the dopant. As discussed previously, increasing the oxygen affinity facilitates adsorption of the acetaldehyde molecule. If (as mentioned earlier), we were to increase the reactivity too much then we would see the activity dropping off due to the ethoxy species being bound too strongly. Figure 10 shows the projected density states for the surface Ni and dopants (Cr, Mo, and Zr), as can be seen there is a weak overlap between the early transition metal d-band and the Ni d-band, which again, rather like Al leads to an upshift of the Ni d-band.

Discussion

When considering the physical origin of differences in reactivity due to the presence of dopants we can consider two primary influences. The first is the change in lattice parameter when a dopant is present (hereafter nominally called the geometric effect) the second is the intrinsic electronic influence of the dopant atom (hereafter nominally called the electronic effect). The geometric effect can be modeled using stepwise expansions and contractions in the lattice parameter and noting how the adsorption energy of a C or O atom changes. In so doing it will be seen how a change in the lattice parameter is related to a change in catalytic activity. The electronic effect can be probed by replacing the Al atom in the surface layer with a dopant element, this time with the lattice parameter fixed at the equilibrium Ni value. Again calculating the change in adsorption energy of a C or O atom (when compared to pure Ni) will provide an indication on the size of the influence. The results presented in Figs. 11 and 12 are for a {211} surface and show that, in general, the electronic influence is dominant for the doped Ni structures considered within this work.

Experimentally observed (X-ray and neutron diffraction) changes in lattice parameter due to doping with either Al and/or Cr are found to be on the order of 0.1–0.5%, with an estimated 0.14% error.[14] This change is clearly a rather small and one could ask the question, how much does it influence reactivity. The following discussion shows that indeed the geometric influence on adsorption energies is much less significant than the electronic influence brought about by the substitution of a Ni atom with another element for the case of O adsorption (and by extension O containing species). However, the differences are not as great for C adsorption (and again, by extension C containing species), although the electronic influence is still greater.

Stepwise expansion and contraction of the Ni lattice was made between ±10% of the equilibrium lattice parameter (Ni $a = 3.56$ Angstrom). From this plot a linear fit was made to obtain a relationship between lattice expansion and change in

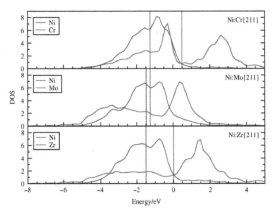

Figure 10 Sum of spin up and spin down density of states projected onto atomic of surface atoms of Cr, Mo and Zr doped Ni-{211} (analogous to structure NiAl_1, Fig. 1), d-band center of top layer Ni and dopant atoms are depicted by vertical line

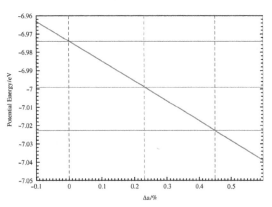

Figure 11 Variation of calculated C adsorption energy on pure Ni {211} surface with varying lattice parameter a, in red. To guide eye; blue corresponds to equilibrium position, green corresponds to average experimental expansion of Ni and Al containing alloys, magenta corresponds to average experimental expansion of Cr doped Ni Al alloys

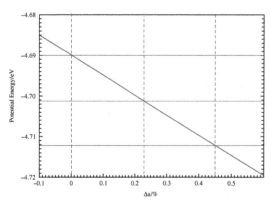

Figure 12 Variation of calculated O adsorption energy on pure Ni {211} surface with varying lattice parameter *a*, in red. To guide eye; blue corresponds to equilibrium position, green corresponds to average experimental expansion of Ni and Al containing alloys, magenta corresponds to average experimental expansion of Cr doped Ni Al alloys

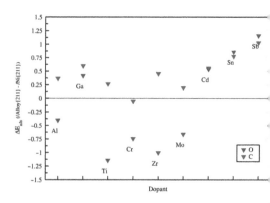

Figure 13 Plot showing deviation of C or O adsorption i doped Ni alloys (these alloys all have Ni in top layer sub stituted by guest element)

adsorption energy (Supplementary Material Fig. SM3). Figures 11 and 12 show the change in C and O adsorption energy on pure Ni{211} adsorption energy zoomed into the region where the experimental lattice changes are observed. At the extreme of the experimentally observed lattice expansion the changes in adsorption energy are of the order of ~ 0.04 eV (C) and ~ 0.02 eV (O). This is very close to the limits of DFT uncertainty and also subject to the errors present in the experiment (the expected error in DFT is expected to be less than 0.1 eV, systemic ways to get a true theoretical error bar are the subject of active research[22]).[†] What this illustrates is that given the size of the changes in lattice parameters the intrinsic reactivity of Ni is not being modulated greatly through this effect. However, looking at Fig. 13 (which depicts the change in C and O binding energy for a number of doped Ni surfaces) the influence of the 'electronic' effect from the elemental substitution can be seen. This in general, is significantly greater than the geometric effect, the difference from pure Ni and between different elements being outside the theoretical noise of the DFT method (values generally being between a tenth and whole eV different).

One notable exception to this observation is for Cr the electronic influence on C adsorption is ~ 0.05 eV, of a similar magnitude to the geometric influence. This can be contrasted with Mo, which of the elements depicted in Fig. 13, most closely resembles Cr and has a change in adsorption energy of 0.2 eV. The geometric influence of Al on the adsorption of C is estimated to be ~ 0.02 eV whereas the electronic influence almost 0.4 eV. Therefore it seems that Cr has a relatively small change on the intrinsic reactivity of Ni towards C, however, the story for O, is much more clear-cut[‡].

It has now been shown that the first order effect and the mos important aspect to understand in the reactivity of the doped N materials presented here, is the role of the electronic influenc of the dopant material. This effect is noted to be particularl significant when considering reactants that contain oxygen. should however, be noted that this is not to say that the reac tivity cannot be more greatly influenced by geometric change were ways to make greater changes to the lattice constan possible. Or even, that fine-tuning of a catalyst could not b achieved through these means.

Experimentally it is indeed found that Cr, Ti, and Mo are goo for the promotion of butyraldehyde reduction, which is i qualitative agreement with the findings of this work. Howeve we find in the work presented herein that Ti should possess th greatest intrinsic activity followed by Mo and Cr. For practica application it is anticipated there could be some deviation from this trend, for the reasons outlined below; however, it is pre dicted that all of these dopants would provide a systemati improvement over the standard sponge Ni catalyst. Despite th limitations of a theoretical screening approach, it has bee shown that a significant amount of insight about catalys composition and the origin of catalytic activity is possible Furthermore, elemental phase space of catalyst materials ha been reduced by the theoretical insight allowing rationa development of practical catalysts.

Other deviations from experiment can occur, not only from intrinsic errors, but also from the issue of 'are we comparing lik with like?' For instance the surface morphology (i.e. nature of th active site) plays an important role in the activity. Herein we hav considered binary systems, whereas real catalysts may in fact b ternary systems. This is an important consideration as it is poss ible that experimentally the different active catalysts posses different morphologies. In turn this could lead to the presence o active sites not considered herein. Further research direction that can also consider the macroscopic structure of the catalys particle include how the size and shape of the pores presen could potentially lead to transport limitations. This woul provide insight into how the surface chemistry (presente herein) interplays with the diffusion of product and reactants to and from the active sites. These transport limitations can i principle be probed by both experiment and theory, future wor will involve studying transport models using multiscal models.[23,24]

[†] It should also be remembered that with the scaling model the difference in adsorption energies of a given molecule from one metal to another is given by a fraction of the change in C or O binding energy. Therefore the changes in binding of the larger molecules as one stretches or compresses the material will be even smaller.
[‡] This also clearly illustrates that we would expect the dopants to have a different influence when we carry out a C = O re-duction compared to a C = C reduction.

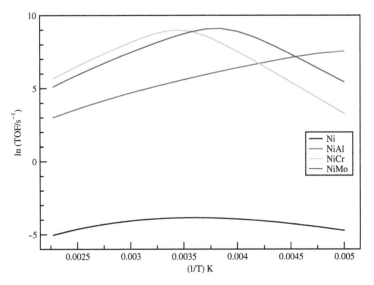

Figure 14 Theoretical temperature dependence of acetaldehyde reduction. It can clearly be seen that doped alloys show improved activity over pure Ni. Effective activation barriers can be obtained by linear fit over small temperature range for direct comparison to experiment

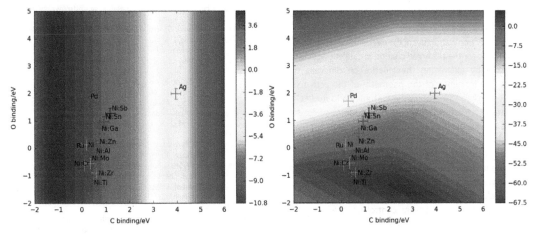

Figure 15 Comparison of turn over frequency of ethylene hydrogenation (left) vs hydrogenation of acetaldehyde (right). Modeled under same conditions as Fig. 8, color bar shows TOF per active site/s^{-1}

Along with well defined analytical characterization, rates and adsorption properties of elementary steps can be investigated in a controlled and consistent fashion and importantly offer the chance for direct quantitative comparison to theory. An example of where theory and experiment can meet is the so called Arrhenius plot, where the rate can be measured as a function of temperature. Under the assumption that the rate is first-order, a linear fit can be made, the slope of which gives the effective activation barrier. Such a plot obtained from theoretical data is shown in Fig. 14. Upon examination this plot serves to highlight a further consideration in the theoretical prediction catalysts that is often overlooked. It can clearly be seen that the optimal catalysts is temperature dependent illustrating that optimizing catalytic activity is not simply a case of optimal material prediction at a single set of conditions, but the optimal material can also depend on the reaction conditions.

The final section of our discussion focuses on the potential to use our screening model to explore the selectivity of catalysts to either C = C or C = O hydrogenation. Figure 15 shows a 2D volcano plot for the ethylene hydrogenation (the supporting material should be consulted for explanation of the kinetic model). If we consider the calculated turn-over frequency of the catalyst, we can see the linear dependence of the activity on only one descriptor leading to a simple case of more or less reactivity depending upon that descriptor. However, if we contrast the two reactions, we can see that the inclusion of even one more descriptor opens up the possibility to tune our catalyst selectivity. For instance, the dopants used in this study enable us to increase the O affinity, whilst maintaining or decreasing the C affinity. Thereby providing the opportunity to get an appreciable turn-over frequency for O contain species, but reducing that of C carbon only containing species. Analysis such as this opens up the possibility for prediction of selective catalysts for multi-functional, hetero atom containing reagents.

Conclusion

In an attempt to understand the catalytic activity of sponge Ni and doped variants, a simple theoretical kinetic model for the reaction has been constructed. The kinetic model shows firstly that the presence of Al does improve the catalytic properties towards carbonyl hydrogenation. This improvement of activity is achieved by increasing the reactivity of Ni towards O containing molecules, whilst decreasing the affinity of Ni towards C. This point is important should one wish to carry out selective hydrogenation of C=O in the presence of C=C where the absence of O means that it is the C affinity that governs adsorption properties. The model presented herein, suggest that should one desire to selectively hydrogenate a carbonyl group in the presence of an alkene then a strategy would be to use a dopant that is able to reduce the affinity for C but maintains a high O affinity, e.g. Ti, Mo, Cr, Zr.

Secondly the influence of dopant metal on geometry and intrinsic electronic structure has been investigated and the results correlated to theoretical activities. The results show that in the case of low dopant concentrations, as is present in the experimental activated sponge Ni catalyst, it is the intrinsic electronic structure of the elemental substitution that is responsible for the observed change in catalytic activity.

Thirdly by examining the kinetics on different surface facets, the influence of the geometry of the active site on activity can be understood. Kinetic Monte Carlo studies[4] of the activated Ni:Al particle indicate that a range of fcc surface facets are exposed as active sites for the reaction. Through DFT simulation of two of these different facets it has been found that the carbonyl reaction is surface sensitive, this is because it requires a low coordinate surface site to adsorb the carbonyl species. This knowledge is important in the control of the chemistry. However to exploit this fact, the technological barrier of selectively controlling the surface morphology under reaction conditions needs to be overcome.

Conflicts of interest

The authors declare that there are no conflicts of interest.

Acknowledgements

The author wishes to acknowledge collaborators on the European Union project IMPRESS, for helpful discussions (Intermetallic Materials Processing in Relation to Earth and Space Solidification).[25] The author would also like to acknowledge Johnson Matthey and the EU for financial support through the IMPRESS integrated project (Contract NMP3-CT-2004-500635), co-funded by the European Commission in the sixth Framework Programme and the European Space Agency.

References

1. M. Raney: US Patent 1628190, 1927.
2. P. Fouilloux, G. A. Martin, A. J. Renouprez, B. Moraweck, B. Imelik and M. Prettre: *J. Catal.*, 1972, **25**, 212.
3. A. J. Smith and D. L. Trimm: *Annu. Rev. Mater. Res.*, 2005, **35**, 127.
4. N. C. Bernard, S. G. R. Brown, F. Devred, J. W. Bakker, B. E. Nieuwenhuys and N. J. Adkins: *J. Catal.*, 2001, **281**, (2), 25.
5. F. Abild-Pedersen, J. Greeley, F. Studt, J. Rossmeisl, T. R. Munter, P. G. Moses, E. Skulason, T. Bligaard and J. K Nørskov: *Phys. Rev. Lett.*, 2007, **99**, 016105.
6. B. Hammer, L. B. Hansen and J. K. Norskov: *Phys. Rev.*, 1999, **59B**, 7423.
7. G. Jones, T. Bligaard, F. Abild-Pedersen and J. K. Nørskov: *J. Phys.: Condens. Matter*, 2008, **20**, 064239.
8. G. Jones, J. Geest Jakobsen, S. S. Shim, J. Kleis, M. P. Andersson, J. Rossmeisl, F. Abild-Pedersen, T. Bligaard, S. Helveg, B. Hinnemann, J. R. Rostrup-Nielsen, I. Chorkendorff, J. Sehested and J. K. Nørskov, *J. Catal.*, 2008, **259**, 147–160.
9. L. Liu, M. Krack and A. Michaelides: *J. Am. Chem. Soc.*, 2008, **130**, 8572.
10. G. Hess: *Bulletin Scientifique, Lacademie Imperiale Des Sciences*, 1840, **174 VIII**, 81–96.
11. NIST Chemistry Webbook, available at: http://webbook.nist.gov/chemistry/.
12. D. J. Wesolowski: *Geochim. Cosmochim. Acta*, 1992, **56**, 1065.
13. F. J. Peryea and J. A. Kittrick: *Clay Clay Miner.*, 1988, **36**, 391.
14. G. Reinhart: private communication. December 2008.
15. W. Pfeiler: 'Alloy physics', 2007, Weinheim, Wiley-VCH Verlag GmbH & Co. KGaA.
16. G. Jones, F. Studt, F. Abild-Pedersen, J. K. Nørskov and T. Bligaard: *Chem. Eng. Sci.*, 2011, **66**, (24), 6318–6323.
17. E. M. Fernández, P. G. Moses, A. Toftelund, H. A. Hansen, J. I. Martínez, F. Abild-Pedersen, J. Kleis, B. Hinnemann, J. Rossmeisl, T. Bligaard and J. K. Nørskov: *Angew. Chem. Int. Ed.*, 2008, **47**, 4683–4686.
18. S. Wang, V. Petzold, V. Tripkovic, J. Kleis, J. G. Howalt, E. Skúlason, E. M. Fernández, B. Hvolbæk, G. Jones, A. Toftelund, H. Falsig, M. Björketun, F. Studt, F. Abild–Pedersen, J. Rossmeisl, J. K. Nørskov and T. Bligaard, *Phys. Chem. Chem. Phys.*, 2011, **13**, 20760–20765.
19. S. Wang, B. Temel, J. Shen, G. Jones, L.C. Grabow, F. Studt, T. Bligaard, F. Abild-Pedersen, C.H. Christensen and J.K. Nørskov, *Catal. Lett.*, 2011, **141**, 370–373.
20. J. K. Nørskov, T. Bligaard, A. Logadottir, S. Bahn, L. B. Hansen, M. Bollinger, H. Bengaard, B. Hammer, Z. Sljivancanin, M. Mavrikakis, Y. Xu, S. Dahl and C. J. H. Jacobsen: *J. Catal.*, 2002, **209**, (2), 275–278.
21. T. Bligaard, J. K. Nørskov, S. Dahl, J. Matthiesen, C. H. Christensen and J. Sehested: *J. Catal.*, 2004, **224**, (1), 206–217.
22. J. J. Mortensen, K. Kaasbjerg, S. L. Frederiksen, J. K. Nørskov, J. P. Sethna and K. W. Jacobsen: *Phys. Rev. Lett.*, 2005, **95**, 216401.
23. V. Novak, P. Koci, M. Marek, F. Stepanek, P. Blanco-Garcia and G. Jones: *Catal. Today*, 2012, **188**, (1), 62–69.
24. M. Dudak, V. Novak, P. Koci, M. Marek, P. Blanco-Garcia and G. Jones: *Appl. Catal. B: Environ.*, 2014, **150–151**, 446–458.
25. D. J. Jarvis and D. Voss: *Mater. Sci. Eng. A*, 2005, **A413–A414**, 583–591.

21

Synthesis of crystalline Mo−V−W−O complex oxides with orthorhombic and trigonal structures and their application as catalysts

Chuntian Qiu[1], Chen Chen[1], Satoshi Ishikawa[1], Zhenxin Zhang[1], Toru Murayama*[1] and Wataru Ueda[1,2]

[1]Catalysis Research Center, Hokkaido University, N-21, W-10, Sapporo 001-0021, Japan
[2]Facility of Engineering, Kanagawa University, 3-27-1, Rokkakubashi, Kanagawa-ku, Yokohama, Kanagawa 221-8686, Japan

Abstract Crystalline Mo−V−W−O complex oxides with the orthorhombic or trigonal structure were synthesized by a hydrothermal method. Those Mo−V−W−O samples with various amounts of tungsten were characterized by inductively coupled plasma atomic emission spectroscopy, TEM, STEM−EDX, X-ray diffraction, Rietveld analysis and a N_2 adsorption method. It was found for the first case that an additional metal such as W can be successfully incorporated into the trigonal Mo−V−O structure by using $(CH_3CH_2NH_3)_2Mo_3O_{10}$. The alkylammonium

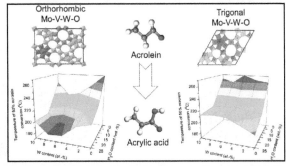

cation acted as a structural stabilizer that was requisite for the formation of a trigonal structure when additional metal ions were present. For the orthorhombic Mo−V−W−O structure, introduction of W into the orthorhombic structure caused a rod segregation effect by which nanoscale crystals formed and the external surface area greatly increased. Additionally, these Mo−V−W−O materials were applied as catalysts for the gas phase selective oxidation of acrolein to acrylic acid. The best catalyst was assigned to the orthorhombic Mo−V−O−W7.5, which possessed an ordered arrangement of heptagonal and hexagonal channels and a large external surface area.

Keywords Crystalline Mo−V−W−O, Selective oxidation, Acrolein, Acrylic acid

Introduction

Mo−V based complex oxides catalysts have a long history of use in selective oxidation and ammoxidation of light alkanes since the 1960s.[1−5] Numerous metal components such as Te, Nb, Cu, W and Fe have been applied to optimize the structure and catalytic performance.[6−8] One of the representative catalysts is MoVTe(Sb)NbO, which was developed by Mitsubishi Chemical Corp. It showed good catalytic performance for oxidation of propane to acrylic acid and ammoxidation of propane to acrylonitrile.[9,10] Additional metals modified the structure and improved the catalytic activity; however, multiple components increased the complexity of structure and composition.

Recently, we have reported the single crystalline Mo−V−O with an orthorhombic or trigonal structure.[11−13] These

Mo−V−Os with microporosity were layered in the direction of the c axis and contained pentagonal $\{Mo_6O_{21}\}$ building units and $\{MO_6\}$ (M = Mo, V) octahedra that were arranged to form heptagonal and hexagonal channels but with different arrangements in the a−b plane. The simplicity of the component and uniformity of the crystalline structure provide a suitable opportunity to investigate the effects of additional metals and the structure−function relationship.

For molecular sieve type porous materials, the incorporation of a transition metal into the framework could modify properties such as acidity, thermal stability and especially catalytic activity.[14,15] However, introduction of heteroatom usually affects the normal crystalline growth process, and it is a challenge to introduce a heteroatom metal while maintaining the crystal structure. In this work, we succeeded in introducing tungsten into the framework of the orthorhombic and trigonal structure. We obtained a trigonal structure with an additional metal for the first time by using

---

*Corresponding author, email: murayama@cat.hokudai.ac.jp

Table 1 Chemical compositions, external surface areas and lattice parameters of Mo–V–W–O

Sample	Composition*/at.-%			External surface area†/m^2 g^{-1}	Lattice parameter/nm		
	Mo	V	W		a	b	c
Amor-MoVO–W0	76.1	23.9	0.0	4.7	...	...	0.3998
Amor-MoVO–W2.5	74.0	23.3	2.7	4.7	...	...	0.3993
Amor-MoVO–W5.0	70.5	23.6	5.9	4.7	...	...	0.3991
Amor-MoVO–W7.5	70.1	21.9	8.0	3.9	...	...	0.3990
Amor-MoVO–W10	67.9	21.8	10.3	4.0	...	...	0.3989
Orth-MoVO–W0	71.5	28.5	0.0	5.0	2.1279	2.6634	0.4002
Orth-MoVO–W2.5	71.8	25.6	2.6	15.6	2.1205	2.6631	0.3999
Orth-MoVO–W5.0	70.0	25.2	4.8	23.3	2.1203	2.6629	0.3998
Orth-MoVO–W7.5	69.7	23.1	7.2	31.4	2.1183	2.6612	0.3994
Orth-MoVO–W10	68.2	21.6	10.2	35.5	2.1053	2.6592	0.3991
Tri-MoVO–W0	74.8	25.2	0.0	17.4	2.1382	...	0.4030
Tri-MoVO–W2.5	74.9	22.1	3.0	17.5	2.1382	...	0.4022
Tri-MoVO–W5.0	73.2	21.4	5.4	14.4	2.1381	...	0.4018
Tri-MoVO–W7.5	71.6	20.7	7.7	18.3	2.1381	...	0.4014
Tri-MoVO–W10	69.8	20.1	10.1	18.9	2.1380	...	0.4011

* Determined by ICP–AES.
† Measured by N$_2$ adsorption and determined by t plot method.

an organic molybdenum source. Moreover, it was found that the introduction of W into the orthorhombic structure caused crystal splitting that contributed to exposure of more active phase. Activity for oxidation of acrolein to acrylic acid over those M–V–W–Os was further investigated.

Experimental

Synthesis of three distinct Mo–V–O complex oxides

The catalysts were synthesized according to a previous report.[13,16] For orthorhombic Mo–V–O (denoted as Orth-MoVO–W0), a solution of VOSO$_4$ (3.28 g, Mitsuwa Chemicals) in deionized water (120 mL) was added to a solution of (NH$_4$)$_6$Mo$_7$O$_{24}$ (8.82 g, Wako) in deionized water (120 mL) with stirring. The mixture was stirred for 10 min and then transferred into an autoclave with a Teflon inner tube and a Teflon sheet enough length for filling about half of Teflon inner tube space. This sheet is necessary for formation of well crystallized sample, because solids are formed on the sheet. The mixture was purged with N$_2$ for 10 min (pH = 3.2) and then hydrothermally treated at 175°C for 48 h. The procedure for synthesis of trigonal Mo–V–O (denoted as Tri-MoVO–W0) was the same as that for Orth-MoVO–W0 except that the pH value of the mixture was adjusted to 2.2 with H$_2$SO$_4$. As synthesized Orth-MoVO–W0 and Tri-MoVO–W0 were purified by treatment in a solution of oxalic acid (0.4 mol L^{-1}) at 60°C for 30 min to remove amorphous impurities. Amorphous Mo–V–O (denoted as Amor-MoVO–W0) was obtained by the same procedure as that for Orth-MoVO–W0 synthesis but with twofold higher precursor concentrations, without the use of a Teflon sheet, and no N$_2$ bubbling.

Synthesis of amorphous and orthorhombic Mo–V–W–O complex oxides

Orthorhombic Mo–V–W–O catalysts (denoted as Orth-MoVWO) with various contents of W were synthesized with (NH$_4$)$_6$Mo$_7$O$_{24}$, VOSO$_4$ and (NH$_4$)$_6$[H$_2$W$_{12}$O$_{40}$].6H$_2$O under a hydrothermal condition. Firstly, solution A was obtained

with 8.82 g of (NH$_4$)$_6$Mo$_7$O$_{24}$ and a certain amount of (NH$_4$)$_6$[H$_2$W$_{12}$O$_{40}$].6H$_2$O (with adjustment of W content to 2.5, 5.0, 7.5 and 10 at.-% in the synthesized Mo–V–W–O) being dissolved in 120 mL of deionized water and solution B was obtained with 3.28 g of VOSO$_4$ being dissolved in another 120 mL of deionized water. Secondly, solution B was poured into solution A under a stirring condition. The obtained mixed solution was introduced into an autoclave with a Teflon sheet in the Teflon inner vessel. The reaction mixture was purged with N$_2$ for 10 min (pH = 3.2) and then hydrothermally treated at 175°C for 48 h. As synthesized Orth-MoVWO catalysts (denoted as Orth-MoVO–W2.5, Orth-MoVO–W5.0, Orth-MoVO–W7.5 and Orth-MoVO–W10, respectively) were treated with oxalic acid solution (0.4 mol L^{-1}) 2 times at 60°C. The treatment was maintained for 30 min each time.

Amorphous Mo–V–W–O catalysts (denoted as Amor-MoVWO) with various contents of W (denoted as Amor-MoVO–W2.5, Amor-MoVO–W5.0, Amor-MoVO–W7.5 and Amor-MoVO–W10, respectively) were also synthesized with the same precursor at a twofold higher concentration and by the same procedure but without a Teflon sheet.

Synthesis of trigonal Mo–V–W–O complex oxides

A trigonal Mo–V–W–O (denoted as Tri-MoVWO) catalyst was obtained by the same procedure as that for Orth-MoVWO synthesis except that ethylammonium trimolybdate (EATM: (CH$_3$CH$_2$NH$_3$)$_2$Mo$_3$O$_{10}$) was used as an Mo source instead of (NH$_4$)$_6$Mo$_7$O$_{24}$. The synthesis procedure for EATM was as follows. MoO$_3$ (0.15 mol, 21.594 g, Kanto) was dissolved in 28.0 mL of 70% ethylamine solution (ethylamine: 0.30 mol, Wako) diluted with 28.0 mL of deionized water. After being completely dissolved, the solution was evaporated under a vacuumed condition at 70°C and then solid powder was obtained. The powder was dried in air at 80°C overnight. As synthesized materials were denoted as Tri-MoVO–W2.5, Tri-MoVO–W5.0, Tri-MoVO–W7.5 and Tri-MoVO–W10, respectively.

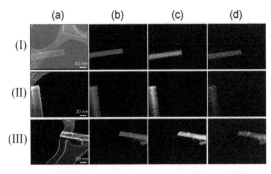

Figure 1 *a* STEM images and element mapping of *b* Mo, *c* V and *d* W of Amor-MoVO–W2.5 (I), Orth-MoVO–W2.5 (II) and Tri-MoVO–W2.5 (III)

Characterization

X-ray diffraction patterns were measured by using an X-ray diffractometer (RINT-Ultima III, Rigaku) with Cu K_α to study their crystalline structure. Crystallite size broading analysis of Rietveld program with Powder Reflex (Material Studio 5.5.3, Accelrys) was utilized to fit the experimental X-ray diffraction (XRD) pattern of Orth-MoVWO. Raman spectra (inVia Reflex Raman spectrometer, RENISHAW) were taken in air on a static sample with Ar laser power. Scanning transmission electron microscopy (STEM) images and metal element mapping of Mo, W and V were obtained on an HD-2000 (HITACHI). Transmission electron microscopy (TEM) images were taken with a 200 kV transmission electron microscope (JEOL JEM-2010F). The chemical composition of the catalysts was determined by the inductively coupled plasma atomic emission spectroscopy (ICP–AES) method with a VISTA–PRO apparatus (Varian). N_2 adsorption–desorption isotherm measurements were carried out on an auto-adsorption system (BELSORP MAX, Nippon BELL) to obtain the external surface area and micropore volume using the *t*-plot method. Before adsorption measurements, the samples were treated at 400°C for 2 h under air and out gassed at 300°C under vacuum for 2 h.

Gas phase catalytic oxidation

Gas phase catalytic oxidation of acrolein to acrylic acid was performed using a fixed bed stainless tubular reactor at atmospheric pressure. The catalysts (0.25 g) were firstly ground for 5 min, then diluted with 2.5 g Carborundum and pretreated at 400°C under N_2 of 50 mL min^{-1} for 2 h. Reactant gas was conducted with change in the water content from 0 to 25.2 vol.-% while keeping other feeding gases constant: acrolein = 2.5 mL min^{-1}, O_2 = 8 mL min^{-1} and N_2 balance (total: 107.6 mL min^{-1}). Quantitative analysis was performed using three on-line gas chromatographs with columns of Molecular Sieve 13X, Gaskuropack 54 and Porapak Q. Blank runs showed that no reaction took place without catalysts under the experimental conditions. Carbon balance was always over 95% and selectivity was calculated on the basis of products of sum.

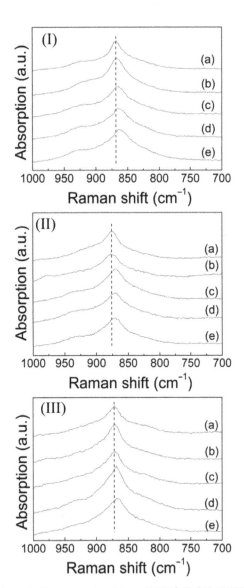

Figure 2 Raman spectra of Amor-MoVWO (I), Orth-MoVWO (II) and Tri-MoVWO (III) with different W contents. *a* W0; *b* W2.5; *c* W5.0; *d* W7.5; *e* W10

Results and discussion

Structural characterization of Mo–V–W–O catalysts

Location of W in Mo–V–W–O

Table 1 shows the chemical compositions of the Mo–V–W–O complex oxides determined by the ICP–AES method. Atomic ratios of tungsten in the Amor-, Orth- and Tri-MoVWO groups were adjusted approximately to 2.5, 5.0, 7.5 and 10 at.-%, respectively. Different Mo and V atomic ratios derived from the distinction of structure requisite. Element mapping of Mo, V and W of Amor-, Orth- and Tri-MoVO–W2.5 showed that metal elements were distributed evenly along the rod shaped materials (Fig. 1). In the Raman spectra (Fig. 2), the main band

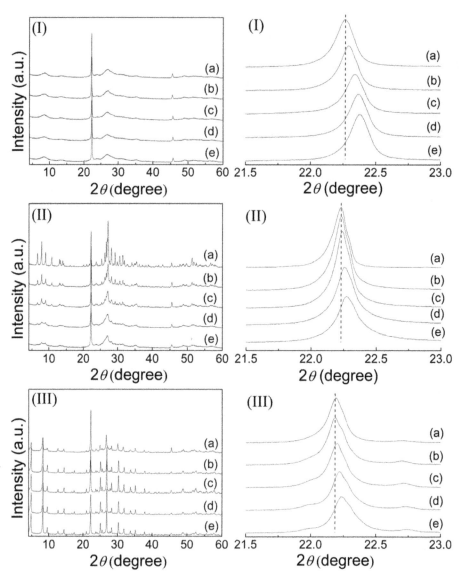

Figure 3 X-ray diffraction patterns of Amor-MoVWO (I), Orth-MoVWO (II) and Tri-MoVWO (III) with different W contents. *a* W0; *b* W2.5; *c* W5.0; *d* W7.5; *e* W10

at 872 cm^{-1} that was ascribed to pentagonal units[16] gradually shifted to a low wave number. All of the results indicated that W was successfully introduced into the three different structures.

The crystalline structure of the materials was investigated by XRD characterization (Fig. 3). Diffraction peaks at 22.2 and 45.4° were ascribed to (001) and (002) plane reflections, respectively. Concentrated at the (001) peak, an obvious peak shift in a high angle direction was observed in all of the three different types of Mo–V–W–O, after being calibrated with silicon as an internal standard (Fig. 2). With increasing W content, the lattice parameter decreased gradually (Table 1), implying that lattice contraction happened with the incorporation of W. Based on the above results, W was considered to be incorporated into the framework. Orthorhombic, trigonal and amorphous Mo–V–O contained the same

pentagonal {Mo$_6$O$_{21}$} units and {MO$_6$} (M = Mo, V) octahedra. The pentagonal unit was constructed with Mo only and octahedra contained V as well as Mo.[16,17] It is notable that, as shown in Table 1, with an increase in tungsten content, the ratio of V in Mo–V–W–O decreased gradually, strongly suggesting that tungsten replaced not only Mo but also the V ions. Therefore, there was a high probability that W formed {WO$_6$} octahedra and acted as linkers connecting {Mo$_6$O$_{21}$} pentagonal units.

Influence of W on Mo–V–O structure

In the XRD pattern of Amor-MoVWO (Fig. 2-I), there were two broad peaks centered at 8 and 27°, implying that Amor-MoVWO was only crystalline along the *c* axis, while a disordered arrangement of pentagonal {Mo$_6$O$_{21}$} and {MO$_6$} octahedra was formed in the *a–b* plane. This is the most

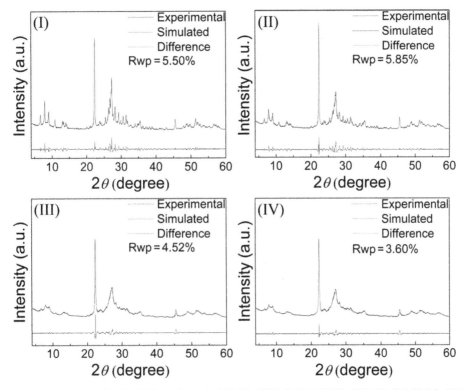

Figure 4 Rietveld analysis of Orth-MoVWO catalysts: Orth-MoVO–W2.5 (I), Orth-MoVO–W5.0 (II), Orth-MoVO–W7.5 (III) and Orth-MoVO–W10 (IV)

distinct difference from the orthorhombic and trigonal structures. With an increase in W content, the XRD patterns of Amor-MoVWO barely changed except for a shift of the (001) peak in a higher angle direction. This revealed that although lattice contraction occurred, the disordered arrangement in the a–b plane was not affected by the introduction of W. Therefore, crystal size (see Fig. S2-I and Table S1 in Supplementary Material on ManeyOnline here http://dx.doi.org/10.1179/2055075814Y.0000000009) and external surface area (Table 1) of Amor-MoVWO slightly changed.

In the XRD pattern of Orth-MoVWO (Fig. 3-II), main diffraction peaks corresponding to the orthorhombic structure emerged at 6.6, 7.9, 9.0 and 27.3°, etc., which were ascribed to the planes of (020), (120), (210) and (630), respectively.[18] The emergence of diffraction peaks at a low angle below 10° indicated that Orth-MoVWO was well crystallized along the a and b axes. However, with an increase in W content, the rod segregation effect proceeded. That could be clearly observed from the TEM image (Fig. S1 in Supplementary Material). Orth-MoVWO crystals partially split into smaller rods and those nanoscale rods had sizes of only several tens of nanometers or even smaller (Average crystal size was calculated and is shown in Fig. S2-II and Table S1 in Supplementary Material). For crystalline materials, lattice expansion or contraction usually results in an unstable structure. Lattice contraction occurred with the incorporation of W and that might have caused cleavage of Mo–O and V–O bonds and decrease in the long range order of the a–b plane, thus facilitating dehiscence. The diffraction peaks

(especially below 10°) of Orth-MoVWO decreased and broadened gradually with increasing W content. Rietveld analysis using the crystallite size broadening procedure was carried out and results are shown in Fig. 4. It was confirmed that broadening of diffraction peaks was not caused by the formation of an amorphous phase but by the dehiscence of orthorhombic crystals. Although the addition of W affected the crystallinity of the orthorhombic structure, a decrease in crystal size contributed to a larger external surface area (Table 1) and more active sites were exposed, which is particularly important for the activity of acrolein oxidation to acrylic acid.[13] When W content rose to 10 at.-%, external surface area increased to 35.5 m^2 g^{-1}, almost 6 times larger than that of the well crystallized Orth-MoVO–W0.

Different from the synthesis process of Amor- and Orth-MoVWO, an organic Mo source $(CH_3CH_2NH_3)_2Mo_3O_{10}$ (EATM) was used to synthesize Tri-MoVWO (the section on 'Synthesis of trigonal Mo–V–W–O complex oxides'). As noted above, additional metal ions usually affect the self-assembly process. In the trigonal structure, there is a linker unit of triple-octahedron and the occupancy of this triple-octahedron is only 0.6, while in the orthorhombic structure, the occupancy of quintuple-octahedron (instead of triple-octahedron) is 1.0.[11,12] This revealed that the trigonal structure is not as stable as the orthorhombic structure. Therefore, simply adding W and other metal precursors into the hydrothermal process only resulted in amorphous phase instead of trigonal structure. The diameter of the heptagonal channel is about 0.4 nm, thus being limited to

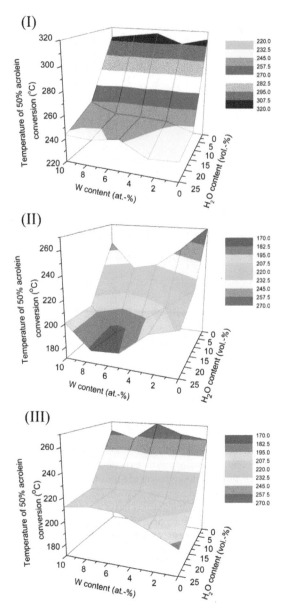

Figure 5 Temperatures of 50% acrolein conversion over Amor-MoVWO (I), Orth-MoVWO (II) and Tri-MoVWO (III) catalysts with different tungsten contents and water partial pressures

Table S1 in Supplementary Material). Due to the good crystallinity of Tri-MoVWO, only a slight difference in the external surface area of Tri-MoVWO was observed (Table 1).

Catalytic activity for oxidation of acrolein to acrylic acid over Mo–V–W–O

Acrylic acid has become a widely used chemical in recent years because of its extensive applications in super absorbent materials, coatings and additives in textile production.[19–21] It can be acknowledged that the crystalline structure was maintained and a much larger external surface was achieved for Orth-MoVWO, which was expected to show excellent performance in the selective oxidation of acrolein to acrylic acid.

Catalytic performance without H_2O feeding was investigated in order to eliminate the effect of H_2O (Figs. S3, S4, S5 and S6 in Supplementary Material). Crystalline Mo–V–W–O catalysts had more active sites because the $a–b$ plane of orthorhombic and trigonal structures was constructed into a high degree of ordered arrangement. Therefore, for acrolein oxidation, crystalline Mo–V–W–O catalysts showed much higher activity than amorphous catalysts did, while selectivity to acrylic acid was almost the same. Figure 5 shows reaction temperatures of 50% acrolein conversion over Amor-, Orth- and Tri-MoVWO catalysts under different H_2O partial pressure and W content conditions. There is no doubt that water acts as an important promoter for the conversion of acrolein. Under the same water feeding condition, Orth- and Tri-MoVWO catalysts always showed higher catalytic activity than that of Amor-MoVWO. Moreover, a positive effect on acrolein conversion was observed when W was incorporated into the orthorhombic structure. However, in the case of trigonal and amorphous structures, W did not show any positive effect because of the slight changes in structure and morphology. The highest catalytic activity was achieved over the Orth-MoVO–W7.5 catalyst, which had a crystalline structure and large external surface area. Therefore, crystalline structure is the guarantee for good catalytic performance and external surface area is another important factor for oxidation of acrolein to acrylic acid.

Conclusion

We succeeded in introducing W into the trigonal structure using EATM as Mo source. For Orth-MoVWO, tungsten acted as a structural promoter that resulted in a rod segregation effect and increase in external surface area. Crystalline Orth- and Tri-MoVWO achieved showed very high catalytic activity compared with that of Amor-MoVWO. It was found that crystalline structure and external surface area were responsible for good catalytic activity.

counter cation with a small molecular size such as an ammonium cation with a size of 0.28 nm. In the self-assembly process, an ethylamine cation with a larger size than that of an ammonium cation balanced the trigonal framework and played an important role as a structure directing agent and stabilizer for the trigonal structure and it could be easily removed by calcination in air. In the XRD pattern of Tri-MoVWO (Fig. 3-III), peak intensity of W-containing trigonal Mo–V–W–O was higher than that of Tri-MoVO–W0, suggesting that the use of EATM as Mo source can provide trigonal Mo–V–W–O crystals with better crystallinity (average crystal size shown in Fig. S2-III and

Conflicts of interest

The authors declare that there are no conflicts of interest.

Acknowledgements

This work was supported by JSPS KAKENHI Grant-in-Aid for Young Scientists (24760635).

References

1. X. L. Tu, N. Furuta, Y. Sumida, M. Takahashi and H. Niiduma: 'A new approach to the preparation of MoVNbTe mixed oxide catalysts for the oxidation of propane to acrylic acid', *Catal. Today*, 2006, **117**, (1–3), 259–264.

2. R. N. d'Alnoncourt, L. I. Csepei, M. Havecker, F. Girgsdies, M. E. Schuster, R. Schlogl and A. Trunschke: 'The reaction network in propane oxidation over phase-pure MoVTeNb M1 oxide catalysts', *J. Catal.*, 2014, **311**, 369–385.

3. K. Amakawa, Y. V. Kolen'ko, A. Villa, M. E. Schuster, L. I. Csepei, G. Weinberg, S. Wrabetz, R. N. d'Alnoncourt, F. Girgsdies, L. Prati, R. Schlogl and A. Trunschke: 'Multifunctionality of crystalline MoV (TeNb) M1 oxide catalysts in selective oxidation of propane and benzyl alcohol', *ACS Catal.*, 2013, **3**, (6), 1103–1113.

4. J. M. M. Millet, H. Roussel, A. Pigamo, J. L. Dubois and J. C. Jumas: 'Characterization of tellurium in MoVTeNbO catalysts for propane oxidation or ammoxidation', *Appl. Catal. A-Gen.*, 2002, **232**, (1–2), 77–92.

5. A. Drochner, P. Kampe, N. Menning, N. Blickhan, T. Jekewitz and H. Vogel: 'Acrolein oxidation to acrylic acid on Mo/V/W-mixed oxide catalysts', *Chem. Eng. Technol.*, 2014, **37**, (3), 398–408.

6. G. Mestl, J. L. Margitfalvi, L. Vegvari, G. P. Szijjarto and A. Tompos: 'Combinatorial design and preparation of transition metal doped MoVTe catalysts for oxidation of propane to acrylic acid', *Appl. Catal. A-Gen.*, 2014, **474**, 3–9.

7. Y. Nakazawa, S. Matsumoto, T. Kobayashi and T. Kurakami, Nippon-kataku Inc.: 'Catalyst and method for producing acrylic acid', US Patent 20130217915 A1, 2013.

8. C. A. Welker-Nieuwoudt, A. Karpov, F. Rosowski, K. J. Mueller-Engel, H. Vogel, A. Drochner, N. Blickhan, N. Duerr, T. Jekewitz, N. Menning, T. Petzold and S. Schmidt: 'Process for heterogeneously catalyzed gas phase partial oxidation of (meth)acrolein to (meth)acrylic acid', Patent 20140018572; 2014.

9. T. Ushikubo, K. Oshima, A. Kayou, M. Vaarkamp and M. Hatano: 'Ammoxidation of propane over catalysts comprising mixed oxides of Mo and V', *J. Catal.*, 1997, **169**, (1), 394–396.

10. H. Tsuji and Y. Koyasu: 'Synthesis of MoVNbTe(Sb)O-x composite oxide catalysts via reduction of polyoxometales in an aqueous medium', *J. Am. Chem. Soc.*, 2002, **124**, (20), 5608–5609.

11. M. Sadakane, N. Watanabe, T. Katou, Y. Nodasaka and W. Ueda: 'Crystalline Mo3VOx mixed-metal-oxide catalyst with trigonal symmetry', *Angew. Chem.-Int. Edit.*, 2007, **46**, (9), 1493–1496.

12. M. Sadakane, K. Kodato, T. Kuranishi, Y. Nodasaka, K. Sugawara, N. Sakaguchi, T. Nagai, Y. Matsui and W. Ueda: 'Molybdenum-vanadium-based molecular sieves with microchannels of seven-membered rings of corner-sharing metal oxide octahedra', *Angew. Chem.-Int. Edit.*, 2008, **47**, (13), 2493–2496.

13. C. Chen, N. Kosuke, T. Murayama and W. Ueda: 'Single-crystalline-phase Mo3VOx: an efficient catalyst for the partial oxidation of acrolein to acrylic acid', *ChemCatChem*, 2013, **5**, (10), 2869–2873.

14. M. Hartmann and L. Kevan: 'Transition-metal ions in aluminophosphate and silicoaluminophosphate molecular sieves: location, interaction with adsorbates and catalytic properties', *Chem. Rev.*, 1999, **99**, (3), 635–663.

15. J. M. Thomas, R. Raja, G. Sankar and R. G. Bell: 'Molecular-sieve catalysts for the selective oxidation of linear alkanes by molecular oxygen', *Nature*, 1999, **398**, (6724), 227–230.

16. T. Konya, T. Katou, T. Murayama, S. Ishikawa, M. Sadakane, D. Buttrey and W. Ueda: 'An orthorhombic Mo3VOx catalyst most active for oxidative dehydrogenation of ethane among related complex metal oxides', *Catal. Sci. Technol.*, 2013, **3**, (2), 380–387.

17. M. Sadakane, K. Yamagata, K. Kodato, K. Endo, K. Toriumi, Y. Ozawa, T. Ozeki, T. Nagai, Y. Matsui, N. Sakaguchi, W. D. Pyrz, D. J. Buttrey, D. A. Blom, T. Vogt and W. Ueda: 'Synthesis of orthorhombic Mo-V-Sb oxide species by assembly of pentagonal Mo(6)O(21) polyoxometalate building blocks', *Angew. Chem.-Int. Edit.*, 2009, **48**, (21), 3782–3786.

18. W. Ueda, D. Vitry, T. Kato, N. Watanabe and Y. Endo: 'Key aspects of crystalline Mo-V-O-based catalysts active in the selective oxidation of propane', *Res. Chem. Intermed.*, 2006, **32**, (3–4), 217–233.

19. E. Saarikoski, H. Rautkoski, M. Rissanen, J. Hartman and J. Seppala: 'Cellulose/acrylic acid copolymer blends for films and coating applications', *J. Appl. Polym. Sci.*, 2014, **131**, (10).

20. M. Y. Zhang, Z. Q. Cheng, M. Z. Liu, Y. Q. Zhang, M. J. Hu and J. F. Li: 'Synthesis and properties of a superabsorbent from an ultra-violetirradiated waste nameko mushroom substrate and poly (acrylic acid)', *J. Appl. Polym. Sci.*, 2014, **131**, (13).

21. A. Ben Fradj, R. Lafi, S. Ben Hamouda, L. Gzara, A. H. Hamzaoui and A. Hafiane: 'Effect of chemical parameters on the interaction between cationic dyes and poly(acrylic acid)', *J. Photochem. Photobiol. A-Chem.*, 2014, **284**, 49–54.

Permissions

All chapters in this book were first published in CSR, by Taylor & Francis Online; hereby published with permission under the Creative Commons Attribution License or equivalent. Every chapter published in this book has been scrutinized by our experts. Their significance has been extensively debated. The topics covered herein carry significant findings which will fuel the growth of the discipline. They may even be implemented as practical applications or may be referred to as a beginning point for another development.

The contributors of this book come from diverse backgrounds, making this book a truly international effort. This book will bring forth new frontiers with its revolutionizing research information and detailed analysis of the nascent developments around the world.

We would like to thank all the contributing authors for lending their expertise to make the book truly unique. They have played a crucial role in the development of this book. Without their invaluable contributions this book wouldn't have been possible. They have made vital efforts to compile up to date information on the varied aspects of this subject to make this book a valuable addition to the collection of many professionals and students.

This book was conceptualized with the vision of imparting up-to-date information and advanced data in this field. To ensure the same, a matchless editorial board was set up. Every individual on the board went through rigorous rounds of assessment to prove their worth. After which they invested a large part of their time researching and compiling the most relevant data for our readers.

The editorial board has been involved in producing this book since its inception. They have spent rigorous hours researching and exploring the diverse topics which have resulted in the successful publishing of this book. They have passed on their knowledge of decades through this book. To expedite this challenging task, the publisher supported the team at every step. A small team of assistant editors was also appointed to further simplify the editing procedure and attain best results for the readers.

Apart from the editorial board, the designing team has also invested a significant amount of their time in understanding the subject and creating the most relevant covers. They scrutinized every image to scout for the most suitable representation of the subject and create an appropriate cover for the book.

The publishing team has been an ardent support to the editorial, designing and production team. Their endless efforts to recruit the best for this project, has resulted in the accomplishment of this book. They are a veteran in the field of academics and their pool of knowledge is as vast as their experience in printing. Their expertise and guidance has proved useful at every step. Their uncompromising quality standards have made this book an exceptional effort. Their encouragement from time to time has been an inspiration for everyone.

The publisher and the editorial board hope that this book will prove to be a valuable piece of knowledge for researchers, students, practitioners and scholars across the globe.

List of Contributors

Q. Yang, G. M. Hughes, A. Varambhia, M. P. Moody and P. A. J. Bagot
Department of Materials, University of Oxford, Parks Road, Oxford OX1 3PH, UK

D. E. Joyce and S. Saranu
Mantis Deposition Ltd., Thame OX9 3RR, UK

Shu Zhao and Xing-Wu Liu
State Key Laboratory of Coal Conversion, Institute of Coal Chemistry, Chinese Academy of Sciences, Taiyuan 030001, China
National Energy Center for Coal to Liquids, Synfuels China Co., Ltd, Huairou District, Beijing 101400, China
University of Chinese Academy of Sciences, No. 19A Yuquan Road, Beijing 100049, China

Chun-Fang Huo and Yong-Wang Li
State Key Laboratory of Coal Conversion, Institute of Coal Chemistry, Chinese Academy of Sciences, Taiyuan 030001, China
National Energy Center for Coal to Liquids, Synfuels China Co., Ltd, Huairou District, Beijing 101400, China

Jianguo Wang
State Key Laboratory of Coal Conversion, Institute of Coal Chemistry, Chinese Academy of Sciences, Taiyuan 030001, China

Haijun Jiao
State Key Laboratory of Coal Conversion, Institute of Coal Chemistry, Chinese Academy of Sciences, Taiyuan 030001, China
Leibniz-Institut für Katalyse e.V. an der Universitä t Rostock, Albert-Einstein Strasse 29a, 18059 Rostock, Germany

A. Torozova
Johan Gadolin Process Chemistry Centre, Åbo Akademi University, 20500 Turku/Åbo, Finland
Tver State Technical University, Tver 170026, Russia

P.Mäki-Arvela, N. Kumar, A. Aho1 M. Stekrova, K.Maduna Valkaj and D. Yu. Murzin
Johan Gadolin Process Chemistry Centre, Åbo Akademi University, 20500 Turku/Å bo, Finland

N. D. Shcherban, S.M. Filonenko, P. S. Yaremov and V. G. Ilyin
L.V. Pisarzhevskii Institute of Physical Chemistry, National Academy of Sciences, 03028 Kiev, Ukraine

P. Sinitsyna
L.V. Pisarzhevskii Institute of Physical Chemistry, National Academy of Sciences, 03028 Kiev, Ukraine Tver State Technical University, Tver 170026, Russia
St. Petersburg State Institute of Technology (Technical University), St. Petersburg 190013, Russia

K. P. Volcho and N. F. Salakhutdinov
N. N. Vorozhtsov Institute of Organic Chemistry, Russian Academy of Sciences, Novosibirsk 630090, Russia

Marco Piumetti, Marco Armandi, Guido Saracco, Edoardo Garrone, Giovanny Esteban Gonzalez and Barbara Bonelli
Department of Applied Science and Technology and INSTM Unit of Torino-Politecnico, Politecnico di Torino, Corso Duca degli Abruzzi, 24, I-10129 Turin, Italy

Francesca S. Freyria
Department of Applied Science and Technology and INSTM Unit of Torino-Politecnico, Politecnico di Torino, Corso Duca degli Abruzzi, 24, I-10129 Turin, Italy
Department of Chemistry, Massachusetts Institute of Technology, 77 Massachusetts Ave, 02139 Cambridge, MA, USA

Kevin Morgan, Robbie Burch, Alexandre Goguet, Christopher Hardacre and David W. Rooney
CenTACat, School of Chemistry and Chemical Engineering, David Keir Building, Queen's University Belfast, Stranmillis Road, Belfast, BT9 5AG, Northern Ireland, UK

Muhammad Daous and Lachezar A. Petrov
Chemical and Materials Engineering Department, King Abdulaziz University, Jeddah, Saudi Arabia

Juan-José Delgado
Departamento de Ciencia de los Materiales e Ingeniería Metalúrgica y Química Inorgánica, Facultad de Ciencias, Universidad de Cádiz. E-11510 Puerto Real, Cádiz, Spain

Y. Hao, M. Li, F. Cárdenas-Lizana and M. A. Keane
Chemical Engineering, School of Engineering and Physical Sciences, Heriot-Watt University, Edinburgh EH14 4AS, Scotland

Simon Jones, Amy Kolpin and Shik Chi Edman Tsang
Wolfson Catalysis Centre, Department of Chemistry, University of Oxford, Oxford OX1 3QR, UK

Tinku Baidya
Solid State and Structural Chemistry Unit, Indian Institute of Science, Bangalore 560012, India

Parthasarathi Bera
Surface Engineering Division, CSIR–National Aerospace Laboratories, Bangalore 560017, India

P. P. Wells
Department of Chemistry, University of Southampton, Highfield, Southampton SO17 1BJ, UK
UK Catalysis Hub, Research Complex at Harwell, Rutherford Appleton Laboratory, Didcot OX11 0FA, UK

E. M. Crabb
Department of Life, Health and Chemical Sciences, The Open University, Milton Keynes MK7 6AA, UK

C. R. King and A. E. Russell
Department of Chemistry, University of Southampton, Highfield, Southampton SO17 1BJ, UK

S. Fiddy
CCLRC, Daresbury Laboratory, Warrington WA4 4AD, UK

A. Amieiro-Fonseca and D. Thompsett
Johnson Matthey Technology Centre, Blounts Court, Sonning Common, Reading RG4 9NH, UK

Eric van Steen and Michael Claeys
Centre for Catalysis Research, Department of Chemical Engineering, University of Cape Town, Private Bag X3,Rondebosch 7701, South Africa

Andrew M. Beale
Department of Chemistry, University College London, 21 Gordon Street, WC1H 0AJ, London, UK
UK Catalysis Hub, Rutherford Appleton Laboratory, Research Complex at Harwell, Harwell, Didcot, OX11 0FA, UK
Finden Ltd, Clifton Hampden, Oxfordshire, OX14 3EE, UK

Ines Lezcano-Gonzalez
Department of Chemistry, University College London, 21 Gordon Street, WC1H 0AJ, London, UK
UK Catalysis Hub, Rutherford Appleton Laboratory, Research Complex at Harwell, Harwell, Didcot, OX11 0FA, UK

Teuvo Maunula
Dinex Ecocat Oy, DET Finland, Catalyst Development, Typpitie 1, FI-90620 Oulu, Finland

Robert G. Palgrave
Department of Chemistry, University College London, 21 Gordon Street, WC1H 0AJ, London, UK

Yufen Hao, Xiaodong Wang, Noémie Perret, Fernando Cárdenas-Lizana and Mark A. Keane
Chemical Engineering, School of Engineering and Physical Sciences, Heriot-Watt University, Edinburgh EH14 4AS, Scotland, UK

M. Dad, H. O. A. Fredriksson, P. C. Thuene and J. W. Niemantsverdriet
Laboratory for Physical Chemistry of Surfaces, Eindhoven University of Technology, P.O. Box 513, 5600 MB, Eindhoven, The Netherlands

J. Van de Loosdrecht
Sasol, Group Technology, P.O. Box 1, Sasolburg 1947, South Africa

Dipak Das and Arup Gayen
Department of Chemistry, Jadavpur University, Kolkata 700032, India

Jordi Llorca and Montserrat Dominguez
Institut de Tècniques Energètiques and Centre for Research in Nanoengineering, Universitat Politècnica de Catalunya, 08028 Barcelona, Spain

Muhammad A. Nadeem and Hicham Idriss
SABIC- Corporate Research and Innovation (CRI) at KAUST, Thuwal 23955, Saudi Arabia

Imran Majeed
Department of Chemistry, Quid-i-Azam University, Islamabad 4200, Pakistan

Geoffrey I. N. Waterhouse
School of Chemical Sciences, University of Auckland, Private Bag 92019, Auckland, New Zealand

Marco Piumetti, Samir Bensaid, Debora Fino, and Nunzio Russo
Department of Applied Science and Technology, Politecnico di Torino, Corso Duca degli Abruzzi, 24, 10129 Torino, Italy

M.A. Khan, M. Al-Oufi, A. Tossef, Y. Al-Salik and H. Idriss
SABIC-Corporate Research and Development (CRD), KAUST, Thuwal, Saudi Arabia

Michael Bowker
Cardiff Catalysis Institute, School of Chemistry, Cardiff University, Cardiff CF10 3AT, UK
UK Catalysis Hub, Research Campus at Harwell (RCaH), Rutherford Appleton Laboratory, Harwell, Oxon, OX11 0FA, UK

Ryan Sharpe
Cardiff Catalysis Institute, School of Chemistry, Cardiff University, Cardiff CF10 3AT, UK

Michael Bowker and Catherine Brookes
Cardiff Catalysis Institute, School of Chemistry, Cardiff CF10 3AT, UK
Rutherford Appleton Laboratory, UK Catalysis Hub, Research Complex at Harwell (RCaH), Harwell, Oxon OX11 0FA, UK

Matthew House and Abdulmohsen Alshehri
Cardiff Catalysis Institute, School of Chemistry, Cardiff CF10 3AT, UK

Emma K. Gibson and Peter P. Wells
Rutherford Appleton Laboratory, UK Catalysis Hub, Research Complex at Harwell (RCaH), Harwell, Oxon OX11 0FA, UK

Glenn Jones
Johnson Matthey Technology Centre, Blounts Court, Sonning Common, Reading RG4 9LJ, UK

Chuntian Qiu, Chen Chen, Satoshi Ishikawa, Zhenxin Zhang and Toru Murayama
Catalysis Research Center, Hokkaido University, N-21, W-10, Sapporo 001-0021, Japan

Wataru Ueda
Catalysis Research Center, Hokkaido University, N-21, W-10, Sapporo 001-0021, Japan
Facility of Engineering, Kanagawa University, 3-27-1, Rokkakubashi, Kanagawa-ku, Yokohama, Kanagawa 221-8686, Japan

Index

Printed in the USA
CPSIA information can be obtained
at www.ICGtesting.com
JSHW051323221024
72173JS00006B/1285

9 781632 406293